Rethinking Networks in Times of Transition

Rethinking Networks in Times of Transition

Ecological, Energy and Social Issues

Edited by

Laure Dobigny
Benoît Robyns

ISTE

WILEY

First published 2026 in Great Britain and the United States by ISTE Ltd and John Wiley & Sons, Inc.

ISTE Ltd
27-37 St George's Road
London SW19 4EU
UK

www.iste.co.uk

John Wiley & Sons, Inc.
111 River Street
Hoboken, NJ 07030
USA

www.wiley.com

Library of Congress Control Number: 2025944876

British Library Cataloguing-in-Publication Data
A CIP record for this book is available from the British Library
ISBN 978-1-83669-018-4

The manufacturer's authorized representative according to the EU General Product Safety Regulation is Wiley-VCH GmbH, Boschstr. 12, 69469 Weinheim, Germany, e-mail: Product_Safety@wiley.com.

Contents

Introduction

Our planet and society are governed by networks, whether biological, ecological, related to energy, transportation, climate or technology. We are continually engaged in interactions with these networks. In the context of energy or ecological and societal transitions, it is essential to examine the positive or negative effects of these interactions. Additionally, it is crucial to explore the influence of the growing role that networks will play and the futures they will shape, open or close.

To address these issues, a multidisciplinary approach is essential, particularly between the technological sciences, the humanities and life sciences. This book is organized around four major themes:

– Ecological networks: biodiversity and ecological transition.

– Energy networks: what socio-technical changes for what futures?

– Social and controversy networks: what dynamics are transforming lifestyles, representations and modes of action? Are multiple transitions/transformations at play?

– Networks of universities in transition: role, impact, difficulties, drivers?

Each theme constitutes a part of this book.

Part 1 deals with the ecological networks for biodiversity and ecological transition. Ecosystems are facing global changes, which is a major challenge for our future. These ecosystems have intrinsic dynamics that we need to integrate into the functioning of our society. It is therefore important to adapt our behaviors in order to preserve biodiversity. Ecosystems function through a multitude of connections

---

Introduction written by Laure DOBIGNY, Sadia BENAMROUZ-VANNESTE and Benoît ROBYNS.

between living organisms and their environment. Their protection thus requires a heuristic understanding of how these ecological networks work.

The five contributions of this part of the book in various domains and contexts converge around a central question: how can we rethink our relationship with the living world in environments transformed by human activity?

They all plead for a deeper reintegration of the living (human and non-human) into our political, social and spatial models. Only then can we start to move toward a more sustainable world. Each contribution challenges dominant paradigms and calls for human activity to be re-embedded within ecological realities. Whether considering freshwater ecosystems, agricultural landscapes, reforested areas, urban interstitial spaces or university institutions, these contributions reveal the richness and complexity of anthropized ecosystems. They are not only spaces to be restored through technical solution, but are also grounds for experimenting and understanding new forms of coexistence between humans' activities and the "living" or biodiversity.

First, in Chapter 1, Rossano Bolpagni, from the University of Parma, highlights the crucial role of macrophytes in freshwater ecosystems, emphasizing their importance as key contributors to both ecological functioning and resilience in the face of global pressures such as climate change and biological invasions. He argues that understanding the multifunctionality expressed by macrophytes is essential for any effective strategy aimed at preserving inland aquatic environments.

In Chapter 2, Florian Kletty, from the Catholic University of Lille, reminds us how intensive farming practices in temperate regions have significantly reshaped landscapes and severely impacted biodiversity, affecting ecosystem services and human well-being. In response to these challenges, alternative agricultural models exist, offering more sustainable approaches across multiple scales. This chapter explores both the potential benefits and the practical barriers to implementing such models, advocating for a holistic rethinking of agriculture that prioritizes sustainability.

In Chapter 3, Kevin Cianfaglione, from the Catholic University of Lille, and his co-authors from Università della Tuscia, Università di Camerino and Nottingham Trent University, examine the large-scale reforestation efforts carried out in Italy from the late 19th to the 20th century, which played a key role in restoring ecological and territorial balance in response to growing environmental degradation. Today, these forest formations are at the center of heated debates, caught between ideological criticisms and economic interests. This chapter argues for moving beyond these polarizations by acknowledging the historical, ecological and social significance of these landscapes, and by rethinking their management through a sustainable perspective.

In Chapter 4, Agathe Douchet and Sadia Benamrouz-Vanneste, from the Catholic University of Lille, examine how individuals perceive and use small urban interstices. Focusing on the sociological dimension, the study aims to assess how citizens and municipal staff perceive these urban areas, considering the temporal dynamics of this perception in relation to their level of knowledge about the biodiversity these places host. Despite their limited size, the findings emphasize the significant social importance of these informal green spaces, providing valuable insights for urban planners and green space managers looking to improve the perception and use of these spaces.

Finally, in Chapter 5, Alice Fouillouze, from the Catholic University of Lille, proposes in a rethink of socio-ecological transition strategies that goes beyond techno-solutionist approaches centered solely on CO_2 emissions to a vivo-centric approach. This contribution defends a model based on reducing unnecessary activities and reconnecting with life, within a paradigm of strong sustainability. The university is seen as a place to experiment with this transition by changing the way it operates, teaches and conducts research. The author advocates a model that calls for action and commitment, while at the same time being ethical, systemic and humanist.

Together, these contributions call for a holistic and transdisciplinary approach to the socio-ecological transition of our human-modified ecosystems, as a necessary condition for building a sustainable world.

Part 2 deals with the energy networks. What socio-technical changes for what futures? The objective of this part is to question, from a transdisciplinary perspective, the evolution and futures of energy networks. Through the increase in local renewable energy production, legislative changes (e.g. energy communities), the growing electrification of consumption (e.g. electric vehicles) and new technological possibilities (smart grids), these networks are moving toward decentralization, with increasing involvement of all stakeholders to enable more flexibility in both production and consumption. These changes cannot occur without considering the role of the user in these networks and the interactions between them. Indeed, by initiating long-term developments, these networks contribute to the (re)construction of social and spatial relationships as well as the representations related to energy. In particular, it is examined how these networks influence choices between alternatives, thus contributing to shaping possible futures.

In order to address both the technical and social challenges of the energy transition, the development of transdisciplinary approaches and indicators is therefore fundamental to make the transition sustainable both environmentally and socially, and to chart directions for the future of these networks.

In Chapter 6, Lucas Lopez, from Bordeaux Montaigne University, discusses the fact that electricity transmission networks are the keystone of neo-electrification and are at the crossroads of multiple issues, including nuclear electricity production and conflicting European and French energy policies.

In order to reconcile the power grid with a form of human scale that allows us to think collectively about sufficiency, Jonathan Coignard, from Grenoble Alpes University, questions the size of power grids in Chapter 7. He highlights the relevance and lessons to be learned from scale issues outside of the strict domain of power grids. Then, the concept of scale is defined specifically for the power grid. The implications of changing scales in the different dimensions of a power system are analyzed. Several research avenues are explored to imagine what a human-scale power grid might look like.

In Chapter 8, Baptiste Beguinet, from INSA Lyon, analyzes the characteristics and role of user networks in the appropriation of self-built complex technologies, using wind turbines manufactured locally in France as a case study. This characterizes French users constituting a network of people engaged in the manufacture of self-built wind turbines, by creating a typology with four profiles. Then, he underlines the core role of the network in the appropriation, through three key moments of a small wind turbine of the Piggott-type life cycle. Particular attention is given to the common representations of users and how new practices emerge, interwoven with the technology and infrastructure specific to the Piggott wind turbine.

Self-production and self-consumption are becoming more common in renewable energy use. In Chapter 9, Jessica Zaphiropoulo, from Grenoble Alpes University, considers collective self-consumption (CSC) and its local limits from a socio-technical approach to energy flexibility. She examines the challenges posed at a transcalar level by renewable energy in order to grasp the flexibility script embedded in the CSC framework. Then, two CSC projects are presented to understand how limits of renewable energy are "scripted" in the relation between demand and supply, giving a specific sense of local flexibility for each project.

The full innovative potential of smart grids can only be released by promoting interaction between the different actors involved in the electric system (producers, consumers, storage and network managers), increasing their "electrical intelligence". These actors may have very different consumption and production profiles, with very varied economic and social objectives and/or constraints. New types of actor may emerge alongside new economic models, all of which may contribute to solving energy and climate issues, promoting the development of renewable energy sources. It is important that all actors should benefit in these cases, including those in situations of “energy poverty”. These questions and issues will be discussed in Chapters 10 and 11, which present several case studies involving very different actors.

In Chapter 10, Adrien Bossu et al., from the Junia Graduate School of Engineering, examine a novel demand-side management approach based on a cooperative game. This chapter demonstrates how prioritising local energy production for community self-consumption can lower individual costs while mitigating environmental impacts. Particular attention is given to the motivations of stakeholders engaged in collective self-consumption initiatives. To evaluate the environmental implications, the authors employ an innovative methodology based on a life-cycle assessment (LCA). Furthermore, the chapter addresses the ethical allocation of benefits through game-theoretic frameworks, raising critical philosophical questions such as whether rewards should favour the most active participants – who are often the wealthiest – or prioritise support for the most vulnerable.

A dynamic energy allocation in a collective self-consumption community integrating user preferences is proposed in Chapter 11, by Amira Dhorbani et al., from the Junia Graduate School of Engineering. The case study, based on a demonstrator of the Catholic University of Lille, provides a detailed analysis of the interaction between buildings, electric vehicles and battery storage. A utility-based framework is introduced to quantify user priorities (maximizing cost effficiency, or consuming local photovoltaic production, or ensuring comfort), enabling an adaptive energy distribution strategy. A rule-based decision-making approach is developed to optimize energy flow while considering technical constraints and economic factors.

Finally, in Chapter 12, Hanene Bettaied et al., from the Junia Graduate School of Engineering, focus on awareness and perception of collective self-consumption in university communities. Through surveys, they show that this new mode of production and consumption is still too little known and understood, except, for example, by a specialized audience of engineers. A communication and training effort for the general public is therefore necessary.

Part 3 deals with social and controversy networks. What dynamics are transforming lifestyles, representations and modes of action? Are several transitions and/or transformations at work?

Whether out of concern for the environment, financial opportunism or even restriction, consumer practices are changing, with ever greater consideration being given to the effects of lifestyle choices in terms of environmental impact and greenhouse gas emissions. Representations, attitudes and practices have also evolved as a result of awareness-raising campaigns, notably toward sufficiency. At the same time as entering the political agenda, the semantic shift in France and elsewhere from militant to political discourse has emptied the term of its radical symbolic and political significance. What representations, practices and dynamics of change are behind the calls for and practices of "simplicity" or "sufficiency"? What distinctions (social, symbolic, political) need to be made between these terms and what they cover? What are the favorable factors or necessary conditions for anyone to embark on the path of lifestyle change? What sociological profiles are to be found in the different degrees of belief and commitment?

The networks here are essentially interpersonal in nature, and take several forms: individuals influencing and recommending to other individuals/collectives that are more or less geographically close, but which have a common project and on which they exchange, etc. And since micro- and macro-dimensions influence each other, we must not underestimate the impact of diplomatic networks, civil society and international economic players (multinationals), which play a strategic and geopolitical role, exerting their influence on major environmental decisions and projects. In this part, the chapters examine the role of networks in current processes of transition, while analyzing the social metamorphoses, resistance mechanisms and justifications they generate.

There is no real consensus about the solutions to environmental issues within societies. While the effects of climate change are increasingly attested and noticeable, climate skepticism has never been so widespread. From promoting sufficiency and low tech to encouraging geoengineering and "techno-solutionism", the paths envisaged to meet the challenges are divergent, as are the models of society on which they work. How, for example, can we analyze the growing mistrust of science and governments (climate skepticism, Covid-19, conspiracy theories, etc.)? What do these changes tell us about the transformations that are taking place and possible futures? Are controversies and mistrust obstacles to the necessary changes or, on the contrary, do they open up alternative avenues that lead to democratic debates? The aim of this part is to understand the controversies and antagonistic positions underpinning current and future transformations.

First, in Chapter 13, Matthew O'Meara Wallis, from Nottingham Trent University, presents a qualitative study of fracking in the East Midlands. He explores the concept of the Social Licence to Operate (SLO) in contested environments, examining the dynamic and interdependent relationships among stakeholders and emphasizing their ethical and relational dimensions.

This interrogation is pursued in Chapter 14, authored by Sébastien Carcelle, from the Center for International Cooperation in Agricultural Research (CIRAD in

France), and Mariana Benitez, from Instituto de Ecologia, Universidad Autónoma de México. The authors propose the notion of "articulation" as a means of analyzing and conceptualizing the Brazilian agroecological movement, with a particular focus on agroecological coffee production. This approach uses the principles of life sciences and recent contributions from the sciences of complexity, which are applied to agroecosystems, to renew our approach and understanding of alternative social movements.

The analysis of alternative movements continues in Chapter 15, with Hector Docarragal Montero, from the Ecole Nationale Supérieure d'Architecture Paris-La Villette in France, who explores innovative and often "out-of-standard" architectural practices developed in response to environmental challenges. Montero proposes an original ethnographic methodology, termed CAST (Context, Actors, Strategies, Theme), which is designed to facilitate a multidimensional reading of architectural practices. By developing the notion of "out-of-standard" experimentation in architecture, the author advances the proposition of cultivating an "*an ecology of the project* itself", as a method of acknowledging the intricate interplay between human and non-human actors within the context of architectural projects.

In Chapter 16, Albin Wagner, from the Catholic University of Lille, proposes to analyze social media during Emmanuel Macron's presidency in France (2017–2022) and the evolution during this five-year term of narratives related to climate and environmental issues. This discourse analysis emphasizes the importance of a lexicological approach in revealing the diversity and complexity of social imaginaries and representations concerning the climate crisis. Furthermore, it exposes the plurality and frequently discordant nature of these discourses, which impede the implementation of coherent systemic actions, as well as the importance of positive narratives in anticipating and adapting to upcoming changes in open ways.

This point is developed in greater depth in Chapter 17 by Cristina Longo, from the University of Lille, and Ishan Jalan, from Nottingham Trent University, who focus on the importance of emotions in sustainable choices and lifestyle. Emotions are frequently underestimated in the understanding of why individuals adopt or resist sustainable practices. However, as demonstrated by Longo and Jalon, the rational decision-making models have several limits. In considering emotions in a non-dichotomous way, Longo and Jalon emphasize the emotional complexities that influence consumer choices and the nuanced emotional and experiential dimensions of living sustainably. The development of this complexity is both important to advance theoretical insights and to offer practical strategies for fostering enduring behavioral change. In conclusion, it is important to note that the role played by emotions may have a significant impact on the transition to sustainable campuses[1]. This issue will be explored in greater detail in the final part of this book.

Part 4 deals with networks of universities in transition. Which role, impact, difficulties, drivers for the universities? One of the missions of the university is to train students for the socio-ecological transition, and to help them reflect on possible technological and social solutions to respond to the climate and environmental crisis. Many questions are addressed within the framework of this theme, including:

– What institutional approaches may be defined to integrate sustainable development priorities, especially in the field of climate action, in training and research?

– What are the current issues and trends in fostering transformative education for sustainable development?

Chapter 18 deals with schools in synergy for sustainability, and the experiences and perspectives of university networks in socio-ecological transitions. Emmanuel D. Delocado et al., from Ateneao de Manila University, highlight that universities or higher education institutions play a key role in knowledge generation through research and knowledge dissemination through teaching. The work in sustainable development requires expertise in multi-disciplinary fields, which all universities hold monopoly of, and so this necessitates constant collaboration and partnerships. These allow for the sharing of knowledge and resources that can help in expanding capacity and building collective action. The authors illustrate this by the progression of sustainability strategies within Philippine higher education institutions.

In Chapter 19, Richard Bull, from Nottingham Trent University, points out that universities play a huge role in their local economies and their impact on the environment, both positively and negatively, is substantial. He defines the concept of Knowledge Exchange (KE) and situates it within wider understandings of partnership working and collaboration with small and medium companies (SMEs). The research findings and case studies in the UK illustrate this concept.

In Chapter 20, Benoît Robyns, from the Catholic University of Lille, focuses on the research for energetic and societal transition. As scientific research, especially of a fundamental nature, does not ensure that it will have a positive contribution to meeting global challenges induced by energy and societal transition, we have to ask

1 As developed, for example, in Aulanier, A. and Dobigny, L. (2025). Une transition écologique sans expérience. Les émotions des usagers des smart-buildings universitaires, ou l'histoire d'un rendez-vous manqué avec les enjeux écologiques. *Émulations, Revue de sciences sociales*, 50.

the question “should we continue scientific research?”. It emerges that research must be increasingly interdisciplinary, that an analysis of the contributions of the results of each research work to the UN SDGs is recommended, as well as the development of low-tech approaches in all areas where this is possible, by encouraging research work on these approaches, which are also a source of innovation as well as of change in lifestyle.

Through its mission of training and education, the university has a special responsibility and opportunity to raise awareness among its students and staff so that they contribute to reducing the university’s carbon footprint, as well as to reduce their personal carbon footprint in their daily activities in the process. To mobilize students and staff, the Catholic University of Lille organized in the fall of 2022 a university convention for the climate, the aim of which was to allow students and staff to propose actions that contribute to the development of the university’s carbon trajectory, which aims for neutrality in 2050, involving actions and visions that transform habits and behaviors, as well as ways of seeing and being in the world. In Chapter 21, Kokou Kouzouahin Somabe and Hervé Barry, from the Catholic University of Lille, propose a socio-ecological analysis of the impact of this university climate convention, and introduce the concept of eco-campus.

Finally, in Chapter 22, Laure Dobigny et al., from the Catholic University of Lille, question the inclusion of users in universities’ ecological transitions. To explore this issue, the authors particularly focus on the decision to build and renovate university buildings as ‘smart’ buildings. What vision of the ecological transition do these technical choices reflect? And on what vision of users and their role in the ecological transition? Through a qualitative study of the decision-making processes behind the construction and renovation of four smart university buildings in the Lille region, the authors highlight both the difficulty and necessity of including users to achieve energy and ecological objectives. This necessitates rethinking the role of users in the ecological transition, as well as developing methods and tools to achieve this technical inclusion.

Acknowledgments

Most of the topics covered in the chapters of this book were debated at the ECOPOSS 2024 conference, *Between Preservation and Transformation, Reinventing our Wolrd*, organized by the Catholic University of Lille.

The editors thank all chapter reviewers who contributed to the quality of this book: Dhaker Abbes, Herinaina Andriamandroso, Audran Aulanier, Hervé Barry, Richard Bull, Camilla Cabral Salles, Arnaud Davigny, Emmanuel Delocado,

Benoit Durillon, Alice Fouillouze, Margot Gaetani-Liseo, Hugo Helbling, Anne-Sophie Loison, Thierry Magnin, Benoit Vanhee, Rémi Platel, Christophe Saudemont, Monica Scarano and Ali Siah.

PART 1

Ecological Networks

1

Climate Crisis, Invasive Species and Biodiversity Collapse: Issues, Challenges and Opportunities for Macrophytes

1.1. Introduction

The purpose of this brief contribution is to frame the role of macrophytes in the context of the climate crisis, the biological invasions and their mutual repercussions on freshwater resources, especially in terms of biodiversity and functioning. The ultimate goal is, therefore, to offer interpretative tools and systematize available knowledge to better understand how ongoing global processes affect all of the key components of inland waters. This is especially important considering the rates of biodiversity loss associated with inland waters compared to all other ecosystems and habitat types. In this direction, recent data suggests that "freshwater wildlife populations have declined by 84% on average since 1970. Almost one in three freshwater species are threatened with extinction" (WWF 2020).

To provide new tools, or more effective tools that can halt the collapse of inland aquatic biodiversity, it is necessary to fully understand the multifunctionality expressed by macrophytes.

One possibility, in this regard, is understanding the responses and adaptations enacted by aquatic plants to serve as ecological pillars of aquatic ecosystems, starting from considering what a macrophyte is and what their peculiarities are.

Chapter written by Rossano BOLPAGNI.

For a color version of all figures in this chapter, see www.iste.co.uk/dobigny/rethinking.zip.

In other words, it is crucial to grasp the ecological exceptionality of this biological component and the irreplaceable role it plays in freshwater, before deepening its responses and adaptations to the global crisis and biological invasions. Particular attention will be devoted to clarifying the essential roles played by macrophytes in sustaining the provision of aquatic ecosystem services and our wellbeing more in general (Ferreira et al. 2023).

1.2. The world of macrophytes

By the term macrophyte, we refer to all of the macroscopic plants that can grow in or near water and in waterlogged sediments and are large enough to be seen with the naked eye (see Figure 1.1) (Murphy et al. 2019). Consequently, macrophytes are found in freshwater and marine environments, such as lakes, rivers, wetlands and estuaries and mediate key processes by providing, for example, oxygen, shelter and food for a variety of aquatic organisms.

Different forms of growth can be identified; the main groups include (i) emergent plant species, i.e. rooted plants, anchored to sediments with much of their culms emerging from the water (e.g. cattails, common reed); (ii) submerged macrophytes, capable of growing completely underwater (e.g. seaweed, pondweed); (iii) floating-leaved macrophytes, which are rooted plants with leaves floating on the water surface (e.g. water lilies); and (iv) free-floating plants, rootless taxa which float on the water surface (e.g. duckweed, water hyacinth) (Sculthorpe 1985; see Figure 1.1).

Regarding the peculiarities of macrophytes, the most recent definition states that they are "plants that live in excess moisture conditions (macrophyte plant life) and grow in water or sediment, or in periodically anaerobic soils" (Cronk and Siobhan Fennessy 2001). It is clear from this definition that macrophytes are intimately associated with water and sediment (i.e. water-saturated soil).

Thus, how they interact (are related) with water and sediment emerges as fundamental in guiding the peculiarities evolved by macrophytes. Considering the classical distribution of plants along a moisture gradient around a water body, it can be observed that macrophytes are arranged in concentric bands (see Figure 1.2). These bands, in turn, correspond to different soil saturation conditions or different flooding periods, or different growth depths relative to the water–atmosphere interface.

One of the key factors regulating the position of a macrophyte with respect to these gradients is oxygen, the availability of which in soils/sediments is strongly influenced by the presence and persistence of water. Oxygen that is virtually never limiting in terrestrial ecosystems or in the uppermost soil layers can become so – with very important consequences – in wetlands or within water bodies. The onset of deoxygenation phenomena (such as hypoxia or anoxia) has dramatic effects on the presence and viability of autotrophic and heterotrophic organisms in aquatic ecosystems. Under these conditions, oxygen becomes a key factor in regulating the diversity and functioning of aquatic ecosystems, triggering limitations and eco-physiological damage to macrophytes.

To avoid this, macrophytes have developed refined adaptations to ensure oxygen availability to all cells in their organs/tissues, especially those in contact with sediments. To do this, macrophytes have evolved different transport mechanisms (e.g. internal gas ventilation) and differentiated a tissue specifically dedicated to this function: the aerenchyma. This is all with the main aim of reducing the toxicity of oxygen-deficient environments.

One of the most relevant implications of this enormous gas/oxygen transfer between different components of the macrophyte body is ROL, the "radial oxygen loss" (Sand-Jensen et al. 1982). This refers to the process of oxygen release from the thin roots to the surrounding sediment. ROL creates very thin oxidized layers/sheaths all around the roots with important physiological and biogeochemical repercussions (Benelli et al. 2020).

Some of the major effects include the prevention of the accumulation of toxic compounds such as hydrogen sulfide (H_2S) and methane (CH_4); enhancement of nutrient uptake through stimulation of beneficial microbial activity and nutrient oxidation, for example, conversion of ammonium (NH_4^+) to nitrate (NO_3^-); and support of root respiration, which enables plant survival in anaerobic environments.

The main ecological implications of root-level oxygen release are a significant increase in sediment quality, which is essential to support wetland restoration efforts; a significant reduction in excess nitrogen by stimulating nitrification processes; and positive modulation of plant–microbe interactions useful in bioremediation programs. Overall, through internal gas transport and mediated release of molecules such as oxygen and methane, macrophytes exert a strong control of the functioning of colonized water.

The widespread presence of macrophytes and macrophyte-dominated stands is critical in maintaining a balance of "local metabolism" with a general progressive burial/accumulation of C. In other words, all of this results in a substantial control of the food web of wetlands/water bodies densely colonized by macrophytes, highlighting a fundamental regulatory role of this type of plant in maintaining freshwater biodiversity (Karus et al. 2022).

Figure 1.1. *Growth forms of macrophytes: from top to bottom, we can observe stands dominated by* Phragmites australis *(emergent plant species),* Vallisneria spiralis *(in the first box; submerged macrophytes),* Nuphar lutea *(floating-leaved macrophytes) and* Trapa natans + Lemnaceae *(free-floating plants) (photo credits, R. Bolpagni)*

Figure 1.2. *Spatial arrangement of macrophytes across ecological gradients, from the aquatic domain (where submerged, floating-leaved and free-floating macrophytes dominate) to riparian sectors (dominated by helophytes) (photo credits, R. Bolpagni)*

1.3. Macrophytes as engineering species

Due to the extraordinary adaptations evolved by macrophytes to thrive in critical environments, these species can be included among the so-called "engineering species" (Bolpagni et al. 2015; Ribaudo et al. 2018). Indeed, macrophytes include organisms that modify their environment in ways that significantly affect other species, as well as exerting direct control over the peculiarities of the environments they colonize. These modifications can change resource availability, alter habitats and affect ecosystem structure and function. Macrophytes are, therefore, capable of generating a myriad of services useful to us. The most relevant services mediated to mention are (i) water purification, nutrient cycling, sediment stabilization and erosion control (among regulatory services); (ii) the provision of habitat and food for aquatic life, raw materials and medicine principles; and (iii) the capability to sustain high rates of primary and oxygen production and exceptional biodiversity hotspots (among supporting services) (Thomaz 2023).

In terms of biological interactions, macrophytes mediate a wide range of relationships. Macrophytes can host many epiphytes and biofilms that aid in nutrient cycling and pollutant control. They are also able, as already reported in the text above, to provide habitat and shelter for insects, mollusks and crustaceans. Several snails graze at the canopy controlling biofilms and pathogens while benefiting from food, whereas the juvenile fish use macrophytes as hiding spots to avoid predation. Other fish species such as grass carp graze on macrophytes controlling their growth, often triggering a strong ecosystem imbalance.

At the same time, macrophytes compete for light, nutrient and space with phytoplankton regulating the biological turbidity of water (Maberly 1981). This is a key point in terms of aquatic ecosystem self-regulation. In fact, macrophytes can also mediate so-called "disservices". Their overgrowth can trigger dystrophic events such as consuming the oxygen present in the water column, which leads to the death of the entire heterotrophic component, and most autotrophic organisms, present in a water body (see Figure 1.3).

Figure 1.3. *Cascading effects of overgrowth of a free-floating macrophyte (in this case,* Trapa natans*), capable of creating floating mats associated with the progressive reduction of oxygenation in the underlying water column; all images refer to the same critical moment (site: Lancone di Po, Piacenza province, North Italy; photo credits, R. Bolpagni, M. Bartoli)*

However, despite the high resistance and resilience shown by macrophytes, even macrophytes are at risk with increasingly high rates of loss globally (Zhang et al. 2017). These authors certified an acceleration in aquatic vegetation loss, especially in lakes with an area larger than 50 km^2 and in terms of submerged aquatic vegetation. The main drivers of this collapse are "eutrophication, algal blooms, land reclamation, aquaculture cultivation and global climate changes". In general, the role of the availability of nutrients and, therefore, of eutrophication is fundamental. As nutrient availability increases, macrophyte communities, especially the submerged ones, undergo profound transformations. Alterations in the nutrient content affect the composition of macrophyte communities, particularly free-floating and rooted. An increase in phosphorus concentration can exacerbate a competition between macrophytes and phytoplankton. This may result in phytoplankton dominance and the disappearance of macrophytes (Declerck et al. 2005).

1.4. Macrophytes, the climate crisis and biological invasions

Freshwater ecosystems, and especially macrophytes, are remarkably exposed to the impacts of the climate crisis. Regarding this, the recent study by Fu and Wu (2024) reported that "climate models indicate that a warmer environment will increase low-level moisture, potentially intensify extreme precipitation". With a focus on tropical zones, this chapter showed a $-9 \pm 2\%$ ($23 \pm 2\%$) change in rain intensity and a $13 \pm 2\%$ ($-24 \pm 1\%$) change in rain frequency. In addition, warming is greater over continents than over oceans, stimulating intense water evaporation that worsens the severity of droughts, taking away room from macrophytes in both lotic and lentic water bodies. A warm climate also supports eutrophication and hence increases the availability of nutrients such as nitrogen and phosphorus for plants (Dhir 2015). Warming can therefore lead to a shift from a clear, macrophyte-dominated state to a turbid, phytoplankton-dominated state (Mooij et al. 2007). The rise in temperature also increases the decomposition processes in aquatic ecosystems, which are much more reactive. All this, combined with the boost in the atmospheric concentration of CO_2, determines fertilizing effects on macrophytes, many of which have been shown to grow faster. However, their excessive growth (such as algal blooms) can disrupt ecosystems leading to increasingly frequent dystrophic phenomena. No less significantly, the increase in temperature leads to a progressive escape of dissolved oxygen from the water masses, establishing conditions that are very critical for the survival of macrophytes and aquatic biota in general. Similarly, a growing frequent appearance of extreme weather events, such as storms, floods and droughts, can physically damage macrophytes, uproot them or increase sedimentation that completely changes the local, edaphic conditions of water ecosystems. At the same time, the intensification of runoff can fuel the development of ruderal or invasive *taxa* associated with eutrophication phenomena. Macrophytes therefore respond quickly and very intensely to the local effects of climate change, often with distorting effects associated with the collapse of communities and their progressive structural simplification. Furthermore, scientific evidence indicates that these processes are likely to be faster in shallow water bodies compared to those in deep lakes (Lind et al. 2022). The reduction in transparency, associated with variations, even rapid ones, in the water levels and an increase in temperature, induces the replacement of vascular macrophytes by algae. Moreover, in remote zones, for example, high-latitude and high-altitude areas, specialized macrophytes highlight the tendency to become rarer following unfavorable interactions with more competitive species.

Finally, the spread of invasive species has a dramatic role in changing macrophyte communities (Bolpagni 2021). Alien species are defined as those species found in habitats or areas placed outside their natural native range, this being due to accidental or deliberate release by humans (Bellard et al. 2015). Over the last decade, species introductions have indeed emerged as one of the most important, least controlled and least reversible drivers of human impact on the world's ecosystems (Strayer 2010); so much so that some scholars have speculated that we are entering a new era called the Homogocene (a term coined by Gordon Orians in the 1990s), in which, all of the world's continents are seen as connected into a single biological Pangaea through global economic trades (Rosenzweig 2001). Recently, this paradigm, with reference to inland aquatic ecosystems and the macrophyte component, has been further investigated by Bolpagni (2021) who introduced the topic of the Exocene: understood as a new global "bio-historical horizon" dominated by alien species. The phenomenon of biological invasion has seen a notable and significant increase in its rates since the industrial revolution. This is due to fundamental scientific advances in the field of transportation which, over the years, have then seen their efficiency (in terms of speed) increase more and more. The emergence of the alien species issue has turned out to be intimately related to a second phenomenon of absolute importance represented by invasive alien species. Invasive alien species are those alien species that, by settling and establishing themselves in the new habitat, are able to spread in an uncontrolled and uncontrollable manner, causing major negative impacts toward biodiversity, ecosystem services, as well as economic activities and human health (Westphal et al. 2008).

Alien invasive species are one of the most rapidly increasing and pressing environmental issues of recent times (see Figure 1.4). Globally, about 6–10% of alien aquatic plants are considered seriously invasive (Houlahan and Findlay 2004). In addition, alien invaders typically have few or no natural herbivores or diseases to limit their spread. Disturbance may play a role in enabling invasive species to penetrate relatively mature communities (Nagasaka et al. 2002). Severe damage caused to aquatic and wetland habitats by invasive plants has been reported worldwide, and the problem is increasing in both extent and severity.

Figure 1.4. *Stands of* Pistia strotioites *(site: Po di Primaro, Ferrara province, North Italy; photo credits, R. Bolpagni)*

Invasive aquatic plants can spread quickly when introduced to suitable habitats beyond their native range. For instance, the small duckweed *Landoltia punctata* spread 2,000 km in North America in less than 70 years, while *Elodea canadensis* expanded 365 km in only five years in Britain (Les and Mehrhoff 1999). Free-floating *Eichhornia crassipes* and *Salvinia molesta* have also spread remarkably quickly in tropical and subtropical regions (Talling and Lemoalle 1998). Bora and Padial (2023), analyzing the traits associated with the invasive behavior of macrophytes and their success, highlighted how these are associated with growth and reproduction indicators. Overall, "invasive species had a higher number of morphological and biotic interaction traits related with invasiveness than native species".

To sum up, the most recent literature suggests a strong interplay between all major climate change-induced issues ("changes in CO_2, temperature, and precipitation patterns") (Lind et al. 2022) toward macrophytes. In this regard, an increase in the abundance and distribution of emerging and floating plant species, and a decrease in the abundance and distribution of submerged macrophytes is expected (Lind et al. 2022). Again, in shallow water bodies, these processes will be much faster than what we observe in deep lakes. Reduced transparency, increased variability of hydrological levels and increasingly marked changes in temperature will favor the establishment and dominance of the algal component. At the same time, macrophytes typical of high altitudes and high latitudes will be in a critical condition, while invasive taxa will gain the upper hand over native species (Lind et al. 2022). Implementing these eco-functional predictions on a spatial scale is very difficult. Local physico-chemical conditions, as well as water resource exploitation rates, can profoundly influence the dispersal potential of macrophytes (native or allochthonous) and their ability to give rise to well-developed populations. Particular attention must always be paid to the conservation status of water bodies in order to hypothesize their macro-ecological dynamics. These dynamics are under the control of trophic relationships that regulate growth rates (nutrient availability, sediment quality, light competition) and herbivory (plant consumption, decomposition) (see Figure 1.5).

1.5. Macrophytes as nature-based solutions

With the goal of adapting to climate change and reducing the risk of disasters, it is crucial to improve the conservation status of aquatic habitats/ecosystems while also strengthening their functioning. With this perspective, the so-called nature-based solutions (NbS) have been developed and are finding increasing application globally (*nature-based solutions – European Commission*). NbS are solutions inspired and sustained by nature. This peculiarity makes these tools extremely effective and sustainable in economic terms, due to their ability to regenerate natural capital, protect biodiversity and provide irreplaceable ecosystem benefits. They guarantee, in fact, simultaneous environmental, social and economic benefits, as well as helping to strengthen the resilience of natural and human-made systems to the climate crisis.

NbS are born and must be implemented to conserve and enhance ecosystems – and the biodiversity they host – under the assumption that our socioeconomic systems rely "almost" exclusively on them. In reality, ecosystems generate a plurality of services on which the survival of our communities depends (Dunlop et al. 2024). From water and air purification to the production of biomass, food, active molecules, genetic and energy resources, carbon storage and the reduction of hydrogeological risk, these are among the irreplaceable services generated by ecosystems. As humanity, we bear the burden of protecting, managing and restoring ecosystems (natural and otherwise), moving beyond the formalism of "merely protecting nature" and turning our attention to the need to rely on natural capital for our survival. Nature can provide solutions to best and most effectively address the challenges posed by the climate crisis. However, it is necessary to listen to and observe nature. Ecosystem-based approaches, green infrastructure, agriculture and aquaculture and farming practices that can ensure reduced production and impacts on the environment, and urban greening become "the go-to for protecting nature while improving societies' response to climate change" (*Nature-based solutions for climate; IUCN*).

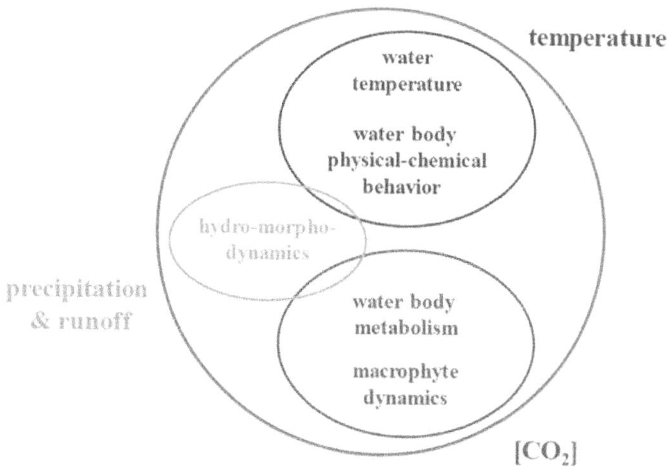

Figure 1.5. *The three spheres that govern the responses of macrophytes to climate change, temperature, precipitation and runoff and the rising of CO_2 availability (modified by Lind et al. (2022))*

With all this in mind, and considering the extraordinary ecological peculiarities of macrophytes, it is possible to think of using macrophytes as a tool for restoring nature and strengthening the provision of ecosystem services (Szoszkiewicz et al. 2025). As widely illustrated in the text, the fundamental role played by macrophytes in aquatic ecosystems in terms of engineering species is widely recognized. This is possible thanks to their refined adaptations to aquatic life, capable of providing a multitude of services. At the same time, working with macrophytes is essential, considering the enormous rates of loss that affect this biological component at a global level; also, because we need to strengthen the resistance/resilience of the freshwaters to the climate crisis. In fact, macrophytes should be counted among the "umbrella" species capable of creating habitats and providing spaces and niches for a multiplicity of other biological forms. The gradual disappearance of macrophytes is associated with a dramatic loss of functions and benefits (e.g. self-purification of water, provision of food and raw materials, carbon assimilation and nutrient cycling). It is possible, again, to think of macrophytes as an essential tool for the redevelopment and enhancement of aquatic ecosystems, among the most vulnerable to the local and global effects of the climate crisis. It is no coincidence that the loss of freshwater biodiversity is far greater than that of any other ecosystem.

The benefits that can be derived from enhancing macrophyte cenoses are widely established in the management of medium and large lakes and lentic water bodies more generally. In this regard, triggering the recovery of submerged stands (dominated by, for example, species of the genus *Potamogeton* or by charophytes) promotes (i) the uptake of N and P from the water body and sediment; (ii) the mitigation of surface sediment resuspension, reduction of turbidity; (iii) the reduction of P release through sediment oxidation; and (iv) inhibition of phytoplankton growth through competition for resources, creation of refuge zones for herbivorous zooplankton, as well as secretion of allelopathic substances. The expected results are improved water quality (although there is a progressive reduction in removal efficiency with the gradual accumulation of biomass, release of nutrients with the senescence and decomposition of plants).

There is, in addition, clear evidence on the positive impacts of the recovery of macrophyte communities on the overall specific richness (with clear increases) of the colonized water bodies, the consequent stabilization of the aquatic community and the increased resilience to external disturbances (both of the communities and the ecosystems more generally).

1.6. Conclusions

In the study "Linking vegetation patterns, wetlands conservation, and ecosystem services provision: From publication to application" by Bolpagni (2020), the key message underlined:

> The necessity of human-mediated actions to reinforce (aquatic) vegetation diversity in over-exploited lowlands. [...]. This must necessarily imply a clear understanding of the continuing context of climate change, as well as of the local and global implications of economic strategies.

This contribution aimed at strengthening the perspective of action here detailed, offering a snapshot of the key peculiarities of the world of macrophytes, reiterating the need to fully exploit their regeneration potential. All of this knowledge emphasizes the extraordinary potential of macrophytes in adapting to change. In this regard, it is urgent to substantiate the development and application of effective "*macrophyte-based solutions*" (MbS). These should be implemented not with the unrealistic objective to recover aquatic ecosystems, bringing them back to the states of quality they manifested in the pre-industrial era, but envisioning the ecosystems of the future (Parmesan 2006). Currently, freshwater ecosystems manifest conditions that have never been observed; to give just one key example, limiting nutrients' (P and N) availability exceed the natural baseline by orders of magnitude. In this regard, macrophytes represent an essential tool to address the current functional states of freshwater ecosystems (see Scheffer et al. 1993) and the global phenomenon of biological invasions. Exploiting the extraordinary ecological peculiarities of macrophytes means offering a chance of survival to the irreplaceable freshwater biodiversity.

1.7. References

Bellard, C., Cassey, P., Blackburn, T.M. (2015). Alien species as a driver of recent extinctions. *Biol. Lett.*, 12(2), 20150623.

Benelli, S., Ribaudo, C., Bertrin, V., Bartoli, M., Fano, E.A. (2020). Effects of macrophytes on potential nitrification and denitrification in oligotrophic lake sediments. *Aquat. Bot.*, 167, 103287.

Bolpagni, R. (2020). Linking vegetation patterns, wetlands conservation, and ecosystem services provision: From publication to application. *Aquat. Conserv.: Mar. Freshw. Ecosyst.*, 30(9), 1734–1740. doi: 10.1002/aqc.3358.

Bolpagni, R. (2021). Towards global dominance of invasive alien plants in freshwater ecosystems: The dawn of the Exocene? *Hydrobiologia*, 848, 2259–2279.

Bolpagni, R., Laini, A., Soana, E., Tomaselli, M., Nascimbene, J. (2015). Growth performance of *Vallisneria spiralis* under oligotrophic conditions supports its potential invasiveness in mid-elevation freshwaters. *Weed Res.*, 55(2), 185–194.

Bora, L.S. and Padial, A.A. (2023). A global review on invasive traits of macrophytes and their link to invasion success. *Acta Limnol. Bras.*, 35, e20.

Cronk, J.K. and Siobhan Fennessy, M. (2001). *Wetland Plants. Biology and Ecology*. CRC Press, Boca Raton and Taylor & Francis, London.

Declerck, S., Vandekerkhove, J., Johansson, L., Muylaert, K., Conde-Porcuna, J.M., Van Der Gucht, K., Perez-Martinez, C., Lauridsen, T., Schwenk, K., Zwart, G. et al. (2005). Multi-group biodiversity in shallow lakes along gradients of phosphorus and water plant cover. *Ecology*, 86(7), 1905–1915.

Dhir, B. (2015). Status of aquatic macrophytes in changing climate: A perspective. *J. Environ. Sci. Technol.*, 8, 139–148. doi: 10.3923/jest.2015.139.148.

Dunlop, T., Khojasteh, D., Cohen-Shacham, E., Glamore, W., Haghani, M., van den Bosch, M., Rizzi, D., Greve P., Felder, S. (2024). The evolution and future of research on nature-based solutions to address societal challenges. *Commun. Earth Environ.*, 5, 132.

Ferreira, V., Bini, L.M., Sagrario G., de Los Angeles, M., Kovalenko, K.E., Naselli Flores, L., Padial, A.A., Padisák, J. (2023). Aquatic ecosystem services: An overview of the Special Issue. *Hydrobiologia*, 850, 2473–2483.

Fu, Y. and Wu, Q. (2024). Recent emerging shifts in precipitation intensity and frequency in the global tropics observed by satellite precipitation data sets. *Geophys. Res. Lett.*, 51, e2023GL107916.

Houlahan, J.E. and Findlay, C.S. (2004). Effect of invasive plant species on temperate wetland plant diversity. *Conserv. Biol.*, 18(4), 1132–1138.

Karus, K., Zagars, M., Agasild, H., Tuvikene, A., Zingel, P., Puncule, L., Medne-Peipere, M., Feldmann, T. (2022). The influence of macrophyte ecological groups on food web components of temperate freshwater lakes. *Aquat. Bot.*, 183, 103571.

Les, D.H. and Mehrhoff, L.J. (1999). Introduction of nonindigenous aquatic vascular plants in southern New England: A historical perspective. *Biol. Invasions*, 1(2), 281–300.

Lind, L., Eckstein, R.L., Relyea, R.A. (2022). Direct and indirect effects of climate change on distribution and community composition of macrophytes in lentic systems. *Biol. Rev.*, 97(4), 1677–1690.

Maberly, S.C. (1981). The nature of competition between macrophytes and phytoplankton in freshwaters. PhD Thesis, University of St Andrews.

Mooij, W.M., Janse, J.H., De Senerpont Domis, L.N., Hülsmann, S., Ibelings, B.W. (2007). Predicting the effect of climate change on temperate shallow lakes with the ecosystem model PCLake. *Hydrobiologia*, 584, 443–354.

Murphy, K., Efremov, A., Davidson, T.A., Molina-Navarro, E., Fidanza, K., Betiol, T.C.C., Chambers, P., Tapia Grimaldo, J., Martins, S.V., Springuel, I. et al. (2019). World distribution, diversity and endemism of aquatic macrophytes. *Aquat. Bot.*, 158, 103127.

Nagasaka, M., Yoshizawa, K., Ariizumi, K., Hirabayashi, K. (2002). Temporal changes and vertical distribution of macrophytes in Lake Kawaguchi. *Limnology*, 3(2), 107–114.

Parmesan, C. (2006). Ecological and evolutionary responses to recent climate change. *Annu. Rev. Ecol. Evol. Syst.*, 37, 637–669.

Ribaudo C., Tison-Rosebery, J. Buquet, D., Jan, G., Jamoneau, A., Abril, G., Anschutz, P., Bertin, V. (2018). Invasive aquatic plants as ecosystem engineers in an oligo-mesotrophic shallow lake. *Front. Plant Sci.*, 9, 1781.

Rosenzweig M.L. (2001). The four questions: What does the introduction of exotic species do to diversity? *Evol. Ecol. Res.*, 3(3), 361–367.

Sand-Jensen, K., Prahl, C., Stokholm, H. (1982). Oxygen release from roots of submerged aquatic macrophytes. *Oikos*, 38, 349–354.

Scheffer, M., Hosper, S.H., Meijer, M.-L., Moss, B., Jeppensen, E. (1993). Alternative equilibria in shallow lakes. *TREE*, 8, 275–279.

Sculthorpe, C.D. (1985). *The Biology of Aquatic Vascular Plants,* 2nd edition. Koeltz Scientific Books, Kšnigstein.

Strayer, D.L. (2010). Alien species in fresh waters: Ecological effects, interactions with other stressors, and prospects for the future. *Freshw. Biol.*, 55(1), 152–174.

Szoszkiewicz, K., Achtenberg, K., Debbaut, R., Carreira, V.D., Gebler, D., Jusik, S., Kałuża, T., Karttunen, K., Lehti, N., Martin Muñoz, S. et al. (2025). Diversification of macrophytes within aquatic nature-based solutions (NBS) developing under urban environmental conditions across European cities. *Ecol. Ind.*, 172, 113331.

Talling, J.F. and Lemoalle, J. (1998). *Ecological Dynamics of Tropical Inland Waters.* Cambridge University Press, Cambridge.

Thomaz, S.M. (2023). Ecosystem services provided by freshwater macrophytes. *Hydrobiologia*, 850, 2757–2777.

Westphal, M.I., Browne, M., MacKinnon, K., Noble, I. (2008). The link between international trade and the global distribution of invasive alien species. *Biol. Invasions*, 10, 391–398.

WWF (2020). *Living Planet Report 2020. Bending the Curve of Biodiversity Loss*, Almond, R.E.A., Grooten M., Petersen, T. (eds). WWF, Gland.

Zhang, Y., Jeppensen, E., Liu, X., Qin, B., Shi, K., Zhou, Y., Thomaz, S.M., Deng, J. (2017). Global loss of aquatic vegetation in lakes. *Earth-Sci. Rev.*, 173, 259–265.

2

Reconciling Productivity, Maintaining Biodiversity and Human Well-being: How Will Tomorrow's Agriculture Solve this Equation?

2.1. Introduction

Arable farming in temperate climates has undergone major changes over the last few decades, particularly in areas dominated by arable crops. The predominantly industrial and intensive farming practices have given rise to a number of problems. Biodiversity in agricultural environments has been particularly affected. A 75% collapse in insect populations in favorable environments close to agricultural plots has been observed over a period of 27 years (Hallmann et al. 2017). Specialist bird species are in decline, and this is most marked for farmland specialists (Rigal et al. 2023). This is not just a question of protecting biodiversity, but also of its capacity to provide services, particularly for agricultural production. For example, the decline in pollinating insects and natural predators of pests leads to a drop in agricultural yields (Dainese et al. 2019).

This decline in invertebrates providing ecosystem services is linked in particular to the simplification of agricultural landscapes, i.e. the proportion of cultivated plots compared to other habitats, particularly semi-natural habitats, such as hedgerows, woodlands, grass strips and meadows. The use of pesticides is also an issue, for instance, reducing pollination services as much as they increase yields (Catarino et al. 2019a). The race for yield in this intensive agriculture is not synonymous with profitability for farmers, whose financial situation is often complicated.

Chapter written by Florian KLETTY.

Moreover, the negative effects of this type of farming are not confined to agricultural plots. The use of pesticides affects air quality, the health of nearby residents and generates conflicts with the farming community. 25% of the French population is directly exposed to pesticide treatments, and 5% is highly exposed, while most of the population live in cities (Nowak and Marliac 2024).

Furthermore, from 20–30% of waters in Europe are subject to diffuse pollution linked to agriculture, particularly nitrogen and pesticides (European Environment Agency 2021). Unhealthy soils are prone to erosion and have more difficulty managing excess or shortages of water, which contributes to mudflows, floods and droughts. The nutritional quality of agricultural production has also declined over recent decades (Davis et al. 2004; Mariem et al. 2020). This has a negative impact on access to good quality food and good human health. Finally, agricultural landscapes have become more uniform, with an increase in the size of plots, a reduction in the diversity of crops and the presence of semi-natural habitats.

In this context, this chapter sets out alternative agricultural practices and models to the intensive industrial agriculture that has now become "conventional", the benefits they can bring at different scales and the limits, obstacles and levers that exist for their development.

2.2. Alternative agricultural practices and models

2.2.1. *Agricultural practices*

2.2.1.1. *Grasslands, crops and species*

The choice of implanted species can alter agricultural practices. Grasslands cover 70% of global agricultural surfaces (Smith et al. 2016). They provide numerous agronomic and environmental services, depending on their type and management (Dumont et al. 2019). Croplands would benefit from diversification through the implementation of grasslands and from crop diversification.

While three crops produce more than half of the calories from agriculture, there are hundreds of species suitable for crop production, and about 7,000 species have been cultivated or collected for food (FAO 1997). Crop rotation diversification and rotation is thus possible and would bring many advantages (Tamburini et al. 2020; Beillouin et al. 2021a). Furthermore, perennial crops can be grown instead of annual ones to improve soil health and optimize resource use (Glover et al. 2010). Legume crops fix atmospheric nitrogen (Hartwig and Ammon 2002). Water, input, pesticide and soil work requirements vary among crops, impacting the ecological performance of the system.

2.2.1.2. *Cover crops*

Annual crops, which make up the majority of arable land, are grown on plots for a limited part of the year. In temperate climates, winter crops are generally grown in fall, winter and spring, while spring crops are grown in spring and summer. This means that agricultural plots can remain without a cash crop for a large part of the year, particularly when a spring crop is grown after a winter crop. Cover crops are plant species that grow during this intermediate period. Various objectives are sought: limit nitrate leaching and pollution to soil and water; provide nitrate or other nutrients to following crop; limit the development of weeds; limit soil erosion; improve soil health and structure; provide resources to feed livestock or for biogas production; provide organic matter to the soil and living organisms; regulate certain pests or favor their predators; favor biodiversity (Blanco-Canqui et al. 2015).

2.2.1.3. *Intercropping*

Plant diversification can be achieved through spatial or temporal association, known as intercropping or several other names such as crop associations or mixed crops. These species may be the main crop (cash crop) with other plant species (service crop) such as those used in cover crops, or several cash crops grown together (Hartwig and Ammon 2002; Gardarin et al. 2022). These crops can be grown synchronously in the field (or not). An example of synchronous intercropping can be a wheat-pea intercropping, both sown in fall, either fully mixed or with alternating strips, and harvested in early summer (Lopes et al. 2015). Performing a cultivar mixture of a given crop can also improve farming practices but the benefits are still reduced in comparison with species diversification (Beillouin et al. 2021b). Asynchronous intercropping can be implemented by the sowing of a cash crop in a service crop, then called living mulch, by the under-seeding of the service crop before the harvest of the cash crop, or by relay cropping (Gardarin et al. 2022). Relay cropping consists of the seeding of a crop in a field where another crop is already growing, so that the two crops will be present simultaneously for only part of their cycle (Tanveer et al. 2017). For instance, a wheat–soybean relay-cropping can be performed by seeding wheat in fall, rows of soybean in spring, harvesting wheat in early summer and soybean in early autumn (Duncan and Schapaugh 1997).

2.2.1.4. *Reduced tillage and no-tillage*

Ploughing and tillage, in general, destroy aggregates, compact the soil, disrupt biological communities, reduce the activity of micro-organisms and the quantity of organic carbon in soil, etc. (Plante and McGill 2002; Bronick and Lal 2005). It is possible to perform reduced tillage which consists of superficial tillage without soil inversion. No-tillage is also possible for many crops, the seeds being directly sown without tillage, and weeds being managed in a different way (herbicides, cover crops).

2.2.1.5. *Chemical pesticide and fertilizer avoidance*

Given the impact of pesticides and fertilizers on biodiversity, environment and human health, stopping their use is desirable. This can be through a total avoidance of chemical molecules, like performed by organic agriculture (see section 2.2.2.2), or by banning the most harmful ones (Geiger et al. 2010). It is also possible to limit their use, by diminishing the doses while maintaining their efficiency (Catarino et al. 2019b). Limitation can also occur through spatial disruption, with untreated field borders then called "conservation headlands" (Stoate et al. 2001).

Alternative practices can also help chemical pesticide and fertilizer avoidance. Manuring can limit or avoid the use of chemical fertilizers. Mechanical weeding can avoid the use of herbicides. Biocontrol, such as releasing beneficial arthropods or entomopathogens, limits or avoids the use of insecticides or fungicides.

2.2.1.6. *Seminatural habitats*

The diversification of habitats, ecosystem structures and ecological conditions in arable landscape is important. This can be done with different ecological structures. i) The promotion of grassed or flowered vegetation through meadows (see section 2.2.1.1), grassed fields and path verges, riverbanks or even the implementation of flowered strips in or around fields; ii) the implementation of trees at the vicinity of crops, i.e. agroforestry. This can be done by developing hedgerows around the fields, as well as by implementing trees inside the fields such as in arable alley-cropping systems (Kletty et al. 2023); iii) the conservation or restauration of rivers, marshes and ponds in the agricultural landscape; iv) the improvement of habitat diversity in existing structures, such as with embankments, dead wood and logs, leaf litter and stones (Lecq et al. 2017).

2.2.2. *Agricultural models*

2.2.2.1. *Agroecology*

Agroecology is a set of practices that aims to improve the agricultural system by creating synergies with the other components of the agroecosystem (biodiversity, local and renewable resources, local farmer's knowledge), thus taking advantage of, and trying to maximize many ecosystem services (Wezel et al. 2018).

It is therefore a model that is particularly desirable but in practice little defined or delimited, and it includes the other models discussed below. Moreover, the term agroecology refers not only to this agricultural model but also to agroecology as a science, and as an innovative farming movement (Wezel et al. 2018).

2.2.2.2. *Organic agriculture*

Organic farming is an agricultural model that operates without the use of chemical inputs such as pesticides or synthetic fertilizers, although a few mineral-based pesticides, such as copper-based products, are allowed. However, it goes far beyond that, with the prohibition of GMOs, some additives and regulation about animal welfare. Beyond a controlled and labeled agricultural model, organic agriculture is a model that aims to foster biodiversity, take care of environment, people and farmers (Migliorini and Wezel 2017).

2.2.2.3. *Conservation agriculture*

Conservation agriculture is an agricultural model whose initial aim is to conserve the soil and optimize its functionality. It is based on three main principles: minimal soil disturbance (no ploughing, reduced or no-tillage), maximization of soil cover by plants and the diversification of crop species (FAO 2017). This is a nature-based agricultural model which promotes biodiversity and related services. Thus, conservation agriculture aims to reduce pesticide use (especially insecticides and fungicides) to maintain ecosystem functions, although this is not part of the three definition pillars.

2.2.2.4. *Mixed crop and livestock farming*

Many farming areas have specialized in certain types of agricultural production, and livestock farming areas are often far from arable farming areas (Dumont et al. 2019). However, this specialization is not always desirable, and it is possible to combine crops and livestock within the same agricultural landscape. In particular, this leads to better management of nitrogen resources, with livestock effluents used as fertilizer for crops, rather than creating dependence on chemical inputs, on the one hand, and nitrate pollution, on the other. Livestock farming favors the use of temporary grassland in rotation with crops, allowing plots to benefit from several years without tillage and with permanent cover. The use or harvest of crop combinations is easier for livestock than for human consumption. In addition, livestock can interact with cultivated plots, for example, by grazing the plant cover. This provides a way to manage the cover crops, as well as a food source in lean periods. Animals can also be used to weed crops. This has traditionally been done with rice and ducks (Men et al. 2002), but is also possible with crops such as wheat using sheep, for example, enabling the plot to be cleaned and fertilized without significant negative impact on yields (Emile et al. 2011).

2.2.2.5. *Permaculture*

Permaculture is also an agricultural model, since it constitutes a set of different agricultural practices, and since it goes beyond the agricultural technic. It is a nature-based solution to live and cultivate in harmony with the territory where it is implemented (Ferguson and Lovell 2014). It requires an ecological coherence through biodiversity, resources and people, full integration in the model. It relies on organic agriculture with few or no mechanizations, and the minimum of perturbation in the production system. It means no or few tillage, crop associations, cover cropping and often association with livestock, use of perennial crops and agroforestry. The scale of such systems is therefore much smaller than other models mentioned. This allows us to go deeper in developing naturality and natural processes, but it significantly diverges from conventional systems.

2.3. Benefits of alternative practices and models

2.3.1. *Biodiversity*

Living organisms need to find food resources and protection in their environment. The agricultural practices and models mentioned above make this possible, through the compositional and structural diversification of plots and the agricultural landscape. Moreover, soil protection and the addition of organic matter favor soil organisms.

A multitude of studies indicate how biodiversity can be enhanced by these agricultural practices and models. In the case of silvoarable agroforestry, for example, existing studies looking at biodiversity conservation have all shown a positive effect (flying and ground invertebrates, plants, birds, small mammals, etc.) (Kletty et al. 2023).

More generally, crop diversification strategies (agroforestry, crop rotation, cover crops, intercropping, cultivar mixture) increase by 24% on average of the crop-associated biodiversity (Beillouin et al. 2021a). Biodiversity benefits can be seen at different trophic levels. For instance, organic agriculture or reduced tillage associated with herbicide reduction increases abundance or activity of vertebrates such as bats and birds (Barré et al. 2018a, 2018b). Thus, it indicates at the same time that such agricultural systems also improve prey species abundance, such as invertebrates.

Permaculture improves abundance or richness of plant, microorganisms, earthworms and birds (Reiff et al. 2024). Despite the knowledge already gathered, many questions remain unanswered: what is the relative impact between taxonomic groups? What parameters are affected (abundance, species diversity, community composition, ecosystem functionality)? What are the effects of the interaction of different practices?

2.3.2. *Global environment*

2.3.2.1. *Carbon storage*

Atmospheric carbon can be recovered by plants and incorporated into biomass or soils, in the form of free or bound organic matter (Jones and Donnelly 2004). Grasslands are among the systems that store the most carbon in soils, particularly when compared with croplands (Guo and Gifford 2002). The presence of trees also contributes to this through storage in the aerial and root parts of trees, as well as by improving carbon storage in soils (Mayer et al. 2022). No (or reduced) tillage improves C storage by more than 40% in four years, in comparison with conventional tillage (Fiorini et al. 2020). Cover crops increase soil organic carbon (SOC) by 0.1–1 ton per hectare and per year (Blanco-Canqui et al. 2015). Therefore, benefits of conservation agriculture for C storage in soils have been recognized in France with the label "Au Coeur des Sols", for farms in conservation agriculture, being recognized as a low carbon label ("Du Carbone Au Coeur des Sols" 2023). Meanwhile, permaculture has already shown on SOC and C storage capacity (Reiff et al. 2024). Enhancing soil carbon storage through improved agricultural practices is the objective of the "4 per 1,000" initiative, considering that an annual increase of 0.4% of SOC in agricultural area would compensate all of the annual anthropogenic emissions.

2.3.2.2. *Water quality*

A reanalysis of seven meta-analyses on 1,399 experiments shows that crop diversification, through diversified practices such as cover-cropping or agroforestry, improves water quality by 51% on average (Beillouin et al. 2021a). Such water quality mostly refers to nitrate and pesticides in water. Obviously, pesticide reduction improves water quality in terms of pesticide pollution, but soil health can also improve it through better pesticide retention or degradation. Both tillage reduction and pasture integration in crop rotation reduce nitrogen losses without impacting crop production (Walkup et al. 2020; Zhang et al. 2020).

2.3.3. *Farmers*

Agricultural production relies on the ability of crops to grow on soil. Avoiding soil loss and improving soil health is therefore necessary and sometimes actively targeted by farmers. Soil erosion and runoffs are greatly reduced with no-tillage (Soane et al. 2012; Carretta et al. 2021) and also with cover cropping (Blanco-Canqui et al. 2015). Meanwhile, in most cases, cover crops have no effect or increase yields of the following crop (Blanco-Canqui et al. 2015). Both tillage reduction and no-tillage seem to improve yields over a long period (Fiorini et al.

2020). This improves margins thanks to the economies of time and fuel brought by reduced or no-tillage.

Farmers can also benefit from increased margins and subsequent profits by pesticides or fertilization reduction. Indeed, these inputs do not necessarily improve profit, since their cost may outweigh their benefits on yields (Catarino et al. 2019b). Nitrogen can be brought by legume crops instead of mineral fertilization, which improves farmers' margins (Carof et al. 2019). Crop rotations including legumes are more financially competitive when comparing against those without (Preissel et al. 2015). While pesticides reduce pollination performance, a better pollination also improves yields (Catarino et al. 2019a).

The introduction of conservation headlands increases the services provided by wildlife and thus yields, leading to similar or higher yields, even when the non-productive areas required for their introduction are included (Pywell et al. 2015). The cost of hedgerow implementation can be balanced in only seven years thanks to insecticide savings and pollination services (Morandin et al. 2016).

In general, the improvement of resources and habitat for biodiversity increases the benefits brought to farmers with ecosystem services such as biocontrol or pollination. This can be illustrated at landscape scale with the meta-analysis of Dainese et al. (2019). This demonstrates that landscape simplification (percentage of cropland in the landscape) decreases both pollination and pest control, through a decrease of abundance and richness of pollinators and natural enemies of pests. It also showed that these pollination and pest control reductions negatively affect subsequent yields. At the field scale, the second-order meta-analysis from Beillouin et al. (2021a) examined 54,500 paired observations from 95 meta-analyses and showed that most of the diversification strategies such as agroforestry, intercropping, crop rotations and cover crops improved ecosystem services and subsequent agricultural production.

The implementation of crop associations can also improve total yields, increasing land use efficiency, along with economic profitability. This is usually measured by the land equivalent ratio (LER), which compares the required area between monocropping and diversified systems, to rich similar yields. For instance, LER of 1.3–1.6 have been found in silvoarable agroforestry system over 20 years, indicating that a 100 ha agroforestry farm with crops and walnuts intercropped produces as much as 130–160 ha of a farm where crops and walnut fields would have been separated (Lovell et al. 2018). Relay-cropping systems also permit a good LER, from 1.4 to 1.8 for three case studies (Tanveer et al. 2017). Not only can a combination of two crops improve land efficiency, but the implementation of high crop diversity – such as in the permaculture model – can as well. LER calculated in permaculture systems are 1.4 in comparison with organic systems (Reiff et al. 2025).

In addition to the services already mentioned, trees and hedges can remobilize deep-seated nutrients to the surface, play a role in regulating erosion and bad weather (wind and cold), serve as natural enclosures for livestock, enable the production of firewood, chipped wood for fertilization, materials for basketry, leaves for livestock or fruit production, etc. (e.g. Jose 2009). They are thus a good example of multifunctionality of green infrastructures.

2.3.4. *Whole society*

Improving water and air quality improves public health. In addition, the cost of treating drinking water due to agricultural pollution is very high. Improving agricultural practices therefore generates significant financial savings for local authorities. In addition, reducing the use of pesticides improves the quality of life of citizens and limits conflicts with farmers. The nutritional quality of food can also be improved. Conservation agriculture increases the micronutrient composition of vegetal foods and even meat (Montgomery et al. 2022).

The aesthetic value of agricultural landscapes must also be taken into account. Grasslands are an essential component of the most remarkable agricultural landscapes (Beudou et al. 2017). Vegetation cover, grassy strips and even flower beds, but above all the presence of trees and hedges, improve users' perception of these spaces. Indeed, the diversity of species observed, and especially the structure and heterogeneity of the landscape, are key elements in human appreciation of our environment (Lindemann-Matthies et al. 2010; Meyer-Grandbastien et al. 2020). Crop diversification, a pillar of conservation agriculture in particular, or mixed crop and livestock farming, are agricultural models that promote the well-being of rural citizens and users.

Considering the negative impact of agricultural practices on society as a whole also makes it possible to realize the benefits of improving practices. For agroforestry, for example, the reduction in the societal cost linked to sediment loss and the benefit linked to carbon sequestration make this model financially attractive, under a variety of climatic conditions (Kay et al. 2019).

2.4. Limits, hinders and levers for transition

2.4.1. *Time and uncertainty*

Changing practices induces benefits, which are expected to be higher than the related costs. However, negative impacts are rapidly occurring (soil compaction, weeds and pest colonization, etc.) while beneficial modifications in the agroecosystem take time. No-tillage soil structuring requires the development and subsequent functioning of a variety of different organisms, themselves dependent on improved environmental conditions. In comparison with ploughing, for example, yields were lower in the first few years after ploughing was stopped, but recovered and even exceeded those of ploughing control after approximately ten years (Fiorini et al. 2020; Nouri et al. 2020). The benefits of creating seminatural habitats such as flowered strips have a similar evolution. Pollination services increase with the age of the infrastructure, and continue to do so for several years (Albrecht et al. 2020). The agricultural transition phase can therefore be a complicated time for farmers, due to a transitory decline in ecosystem services and even yields. A certain amount of time is needed for a balanced and generally favorable system to be established.

In addition, there are major uncertainties surrounding both this transition phase and the viability of the targeted model. A study of various permaculture farms showed that, on average, they were more productive per hectare than usual organic farms (LER = 1.44), but the land equivalent ratio (LER) varied widely, ranging from 0.5 to 2.7, and was less than 1 in half the cases (Reiff et al. 2025). This type of variability has also been observed in other agricultural models, such as conservation agriculture (Chabert and Sarthou 2020), or practices such as cover crops (Blanco-Canqui et al. 2015). Projecting ahead thus seems difficult for farmers motivated by the transition, and many uncertainties and influencing factors are at work. More detailed knowledge of the parameters influencing this variability would help limit the reluctance of some farmers to embark on the transition.

2.4.2. *Financial cost and subsidies*

In addition to the potential losses and uncertainties associated with the transition to new farming practices or models, there are also costs associated with changes in equipment and machinery. Direct seeding or strip-tilling, for example, require specific equipment, while hedge planting can be costly, as can the launch of new activities such as livestock farming to diversify activities and the agricultural landscape. Subsidies for farmers would make such transitions possible more often.

Another parameter to consider concerns negative externalities, i.e. the negative impacts of agriculture on society as a whole. The agricultural practices and models mentioned above help to limit these negative externalities. However, farmers' choices are mostly influenced by a cost–benefit balance that takes into account direct benefits but not the reduction of negative externalities. In the case of agroforestry, for example, the financial benefits are low or non-existent if we consider only labor and machinery costs, but become interesting if we include ecosystem services such as pollution reduction or carbon storage (Kay et al. 2019).

To encourage agricultural practices that limit negative externalities and promote positive ones, three mechanisms have been described (Pretty et al. 2001): i) environmental taxes to internalize the costs associated with agricultural production; ii) the introduction of subsidies and incentive reforms, which may involve making subsidies conditional upon the implementation of certain practices or the achievement of defined results; and iii) communication and learning about the issues associated with agricultural practices, facilitating exchanges between stakeholders and enabling voluntary, joint and participatory decision-making.

2.4.3. *Complexity of ecological systems and knowledge*

The knowledge acquired and available in the scientific community is not always well known and transmitted to farmers. Communication and popularization of information is therefore necessary. This is all the more true as farming systems based on biodiversity are intrinsically more complex than those dominated by chemistry and mechanization. The beneficial effect on biodiversity of the abandonment of tillage can only be seen when combined with the limitation or the absence of pesticide use, and this for different taxonomic groups (Bai et al. 2018; Barré et al. 2018a). In a similar way, agroforestry can in some cases show positive effects when implemented following organic farming, and negative or no effects with conventional farming (Boinot et al. 2019, 2020). A review on cover crops indicates that effects on yield can be positive or negative, as a function of diverse factors such as precipitations, growing season, tillage system or age of the system (Blanco-Canqui et al. 2015). Agricultural models such as soil conservation agriculture thus appear to provide important ecosystem services for agricultural production, but pest–predator relationships are still largely misunderstood to sufficiently limit crop damage (Kletty et al. 2025). In silvoarable agroforestry, a multitude of factors that can influence biodiversity or the services it provides have been identified, but only a limited number have been studied to date, and therefore in a very incomplete way (Kletty et al. 2023). Given the complexity of the networks of interactions within ecosystems, based on the functioning of biodiversity, it is therefore necessary to significantly develop the studies carried out and the knowledge acquired, to better understand the phenomena and better help farmers face the difficulties they encounter.

2.4.4. *Customer support*

The potential power of consumers over agricultural models is significant. However, a first limitation lies in the choice and ability of consumers to pay more for products to support a particular mode of production. To overcome this, it is important to enable more sustainable practices to compete financially (see section 2.4.2). Secondly, it is currently difficult for farmers to valorize the ecological benefits and ecosystem services they promote to consumers (Dumont et al. 2019). The organic farming label is widely recognized and deployed, but this is not the case for other equally virtuous production methods. Better communication and recognition of more sustainable agricultural practices and models would therefore be necessary to encourage their development through informed consumer choice. Another possibility is to forge closer links between producers and consumers. This can be achieved through the development of Community Supported Agriculture, such as the French AMAP (Association for the Maintenance of a Peasant Agriculture) system (Wezel et al. 2018). In such short circuits, consumer choices are passed on directly to the farmer and the practices put in place, making it possible to realize the added value of the choices made.

2.4.5. *Law and politics*

Political will and the implementation of appropriate laws and regulations are needed to support and encourage the necessary agricultural transition. Policies along these lines have already been put in place, such as the greening of the Common Agricultural Policy and pesticide reduction plans. However, concrete achievements have often failed to live up to the ambitions or rhetoric (Pe'er et al. 2014; Hossard et al. 2017). More than a financial issue, the reason is often linked to the policy of allocating resources, which is not sufficient (Pe'er et al. 2014). It is now necessary to implement ambitious programs, decisions and regulations that break with the current status quo, to enable the large-scale development of another form of agriculture, beneficial both to farmers and to the environment and society as a whole.

2.5. Conclusion

To deal with the problems raised by today's predominantly industrial agriculture, different agricultural practices and models can be developed. These can bring numerous benefits in terms of protecting biodiversity and the environment, agricultural production and farmers' incomes, as well as the health and well-being of citizens. In this way, the same solutions can address different issues for different stakeholders. Although there are limits to the development of these nature-based solutions, it is important to take into account cost–benefit considerations at a global scale, rather than in a compartmentalized way. This would be a real lever for developing such promising solutions. This also reaches the One Health concept, which stresses the need to have a unified approach of human, animal and ecosystem health (OHHLEP 2021). Thus, it is through a holistic vision linking agricultural production, biodiversity and its associated services, and human well-being, that the fairest, most favorable and most effective transition possible can be achieved.

2.6. Acknowledgments

Many thanks to the reviewer who commented on this work and helped to improve this chapter.

2.7. References

Albrecht, M., Kleijn, D., Williams, N., Tschumi, M., Blaauw, B., Bommarco, R., Campbell, A., Dainese, M., Drummond, F., Entling, M. et al. (2020). Global synthesis of the effectiveness of flower strips and hedgerows on pest control, pollination services and crop yield. *Ecology Letters*. doi: 10.22541/au.158618502.29242370.

Bai, Z., Caspari, T., Gonzalez, M.R., Batjes, N.H., Mäder, P., Bünemann, E.K., de Goede, R., Brussaard, L., Xu, M., Ferreira, C.S.S. et al. (2018). Effects of agricultural management practices on soil quality: A review of long-term experiments for Europe and China. *Agriculture, Ecosystems & Environment*, 265, 1–7. doi: 10.1016/j.agee.2018.05.028.

Barré, K., Le Viol, I., Julliard, R., Chiron, F., Kerbiriou, C. (2018a). Tillage and herbicide reduction mitigate the gap between conventional and organic farming effects on foraging activity of insectivorous bats. *Ecology and Evolution*, 8, 1496–1506. doi: 10.1002/ece3. 3688.

Barré, K., Le Viol, I., Julliard, R., Kerbiriou, C. (2018b). Weed control method drives conservation tillage efficiency on farmland breeding birds. *Agriculture, Ecosystems & Environment*, 256, 74–81. doi: 10.1016/j.agee.2018.01.004.

Beillouin, D., Ben-Ari, T., Malezieux, E., Seufert, V., Makowski, D. (2021a). Positive but variable effects of crop diversification on biodiversity and ecosystem services. *Global Change Biology*, 27, 4697–4710. doi: 10.1111/gcb.15747.

Beillouin, D., Ben-Ari, T., Malezieux, E., Seufert, V., Makowski, D. (2021b). Positive but variable effects of crop diversification on biodiversity and ecosystem services. *Global Change Biology*, 27, 4697–4710. doi: 10.1111/gcb.15747.

Beudou, J., Martin, G., Ryschawy, J. (2017). Cultural and territorial vitality services play a key role in livestock agroecological transition in France. *Agronomy for Sustainable Development*, 37, 36. doi: 10.1007/s13593-017-0436-8.

Blanco-Canqui, H., Shaver, T.M., Lindquist, J.L., Shapiro, C.A., Elmore, R.W., Francis, C.A., Hergert, G.W. (2015). Cover crops and ecosystem services: Insights from studies in temperate soils. *Agronomy Journal*, 107, 2449–2474. doi: 10.2134/agronj15.0086.

Boinot, S., Fried, G., Storkey, J., Metcalfe, H., Barkaoui, K., Lauri, P.-E., Meziere, D. (2019). Alley cropping agroforestry systems: Reservoirs for weeds or refugia for plant diversity? *Agriculture Ecosystems & Environment*, 284, 106584. doi: 10.1016/j.agee.2019.106584.

Boinot, S., Meziere, D., Poulmarc'h, J., Saintilan, A., Lauri, P.-E., Sarthou, J.-P. (2020). Promoting generalist predators of crop pests in alley cropping agroforestry fields: Farming system matters. *Ecological Engineering*, 158, 106041. doi: 10.1016/j.ecoleng.2020.106041.

Bronick, C.J. and Lal, R. (2005). Soil structure and management: A review. *Geoderma*, 124, 3–22. doi: 10.1016/j.geoderma.2004.03.005.

Carof, M., Godinot, O., Ridier, A. (2019). Diversity of protein-crop management in western France. *Agronomy for Sustainable Development*, 39, 15. doi: 10.1007/s13593-019-0561-7.

Carretta, L., Tarolli, P., Cardinali, A., Nasta, P., Romano, N., Masin, R. (2021). Evaluation of runoff and soil erosion under conventional tillage and no-till management: A case study in northeast Italy. *Catena*, 197, 104972. doi: 10.1016/j.catena.2020.104972.

Catarino, R., Bretagnolle, V., Perrot, T., Vialloux, F., Gaba, S. (2019a). Bee pollination outperforms pesticides for oilseed crop production and profitability. *Proceedings of the Royal Society B: Biological Sciences*, 286, 20191550. doi: 10.1098/rspb.2019.1550.

Catarino, R., Gaba, S., Bretagnolle, V. (2019b). Experimental and empirical evidence shows that reducing weed control in winter cereal fields is a viable strategy for farmers. *Scientific Reports*, 9, 1–10. doi: 10.1038/s41598-019-45315-8.

Chabert, A. and Sarthou, J.-P. (2020). Conservation agriculture as a promising trade-off between conventional and organic agriculture in bundling ecosystem services. *Agriculture, Ecosystems & Environment*, 292, 106815. doi: 10.1016/j.agee.2019.106815.

Dainese, M., Martin, E.A., Aizen, M.A., Albrecht, M., Bartomeus, I., Bommarco, R., Carvalheiro, L.G., Chaplin-Kramer, R., Gagic, V., Garibaldi, L.A. et al. (2019). A global synthesis reveals biodiversity-mediated benefits for crop production. *Science Advances*, 5, eaax0121. doi: 10.1126/sciadv.aax0121.

Davis, D.R., Epp, M.D., Riordan, H.D. (2004). Changes in USDA food composition data for 43 garden crops, 1950 to 1999. *Journal of the American College of Nutrition*, 23, 669–682. doi: 10.1080/07315724.2004.10719409.

Du Carbone Au Coeur des Sols (2023). Label bas carbone – ministère de la transition énergétique. [Online]. Available at: label-bas-carbone.ecologie.gouv.fr/projets/du-carbone-au-coeur-des-sols-20232027 [Accessed February 20, 2025].

Dumont, B., Ryschawy, J., Duru, M., Benoit, M., Chatellier, V., Delaby, L., Donnars, C., Dupraz, P., Lemauviel-Lavenant, S., Méda, B. et al. (2019). Review: Associations among goods, impacts and ecosystem services provided by livestock farming. *Animal*, 13, 1773–1784. doi: 10.1017/S1751731118002586.

Duncan, S.R. and Schapaugh Jr., W.T. (1997). Relay-intercropped soybean in different water regimes, planting patterns, and winter wheat cultivars. *Journal of Production Agriculture*, 10, 123–129. doi: 10.2134/jpa1997.0123.

Emile, J.C., Walczak, P., Trillaud, A., Novak, S. (2011). Pâturer une céréale sans trop pénaliser le rendement grain : effet de la date d'exploitation et de l'espèce. *Renc. Rech. Ruminants*, 18, 146 [Online]. Available at: https://journees3r.fr/textes3r/20110528-paturer-une-cereale-sans-trop-penaliser-le-rendement-grain-effet-de-la-date-dexploitation-et-de-lespece/.

European Environment Agency (2021). Drivers of and pressures arising from selected key water management challenges – A European overview [Online]. Available at: www.eea.europa.eu/en/analysis/publications/drivers-of-and-pressures-arising [Accessed 10 March 2025].

FAO (1997). State of the world's plant genetic resources. Report, FAO, Rome.

FAO (2017). Conservation Agriculture, revised version. Report, FAO, Rome.

Ferguson, R.S. and Lovell, S.T. (2014). Permaculture for agroecology: Design, movement, practice, and worldview. A review. *Agronomy for Sustainable Development*, 34, 251–274. doi: 10.1007/s13593-013-0181-6.

Fiorini, A., Boselli, R., Maris, S.C., Santelli, S., Ardenti, F., Capra, F., Tabaglio, V. (2020). May conservation tillage enhance soil C and N accumulation without decreasing yield in intensive irrigated croplands? Results from an eight-year maize monoculture. *Agriculture, Ecosystems & Environment*, 296, 106926. doi: 10.1016/j.agee.2020.106926.

Gardarin, A., Celette, F., Naudin, C., Piva, G., Valantin-Morison, M., Vrignon-Brenas, S., Verret, V., Médiène, S. (2022). Intercropping with service crops provides multiple services in temperate arable systems: A review. *Agronomy for Sustainable Development*, 42, 39. doi: 10.1007/s13593-022-00771-x.

Geiger, F., Bengtsson, J., Berendse, F., Weisser, W.W., Emmerson, M., Morales, M.B., Ceryngier, P., Liira, J., Tscharntke, T., Winqvist, C. et al. (2010). Persistent negative effects of pesticides on biodiversity and biological control potential on European farmland. *Basic and Applied Ecology*, 11, 97–105. doi: 10.1016/j.baae.2009.12.001.

Glover, J.D., Reganold, J.P., Bell, L.W., Borevitz, J., Brummer, E.C., Buckler, E.S., Cox, C.M., Cox, T.S., Crews, T.E., Culman, S.W. et al. (2010). Increased food and ecosystem security via perennial grains. *Science*, 328, 1638–1639. doi: 10.1126/science.1188761.

Guo, L.B. and Gifford, R.M. (2002). Soil carbon stocks and land use change: A meta analysis. *Global Change Biology*, 8, 345–360. doi: 10.1046/j.1354-1013.2002.00486.x.

Hallmann, C.A., Sorg, M., Jongejans, E., Siepel, H., Hofland, N., Schwan, H., Stenmans, W., Müller, A., Sumser, H., Hörren, T. et al. (2017). More than 75 percent decline over 27 years in total flying insect biomass in protected areas. *PLOS One*, 12, e0185809. doi: 10.1371/journal.pone.0185809.

Hartwig, N.L. and Ammon, H.U. (2002). Cover crops and living mulches. *Wees*, 50, 688–699. doi: 10.1614/0043-1745(2002)050[0688:AIACCA]2.0.CO;2.

Hossard, L., Guichard, L., Pelosi, C., Makowski, D. (2017). Lack of evidence for a decrease in synthetic pesticide use on the main arable crops in France. *Science of the Total Environment*, 575, 152–161. doi: 10.1016/j.scitotenv.2016.10.008.

Jones, M.B. and Donnelly, A. (2004). Carbon sequestration in temperate grassland ecosystems and the influence of management, climate and elevated CO_2. *New Phytologist*, 164, 423–439. doi: 10.1111/j.1469-8137.2004.01201.x.

Jose, S. (2009). Agroforestry for ecosystem services and environmental benefits: An overview. *Agroforestry Systems*, 76, 1–10. doi: 10.1007/s10457-009-9229-7.

Kay, S., Graves, A., Palma, J.H.N., Moreno, G., Roces-Díaz, J.V., Aviron, S., Chouvardas, D., Crous-Duran, J., Ferreiro-Domínguez, N., García de Jalón, S. et al. (2019). Agroforestry is paying off – Economic evaluation of ecosystem services in European landscapes with and without agroforestry systems. *Ecosystem Services*, 36, 100896. doi: 10.1016/j.ecoser.2019.100896.

Kletty, F., Rozan, A., Habold, C. (2023). Biodiversity in temperate silvoarable systems: A systematic review. *Agriculture, Ecosystems & Environment*, 351, 108480. doi: 10.1016/j.agee.2023.108480.

Kletty, F., Deville, P., Devigne, C., Machelart, S., Benamrouz-Vanneste, S., Vanhée, B. (2025). Ground beetles in agroforestry system show functional traits and activity-density differences according to distance to trees. *Preprint*. hal-04976621.

Lecq, S., Loisel, A., Brischoux, F., Mullin, S.J., Bonnet, X. (2017). Importance of ground refuges for the biodiversity in agricultural hedgerows. *Ecological Indicators*, 72, 615–626. doi: 10.1016/j.ecolind.2016.08.032.

Lindemann-Matthies, P., Junge, X., Matthies, D. (2010). The influence of plant diversity on people's perception and aesthetic appreciation of grassland vegetation. *Biological Conservation*, 143, 195–202. doi: 10.1016/j.biocon.2009.10.003.

Lopes, T., Bodson, B., Francis, F. (2015). Associations of wheat with pea can reduce aphid infestations. *Neotropical Entomology*, 44, 286–293. doi: 10.1007/s13744-015-0282-9.

Lovell, S.T., Dupraz, C., Gold, M., Jose, S., Revord, R., Stanek, E., Wolz, K.J. (2018). Temperate agroforestry research: Considering multifunctional woody polycultures and the design of long-term field trials. *Agroforestry Systems*, 92, 1397–1415. doi: 10.1007/s10457-017-0087-4.

Mariem, S.B., Gámez, A.L., Larraya, L., Fuertes-Mendizabal, T., Cañameras, N., Araus, J.L., McGrath, S.P., Hawkesford, M.J., Murua, C.G., Gaudeul, M. et al. (2020). Assessing the evolution of wheat grain traits during the last 166 years using archived samples. *Scientific Reports*, 10, 1–13. doi: 10.1038/s41598-020-78504-x.

Mayer, S., Wiesmeier, M., Sakamoto, E., Hübner, R., Cardinael, R., Kühnel, A., Kögel-Knabner, I. (2022). Soil organic carbon sequestration in temperate agroforestry systems – A meta-analysis. *Agriculture, Ecosystems & Environment*, 323, 107689. doi: 10.1016/j.agee.2021.107689.

Men, B.X., Ogle, R.B., Lindberg, J.E. (2002). Studies on integrated duck-rice systems in the mekong delta of Vietnam. *Journal of Sustainable Agriculture*, 20, 27–40. doi: 10.1300/J064v20n01_05.

Meyer-Grandbastien, A., Burel, F., Hellier, E., Bergerot, B. (2020). A step towards understanding the relationship between species diversity and psychological restoration of visitors in urban green spaces using landscape heterogeneity. *Landscape and Urban Planning*, 195, 103728. doi: 10.1016/j.landurbplan.2019.103728.

Migliorini, P. and Wezel, A. (2017). Converging and diverging principles and practices of organic agriculture regulations and agroecology. A review. *Agronomy for Sustainable Development*, 37, 63. doi: 10.1007/s13593-017-0472-4.

Montgomery, D.R., Biklé, A., Archuleta, R., Brown, P., Jordan, J. (2022). Soil health and nutrient density: Preliminary comparison of regenerative and conventional farming. *PeerJ*, 10, e12848. doi: 10.7717/peerj.12848.

Morandin, L.A., Long, R.F., Kremen, C. (2016). Pest control and pollination cost–benefit analysis of hedgerow restoration in a simplified agricultural landscape. *Journal of Economic Entomology*, 109, 1020–1027. doi: 10.1093/jee/tow086.

Nouri, A., Lee, J., Yoder, D.C., Jagadamma, S., Walker, F.R., Yin, X., Arelli, P. (2020). Management duration controls the synergistic effect of tillage, cover crop, and nitrogen rate on cotton yield and yield stability. *Agriculture, Ecosystems & Environment*, 301, 107007. doi: 10.1016/j.agee.2020.107007.

Nowak, B. and Marliac, G. (2024). Assessment of population exposure to pesticide applications with a high-resolution map of pesticide use intensity for mainland France. *Preprint*. doi: 10.1101/2024.11.25.624818.

OHHLEP (2021). Joint Tripartite (FAO, OIE, WHO) and UNEP Statement – Tripartite and UNEP support OHHLEP's definition of "One Health".

Pe'er, G., Dicks, L.V., Visconti, P., Arlettaz, R., Báldi, A., Benton, T.G., Collins, S., Dieterich, M., Gregory, R.D., Hartig, F. et al. (2014). EU agricultural reform fails on biodiversity. *Science*, 344, 1090–1092. doi: 10.1126/science.1253425.

Plante, A.F. and McGill, W.B. (2002). Soil aggregate dynamics and the retention of organic matter in laboratory-incubated soil with differing simulated tillage frequencies. *Soil and Tillage Research*, 66, 79–92. doi: 10.1016/S0167-1987(02)00015-6.

Preissel, S., Reckling, M., Schläfke, N., Zander, P. (2015). Magnitude and farm-economic value of grain legume pre-crop benefits in Europe: A review. *Field Crops Research*, 175, 64–79. doi: 10.1016/j.fcr.2015.01.012.

Pretty, J., Brett, C., Gee, D., Hine, R., Mason, C., Morison, J., Rayment, M., Bijl, G.V.D., Dobbs, T. (2001). Policy challenges and priorities for internalizing the externalities of modern agriculture. *Journal of Environmental Planning and Management*, 44, 263–283. doi: 10.1080/09640560123782.

Pywell, R.F., Heard, M.S., Woodcock, B.A., Hinsley, S., Ridding, L., Nowakowski, M., Bullock, J.M. (2015). Wildlife-friendly farming increases crop yield: Evidence for ecological intensification. *Proceedings of the Royal Society B: Biological Sciences*, 282, 20151740. doi: 10.1098/rspb.2015.1740.

Reiff, J., Jungkunst, H.F., Mauser, K.M., Kampel, S., Regending, S., Rösch, V., Zaller, J.G., Entling, M.H. (2024). Permaculture enhances carbon stocks, soil quality and biodiversity in Central Europe. *Communications Earth & Environment*, 5, 305. doi: 10.1038/s43247-024-01405-8.

Reiff, J., Jungkunst, H.F., Antes, N., Entling, M.H. (2025). Crop productivity of central European permaculture is within the range of organic and conventional agriculture. *Peer Community Journal*, 5. doi: 10.24072/pcjournal.521.

Rigal, S., Dakos, V., Alonso, H., Auniņš, A., Benkő, Z., Brotons, L., Chodkiewicz, T., Chylarecki, P., de Carli, E., del Moral, J.C. et al. (2023). Farmland practices are driving bird population decline across Europe. *Proceedings of the National Academy of Sciences*, 120, e2216573120. doi: 10.1073/pnas.2216573120.

Smith, P., House, J.I., Bustamante, M., Sobocká, J., Harper, R., Pan, G., West, P.C., Clark, J.M., Adhya, T., Rumpel, C. et al. (2016). Global change pressures on soils from land use and management. *Global Change Biology*, 22, 1008–1028. doi: 10.1111/gcb.13068.

Soane, B.D., Ball, B.C., Arvidsson, J., Basch, G., Moreno, F., Roger-Estrade, J. (2012). No-till in northern, western and south-western Europe: A review of problems and opportunities for crop production and the environment. *Soil and Tillage Research*, 118, 66–87. doi: 10.1016/j.still.2011.10.015.

Stoate, C., Boatman, N.D., Borralho, R.J., Carvalho, C.R., de Snoo, G.R., Eden, P. (2001). Ecological impacts of arable intensification in Europe. *Journal of Environmental Management*, 63, 337–365.

Tamburini, G., Bommarco, R., Wanger, T.C., Kremen, C., van der Heijden, M.G.A., Liebman, M., Hallin, S. (2020). Agricultural diversification promotes multiple ecosystem services without compromising yield. *Science Advances*, 6, eaba1715. doi: 10.1126/sciadv.aba1715.

Tanveer, M., Anjum, S.A., Hussain, S., Cerdà, A., Ashraf, U. (2017). Relay cropping as a sustainable approach: Problems and opportunities for sustainable crop production. *Environmental Science and Pollution Research*, 24, 6973–6988. doi: 10.1007/s11356-017-8371-4.

Walkup, J., Freedman, Z., Kotcon, J., Morrissey, E.M. (2020). Pasture in crop rotations influences microbial biodiversity and function reducing the potential for nitrogen loss from compost. *Agriculture, Ecosystems & Environment*, 304, 107122. doi: 10.1016/j.agee.2020.107122.

Wezel, A., Goette, J., Lagneaux, E., Passuello, G., Reisman, E., Rodier, C., Turpin, G. (2018). Agroecology in Europe: Research, education, collective action networks, and alternative food systems. *Sustainability*, 10, 1214. doi: 10.3390/su10041214.

Zhang, Y., Xie, D., Ni, J., Zeng, X. (2020). Conservation tillage practices reduce nitrogen losses in the sloping upland of the three gorges reservoir area: No-till is better than mulch-till. *Agriculture, Ecosystems & Environment*, 300, 107003. doi: 10.1016/j.agee.2020.107003.

3

Conifers Reforestations: An Environmental Problem or a Possible Resource?

3.1. Introduction, aims and scopes

This work explores the current knowledge of conifer forests in Italy, focusing especially on artificial reforestation formations. The aim is to highlight the current critical issues and contradictions, proposing more balanced solutions. The human and ecological opportunities offered by these ecosystems are analyzed, considering global challenges such as climate change, hydrogeological balance, biological invasions, assisted migration, ecosystem restoration, conservation, natural resources use, rewilding, wood supply chain and forest management. This chapter contributes to the debate on human impact, land abandonment and species management, topics with implications and interests extending well beyond Italy's borders.

3.2. Critical appraisal of the present context

3.2.1. *Analysis of the most commonly held beliefs*

Since the 1980s, there has been an ongoing discussion regarding the "renaturalization" of artificial conifer reforestation (Mercurio 2013), even though questions about reforestation and nature conservation had already been raised well before this (e.g. AA.VV. 1962, 1975).

Chapter written by Kevin CIANFAGLIONE, Alessandro BOTTACCI, Silvano LANDI, Bartolomeo SCHIRONE, Franco PEDROTTI and Marcello DI BONITO.

The analysis of scientific literature, protected area plans, and environment and forestry project reports from the last 30 years shows how attention has progressively focused on actions aimed at reducing conifer surfaces, in particular those resulting from artificial reforestations or abandoned plantations. Such interventions are often defined as "renaturalization" (e.g. La Marca et al. 2016; Iovino et al. 2017; AA.VV. 2018) by proposals aimed at harvesting and economic valorization of these formations (e.g. Mercurio 2000; Bagnato et al. 2012); more rarely supporting more heterogeneous approaches (e.g. Ciancio 2000; Bosa and Portoghesi 2002; Nocentini 2006; Borghetti et al. 2024) outside, as well as inside protected areas. Proposals for new plantations for commercial or naturalistic purposes have been much rarer over the past two decades (e.g. Manzi 2017). Biological–ecological research, such as studies on biodiversity, spontaneous dynamics and interactions between ecosystems, remains significantly underrepresented (e.g. Tiberi et al. 2005; Sheffer 2012; Cianfaglione 2014a). In general, most of the papers, projects and field actions are strictly linked to the plans of protected areas and to European and national funding, which tend to favor the exploitation or elimination of artificial reforestations. In addition to these policies, there is a wide and fragmented network of local regulations and administrative competences, which contributes to making the Italian framework particularly complex to follow (Cantiani et al. 2018). An analysis of the literature clearly shows that, over the past 30 years, conifers and artificial reforestation have increasingly been proposed in Italy as problematic elements, or at least as formations to be avoided or removed, often within the framework of a presumed need for renaturalization. In general, the following points can be highlighted:

A) There is a need to act in these formations to eliminate them through clear-cutting or other interventions aimed at facilitating their renaturalization (their disappearance).

B) Reference is often made to the fact that the forest tree species used for reforestation (sometimes referred to as plantations or afforestation) are non-native.

C) Publications often refer to the history and motivations behind the national reforestation efforts in Italy, frequently justifying these actions.

D) Frequent judgment in the literature is that, due to the lack of silviculture treatments, aging processes and infestations by insects and fungi, many reforestation sites have reached an unstable biological equilibrium.

E) The need for more clear-cutting, more or less intense thinning, the removal of undergrowth and forest litter, delimbing and clearings is often emphasized as a preventive measure to reduce fire risks.

We have expanded our research by collecting the most widespread opinions of stakeholders (professionals, technicians, environmentalists, forests and protected areas administrators) actively involved in the conifer reforestation management activities to gather shared perspectives on the criticisms about conifers, conifer forests and artificial reforestation. These perspectives help to understand common viewpoints that, although not always reflected in the literature, they largely influence conservation, forests and land management decisions. Twenty-one key points emerged against conifer reforestations:

1) Nothing grows under conifers, and in these environments, conifers discourage the establishment of other species.

2) We need to eliminate these formations, or we need to foster their elimination to enable renaturation and leave space for natural formations.

3) They are ugly and useless; even their wood is of bad quality and useless.

4) They get sick from too many diseases, especially by the pine processionary moths, and therefore they become increasingly dangerous for human and animal health.

5) These are alien species, and therefore their elimination is positive, and it should be considered as an ecological restoration and a support for pollinators.

6) Conifer reforestations should be eliminated to increase biodiversity and activate the seedbank.

7) Conifer reforestations are ugly because they often have unnatural geometric shapes, trees are too often arranged in straight rows and on unattractive artificial terracing (called "gradonature").

8) These are artificial plantations and, as such, should be removed.

9) Conifer reforestations were made to be exploited, or to be eliminated one day, through human intervention.

10) Soils are negatively impacted by leaves and deadwood and acidified by conifers.

11) There is no life in the conifer forests.

12) Conifer reforestation must be cut down because there are political, financial and project-related incentives, as well as financial support, to do so and make it profitable.

13) They are much too susceptible to fire and are consequently dangerous.

14) Conifer reforestation must be exploited or manipulated in various ways to provide local employment and stimulate the national economy; support biomass power plants and forestry supply chains; ensure proper cleaning and management; or reduce, if not eliminate, the risk of fires.

15) Lianas and climbing herbs do not climb on conifers, not even ivy can do it.

16) Conifers used for reforestation are not native species. There are no naturally occurring conifer species in the Apennines.

17) Conifers risk infesting and destroying primary high mountain grasslands.

18) Conifer reforestations are "abiotic", meaning that there are no living organisms that can adapt to the presence of conifers, exploit them or properly decompose their necromass.

19) Birds do not nest in the crowns of conifers.

20) The living or dead biomass of conifers is noxious for soil microorganisms, aquatic living forms, amphibians and other animals and plants.

21) Reforestation efforts have always been carried out with the intention of exploitation, or, in any case, they were destined to be manipulated or removed by humans.

3.3. Discussion

3.3.1. *How accurate are these claims?*

In recent years, both the previous five points (A, B, C, D, E) and the 21 observations have emerged as the principal criticisms widely circulated among experts and other stakeholders. However, most of these criticisms directed at conifer reforestations can be seen as urban legends, exaggerations or partial truths that warrant a more nuanced, case-by-case evaluation, considering the diverse contexts in which they may arise. In any case, such negative views can often be interpreted differently, if not in an opposite light. For this reason, we have developed an initial set of categories for discussion and logical analysis, as outlined in the following.

3.3.2. *Between the necessity of undertaking reforestations and the will to eliminate them lies a profound contradiction*

The species used in artificial reforestation in Italy cannot be considered as invasive species because of their biological characteristics and also because they are among the plants most easily controlled by humans. They were planted for the restoration of forests in Italy, to promote ecological succession and renaturalize areas deeply degraded by the intensive use of the soil. They also had the task of creating protection from environmental risks, such as erosion, dune stability, windbreaks, climate mitigation, water resource protection and anti-avalanche barriers. Some reforestation was instead carried out for experimental scientific purposes, more rarely for productive purposes, especially in the critical periods of the two world wars, and after 1992, following the EEC/2080/92 regulation, which provided subsidies to farmers who interrupted agricultural crops to plant forestry crops on agricultural land. However, even in earlier periods, interventions carried out for productive purposes, such as timber production, cannot be entirely excluded (Eccher et al. 1986). Nevertheless, these cases do not represent the general trend, nor do they account for the majority of reforestation efforts in Italy. After the great deforestation due to the creation of pastures and agricultural areas, in the post-unification period of Italy, 2,300,000 hectares were further deforested with the Majorana-Calatabiano law of 1877, and other deforestations were also linked to the events of the two world wars. The Italian territory resulted in largely bare forest cover (Schirone 1998; Pedrotti 2014); even the grass cover was scarce, and in many areas, absent; this is easily demonstrated in photos, postcards and aerial photographs of the past.

Artificial reforestation was an urgent action to counteract hydrogeological instability and rebuild a productive and protective forest landscape. Many publications from the past extol its necessity, importance, values and the good results obtained (i.e. Liguori 1928; Pavari and Susmel 1961; de Philippis 1961; Vetturini 1989; Conti 1995; Scimia 2012; Sfregola 2017).

Reforestation efforts, particularly those involving conifer species, were important not only for restoring hydrogeological balance and stability, curbing erosion processes and re-establishing national forest cover, but also for contributing positively to climate regulation and fostering a healthier local microclimate (i.e. Di Bérenger 1871–1872; Pavari 1927, 1959; Pavari and Susmel 1961).

Furthermore, in the journal "*L'Alpe*" of the "*Società Pro Montibus et Sylvis*", founded by Count Ranuzzi Segni, the topics of deforestation, erosion, grazing, and the challenges and achievements of reforestation are thoroughly addressed, starting with the earliest volumes from 1903.

In the last 30 years, while recognizing their historical–cultural value and effectiveness, proposals for the elimination of these formations have multiplied (e.g. La Marca et al. 2016; AA.VV. 2018; Borghetti et al. 2024), most of the time in the name of "renaturalization", "forest restoration" or wildfire prevention.

In the past, conifers were often preferred for their rusticity and pioneering capacity: they take root on degraded soils and promote the return of vegetation and soil. These same qualities still make them strategic today, also in relation to climate change challenges. If reforestation has been well done and not strongly altered (e.g. by cutting, burning, grazing, etc.), they still work. For this reason, generalized campaigns for their eradication risk creating unjustified ecological and economic damages.

Each reforestation has its own environmental context, management history, cultural value and origin. Therefore, generalizations are misleading. There are coastal, hilly and mountain reforestations, varying in levels of human impact, management or natural evolution. Some are composed of native species but with an artificial structure, while others feature non-native species that have since naturalized. Some have been completely forgotten, while others are still managed based on fragmented local decisions, often not documented in archives. Additionally, some conifer reforestations have been established in areas where conifer forests previously existed, even if these forests were later reduced or destroyed by human activity (e.g. Cianfaglione et al. 2014). Alternatively, we can consider the case of the alpine spruces' forests, which are often considered as natural forests, when in fact they often represent ancient plantations that replaced beech, oak, ash and even hygrophilous forests.

The idea that those reforestations were necessarily destined to be eliminated by humans is not supported by any historical documents (Schirone 1998) nor by the testimonies of elderly forestry experts. These plantations were generally intended to promote vegetation succession, and provide environmental protection, with initial cultural treatments and protection against grazing, before being left to evolve according to their spontaneous course, which may vary depending on environmental conditions and human impact. In the rarer cases of plantations established for productive purposes, these can be considered like reforestation if they have been abandoned or are no longer subjected to significant human impact. Therefore, each formation should be evaluated on a case-by-case basis, avoiding schematic thinking, preconceived notions and uniform approaches. The issue cannot be reduced to a simple "natural versus artificial" dichotomy. It is necessary to recognize the complexity, evolution and diversity of the ecological trajectories that these forests, born from human choices, can express over time.

Where there is a desire to accelerate their renaturalization, targeted interventions can be implemented. However, in many cases, simply waiting allows for a more effective and less impactful spontaneous evolution. In many instances, natural processes (wind, snow, ungulates, fungi, insects, etc.) already initiate thinning and mortality dynamics, guiding structural evolution without the need for artificial interventions (Cianfaglione 2014a), fostering biodiversity more effectively, with reduced impact, minimal shocks and economic savings for the community.

Over time, these formations tend to mature and become more ecologically complex, even if they appear "unstable" from a productivity point of view, due to the marked presence of necromass, senescent trees, presence of xylophages and saprotrophs. From a biological and ecological point of view, all of this corresponds instead to an acquisition of naturalness and mature forest characters. This favors the arrival of forest specialist species, associated with mature forests rich in dead wood, undergrowth and complex structure (see Sheffer 2012). It is therefore not an ecological imbalance, but the beginning of the renaturalization process hoped for from the beginning (see Cianfaglione 2014a). Reforestations that appear poorly structured, with a scarce undergrowth, or with trees arranged too regularly and closely together, should not be attributed to the nature of conifers, but rather to excessive or overly frequent human impact, or simply because the stands are still too young. In such cases, allowing spontaneous evolution over time can lead to the development of a more complex and naturaliform structure. Similarly, it is not uncommon to hear claims that "nothing grows" beneath beech forests or that these areas are species-poor forests, for similar reasons. The only appropriate interventions should involve monitoring and, if necessary, measures to accelerate natural dynamics (see Bottacci 2020). To act in that way, it should be preferable to consider targeted reintroductions of species once favorable ecological conditions have been established. Conversely, if the objective is to maintain these reforestations, then more specific management practices are required to halt vegetation succession. In any case, even these artificial forest formations, such as other secondary ecosystems, can contribute to the cultural landscape, enriching it both aesthetically and ecologically.

3.3.3. *Native or non-native, is that really the question?*

It is often said that conifers are non-native in Italy, especially in the Apennines; or that the species used in reforestation are non-native. Some claim that conifers are harmful to biodiversity because local species are not suited to coexisting with conifers, or because local species are repelled by conifers. Occasionally, native species are even confused with non-native ones, and non-invasive species with invasive ones. A rather contradictory situation also arises in Italy, when in the Alps, spruces, firs, larches, and other conifers are often regarded as native and worthy of conservation, despite having been repeatedly planted for harvesting purposes, replacing other spontaneous forest types. Conversely, in the Apennines, conifers are frequently considered alien, or even targeted for removal, simply because they were planted, even when not intended for harvesting and in areas where these species should rightly be considered native.

Let us analyze these assertions to determine where the truths lie and whether alternative interpretations are possible.

In truth, no Italian region is devoid of native conifers, from the coastal plains to the highest mountain. Even along the entire Apennines, there are widespread conifer species; see, for example, junipers (*Juniperus* spp.). Additionally, there are other, less widespread species, either as relicts (remnants from past climatic eras) or as residuals (what remains after human elimination). In fact, there are dozens of conifer species that are endemic to Italy, including those along the Apennines.

The distribution of many native conifer species has been altered by humans throughout Italy, with areas where it has been reduced and areas where it has been facilitated. Where conifers that existed before having been replanted, these should be considered as nothing more than a reintroduction, an environmental rehabilitation or a forest restoration, as it happens with many other species, and as is recommendable, for other conifers (e.g. Corona et al. 2024). Certain conifer species may well deserve new purposefully designed plantations for experimental and conservation purposes, including ex situ efforts. Forest management, including reforestation efforts, requires cautious, case-by-case assessment, with a focus on minimal and low-impact intervention, especially in protected areas and those deserving conservation, even if not yet formally designated (Pedrotti 1987).

Many erroneous ideas that appear in the lists of collected criticisms against conifers and conifer reforestation can already fail by this point. Of course, in nature nothing can be fully restored to its original state without consequences. The disappearance of some populations and the modifications observed in others due to human impact demonstrate that we must always proceed with caution and avoid acting carelessly. For this reason, we need to be judicious in our interventions, both in implementing changes and in attempting to reverse them, as some authors propose or practice. After all, renaturalization does not necessarily have to involve erasing every trace of humans, which can remain in various forms (Cianfaglione 2024a). Another myth that must be unequivocally debunked is that the conifers used in Italian reforestation are generally non-native entities. This assertion even targets species such as *Pinus nigra* s.l. and *Pinus halepensis* s.l., two of the most widely used species in Italian reforestation and, therefore, emblematic for our argument in many respects. Primarily, these are between the relict and residual species in Italy, occurring in various intraspecific taxa. Initially, these species were reduced to a few residual stands in Italy, only to later be widely disseminated by artificial reforestation initiatives, alongside other conifer species (particularly within the Pinaceae and Cupressaceae families) that nonetheless remain natural species, or whose native status may be debatable. The best example in this regard is *Pinus pinea*, whose natural range is nearly impossible to distinguish from the areas where it has been introduced and spread by human activity over the centuries. Another important example, from another conifer family, is *Taxus baccata*, for which it is often virtually impossible to separate native stands from those of human origin (e.g. Thomas and Polwart 2003; Cianfaglione 2014b). The same applies to *Cupressus sempervirens* (e.g. Bagnoli et al. 2009, 2020), *Abies alba*, and even *Pinus nigra* s.l. and *Pinus halepensis* s.l., where the origins of certain populations remain uncertain in Italy. The more this issue is examined rigidly and in detail, the more it could easily extend indefinitely to many other species, of various types. To a lesser extent, non-native species have also been used for reforestation, but they are now historicized or naturalized. In any case, these species should now be considered typical of our landscape, even when they are clearly non-native, and therefore, they should be accepted as they are, not destroyed at all costs. Formations and plants that should now be considered in any case, as typical of the Italian landscape and not as abused in planting; at least, no more than many other species that do not get as much attention. In the case of clearly non-native species, they should be accepted like many other alien species that are now typical of our culture, as has always been done by humans, with various species, from the agricultural revolution onwards (for at least 9,000 years). This has already been successfully advocated for by Pavari (Nocentini 2010) and de Philippis (Piussi 2004).

Pinus nigra and *Pinus halepensis* are increasingly species often targeted by conifer reforestation opponents, even in Italian regions where these pines are absolutely native, or where these species have existed for millennia (Cianfaglione et al. 2021). Opponents always try to justify in some way a certain exoticism of them, or the artificiality of their formations, or the susceptibility of fire, or phytosanitary reasons: in short, everything seems to be used against conifers and artificial reforestations.

A recurring contradiction in this context lies in the tendency to focus on the administrative boundaries of Italian regions and provinces rather than on broader ecologically and geographically coherent areas. This perspective results in a clear ecological inconsistency, especially in the context of climate changes and the urgent need to address emerging environmental threats. It often happens that a species native to one region is considered non-native in a neighboring administrative region, even if it is only a few kilometers away. Conversely, species that have been harmed or removed by humans are sometimes labeled as non-native simply because they have been reintroduced or facilitated in other ways by humans. In Italy, there is a strong emphasis on classifying certain trees as "non-native" or "invasive", while other species in similar situations are treated more leniently. Why is this? For example, there is rarely concern over the planting of species that could be considered "out of place", such as holm oaks, chestnut trees, olive trees, cork oaks or downy oaks, which are frequently planted due to the "frenzied fashions" of the time (Cianfaglione 2014a). Another concern related to current views on species distribution areas is that, at times, it seems they are treated too rigidly and arbitrarily, as if they are immutable or must remain fixed as initially established. This approach is not constructive, especially in light of climate change, the growing crisis of forest diseases, the demands of wood supply chains and the cultural heritage of local communities.

Conifers were once far more widespread in Italy, and their decline is not solely attributable to climate change, but also to human impact (e.g. Paganelli 1956, 1982;

Marchesoni 1957, 1958, 1959a, 1959b; Padula 1972; Cianfaglione et al. 2014). For this reason, it becomes even more challenging to justify the tendency to label pines as species introduced by humans, or as alien, or even invasive species that should be eradicated, particularly when natural populations are still present nearby. The return of forests, including conifers, should be seen as part of a historical process and recurrence, extending even to the highest altitudes (Sangüesa-Barreda et al. 2020). This should greatly reduce, if not even eliminate, the concerns of some authors related to the rise of conifers up to the mountains.

If some people argue that the climate has changed and, consequently, these conifer species should no longer exist, then why should we direct attention to other species that have suffered or are suffering the same fate? Is it simply out of antipathy towards certain species and sympathy towards other species?

In many regions of Italy, pines from artificial reforestations are frequently mistaken for alien species, with a non-native status incorrectly attributed to them, even when they were established using locally sourced genetic material (Alpine pines in the north, central Italian pines in the center and southern provenances in the south). These pines consequently are often removed under the assumption that they are non-native, even when natural populations exist just a few kilometers away, or when historical evidence suggests that the species may have been present in the area until relatively recently. In central Italy, it is not uncommon to conflate highly autochthonous taxa, such as the Italian black pine (*Pinus nigra* J.F. Arnold var. *italica* Hochst.), with Alpine provenances such as the Austrian black pine (*Pinus nigra* J.F. Arnold var. *austriaca*). This taxonomic confusion has often been used to justify the large-scale removal or the manipulation of conifer reforestations, even in regions such as Abruzzo, where the Italian black pine is naturally occurring and strictly native (Mercurio 2015; Pedrotti 2024). This confusion has undoubtedly been compounded by the complex and often debated taxonomic history of the Italian black pine, as documented in various studies (see Bruschi et al. 2006). Over time, this taxon has been variably associated by different authors either with the subsp. *laricio* (southern Italian populations) or with the subsp. *nigra* (northern populations), due to its morphological proximity to both. Indeed, individuals may display traits aligning more closely with one or the other, or exhibit intermediate characteristics. For that reason, Gellini (1968) hypothesized that this form may have arisen through hybridization during periods in which the distribution ranges of the two subspecies overlapped, forming a transitional population between northern and southern black pines. Nevertheless, based on current research, the Italian black pine is now generally regarded as a synonym of *Pinus nigra* J.F. Arnold subsp. *nigra*, the same taxon found in the Alps (Bartolucci et al. (2018) and subsequent updates). Contrary to what is often claimed, i.e. that the black pine used for reforestation is of non-local origin, it can be said that the Italian black pine from Abruzzo is the one that should normally be ascribed to the black pine used for reforestation in central Italy. The same goes for reforestation in the north, generally done using seeds from Alpine populations, and the same goes for the south where southern origins have been used. The visions that see the black pine as an allochthonous and invasive entity in Italy remain very common, even if unjustifiable and exaggerated. Taking into account the taxonomy and autoecological characteristics of the various natural black pine populations, even if individuals from different regions were introduced into the range of other populations, they would likely be disadvantaged by differing edaphic and climatic conditions. This would promote negative selection from an early stage, thereby further minimizing the potential risks associated with genetic contamination. This significantly reduces the concerns previously expressed regarding the possible consequences of genetic introgression, particularly in central Italy (see Pellegrini et al. 2016). These formations of artificial origin, especially those located closer to inhabited areas, could function as a kind of "protective outlet", as they can be visited and used for human activities and for maintaining traditions, thereby helping to reduce human impact on more natural formations and less disturbed forests. This taxonomic confusion and misinterpretation in a native/non-native key show the limits and the drifts that are currently affecting the ideas on non-native plant species.

This confusion has been repeatedly leveraged over time as a misleading and convenient premise to undermine the value of pines and conifer reforestation, despite botanical evidence, archival records and testimonies consistently pointing to the contrary. Certainly, the most natural populations of black pine that remain today must be under conservation. However, this does not mean that artificial reforestation should be destroyed or regarded as a threat as a result. Artificial conifer reforestations should be accepted for what they are, consolidated ecosystems, without resorting to logging/tampering projects, promoting open environments (e.g. pastures) or being exploited for woody biomass purposes, neither under the pretext of reducing fire risk, nor to capitalize on funding linked to current environmental, energy, agricultural or forestry policies. Unfortunately, even native conifer taxa that are rare and deserving of greater attention and protection are deliberately removed as part of various ad hoc projects. What is concerning is that, despite acknowledging the native status of these conifers, many people and institutions still consciously support actions to remove related formations (see Mercurio 2015). It is striking that not only artificial reforestations composed of strictly endemic taxa are subjected to forest reduction or tree cutting activities, but also that these practices are promoted in protected areas, even affecting the most natural and emblematic formations (Bisbiglia 2023). Precautions should be valid for artificial reforestation of conifers and therefore should be even more valid for those more natural formations.

All conifer formations now risk becoming victims of the biases and exaggerations associated with biases, hasty judgments or the opportunism linked to businesses revolving around financing policies, incentives and projects that may justify logging, manipulation, tampering or even their elimination (e.g. Biagioni et al. 2014; Masseti 2023). This includes even the oldest, most representative conifer formations, as well as the most natural ones, even those within the core areas (the most protected surfaces) of the oldest national parks (Ursitti 2022; Bisbiglia 2023; Tassi 2024).

Black pine (*Pinus nigra*) in Italy, with special regard to central Italian reforestations, should be regarded as important, primarily for its conservation value, due to the presence of this relict and residual species endemic to central Italy. The same can be said for *Pinus halepensis, Pinus cembra, Pinus heldreichii* among others, in similar contexts, and also for *Pinus mugo* (especially in the Apennines), where attempts at reintroduction have already been made in some cases during the last decades (Fermanelli et al. 2008), though unfortunately, with poor results at the moment. For rarer or threatened species, it would be conceivable to establish new experimental and ex situ conservation plantations, even at high altitudes, as long as they are carried out with due caution and ethics.

Some authors raised issues about the upward expansion of black pine from reforestations in the Apennines (Piermattei et al. 2012, 2013). This raises another issue often mentioned: the potential danger of the expansion of species from artificial reforestation areas into the surrounding open spaces. However, for the same reason, the conifers used in these reforestations should be considered like all other pioneering species, each with their own characteristics to be studied on a case-by-case basis and potentially managed accordingly, without necessarily generalizing against them. In the highest mountains, in addition to junipers, we can find other conifers of the *Pinus* genus, such as *P. mugo*, a species that is present on almost all the large mountain massifs, including the Apennines, where it has been decreasing or disappearing from some mountains, always as a result of human activity (e.g. Sibillini Mountains). This species can largely thrive above 2,000 m asl, demonstrating that grasslands located above the upper limit of beech forests (approximately 1,800–2,200 m asl) should not always be regarded as primary grasslands, but often as secondary ones. Furthermore, with climate changes, the vegetation limits and the tree-line limits have always changed or shifted.

This significantly reduces concerns about *Pinus nigra* and other conifers potentially developing above the treeline or above 1,800 m asl, either as invasive species or as species that could pose a threat to the primary high-mountain grasslands.

Indeed, thanks to the litter produced by various native grasses (which some authors also consider invasive even if they are native species, or indicative of meadows degradation for similar reasons), along with the litter from high-altitude pines and junipers, it may eventually be possible to recreate the favorable conditions for the return of acidophilic, calcifuge and saprophytic species that have either disappeared or become rare in the Apennines, due to the erosion caused by human activities. Moreover, even at lower elevations, some argue that conifers pose a danger or are invasive, preferring more open formations (e.g. Filibeck et al. 2022).

Proposals for the elimination of conifers are often made too easily, even in areas where conifer forests are more natural, because these formations tend to expand as effective pioneer species on rocky substrates, secondary meadows and garriga. This is true even in cases where open vegetation is more influenced by human activity than the conifer forests themselves. Furthermore, rocky formations rich in succulent plants, which are often thought to be threatened by the advance of conifers, can also be found within undisturbed conifer forests, provided the right conditions exist. This highlights numerous contradictions. Conifers have the ability to recreate lost environments, integrate actual vegetation or establish new ones (e.g. Biondi and Vagge 2015; Ottaviani et al. 2015; Bonari et al. 2017, 2018). From both a vegetational and ecological perspective, these environments can be highly interesting to study and observe (Cianfaglione 2024b), and may even be considered original (Ottaviani 2017) with new plant conenoses and associations to be studied (e.g. Pesaresi et al. 2013; Tesei et al. 2021). Not to mention those less impacted artificial formations or in any case older artificial reforestations could be assimilated to the corresponding more natural conifer forests. Instead of seeing artificial reforestations as harmful to the environment, they can be seen in the opposite way, even as community priority habitats, especially when they have been left undisturbed or little disturbed for enough time. This should remove any doubt as to whether these formations should be eliminated at all costs or not. Another doubt arises on the management of *Abies* spp., in particular *A. alba*, resulting from reforestation, especially on the Apennines. Many of these specimens are extremely subject to attacks by wild and domestic herbivores and ungulates. Often, the regeneration of this species is destroyed, after years of profitable growth. This situation can be explained by two main factors. First, the removal of many protective fences has allowed herbivores to access areas where they were previously restricted. Without these barriers, grazing pressure on plants and vegetation increases, potentially affecting their growth and regeneration. Second, there is often an unbalanced focus on animal protection in natural environments, which can sometimes overlook the importance of plant conservation. While animal welfare is vital, it is equally crucial to consider the ecological balance, ensuring that both plant and animal populations thrive together in harmony.

Considerable attention must be given to certain proposals advocating for the removal of *Abies alba* specimens from conifer reforestations, deemed to be of unknown, non-local or Alpine origin, in favor of replacing them with seedlings of supposedly closer provenance. The issue is that the provenance and systematic classification of *Abies alba* in the Apennines remains uncertain (e.g. Giacobbe 1973;

Magini 1973), and in many cases, the populations considered to be of "closer origin" may in fact be no closer than the specimens used in past reforestation efforts.

These so-called "closer formations" of *Abies alba* are often geographically or ecologically more distant than the gap between artificially reforested stands of *Pinus nigra* or *Pinus halepensis* and their naturally occurring counterparts. This series of considerations highlights several contradictions that deserve further reflection. It is known how, in many so-called "natural" stands, the spontaneous regeneration of silver fir (*Abies alba*) often struggles to occur (e.g. Pallotta 1922; Csilléry et al. 2023) due to human impact consequences, or to changes in the environmental conditions of these sites, which are no longer optimal for this species regeneration. Anyway, this could possibly indicate some problems (i.e. a non-native status, poor genetic characteristics or a reduced ability to compete with other species). Why should we replace established firs from reforestations, which are self-regenerating in artificial reforestations, demonstrating a good environmental fitness, with firs of uncertain provenance and behavior? Is it to reduce genetic variability and, consequently, their resilience to climate change and other stressors, all in the name of the "racial purity"? Risking substituting them with firs that have poorer traits and a reduced ability to regenerate spontaneously? To end up with firs that struggle to compete with other species, making them less capable of natural regeneration and expansion? To increase the risks, and thus create conditions that could justify unnecessary actions today and also in the future? This reasoning applies equally to other conifers of special interest, such as *Taxus baccata.* It is important to resist the impulse to act merely for the sake of doing something, avoiding unnecessary manipulation of spontaneous systems, particularly when such interventions may yield intangible or negative effects.

It would be necessary to pursue a softer and more tolerant approach towards these artificial origin formations and towards their eventual non-native/non-local plant species, as is now suggested by authors who come from numerous experiences of study and experimentation of activities in the management and tampering of forests and reforestation, including conifers, even with non-native species (e.g. Mercurio 2024). Even among many botanists and ecologists who were once skeptical about, and critical of these reforestations, there is now a growing understanding and appreciation, especially as they have matured relatively undisturbed over time. It is necessary to move beyond simplistic or misunderstood notions of "natural" and "artificial", "native" or "non-native" (Clauser 2005; Nocentini 2010); this type of stereotyping bias otherwise risks causing problems, since it compromises our ability to make an objective evaluation (C.E.O. 2023). With regard to species that may be considered non-native, their presence could be reframed in any case as a form of ex situ conservation (Cianfaglione 2014a), paying particular attention in the case of taxa that are threatened in their native habitats, such as *Abies nebrodensis* (and other similar species), *Cedrus libani* and

C. atlantica; for example, see: https://www.iucnredlist.org/. In other cases, these presences may serve as a form of ecological compensation for the destruction or reduction of the original populations.

As for the frequently cited concerns about potential hybridization, we could paradoxically argue that the coexistence of hybridizable taxa in artificial reforestations may, over time, represent a potential advantage rather than a cause for concern. It could lead to the spontaneous emergence and natural selection of new taxa, potentially better adapted to future environmental challenges, without requiring further human intervention. Is that necessarily a bad thing?

3.3.4. *Re-evaluating soil changes under conifer reforestation*

The prevailing perception, often supported by selective literature, is that coniferous reforestation tends to have negative effects on soil properties. Common concerns include soil acidification, nutrient depletion and deterioration of physical structure. While such outcomes can occur under specific conditions, this generalization overlooks cases where conifer reforestation contributes positively to soil development and landscape functioning.

Emerging evidence indicates that conifer plantations can enhance several aspects of soil quality, particularly in degraded or erosion-prone landscapes. For example, conifer stands have been shown to increase the water resistance of soil aggregates compared to control areas such as degraded or eroded lands, pointing to improved physical soil structure and enhanced erosion control (Guo et al. 2021). In riparian zones, surface rooting, tree stumps and associated ground vegetation can also reduce runoff and trap suspended sediments.

In terms of soil carbon dynamics, it is well established that conifer litter decomposes more slowly than some species of broadleaf litter, leading to a greater accumulation of organic matter on the forest floor. Although many studies report a reduction in mineral soil organic carbon (SOC) under conifers compared to broadleaf species or previous land uses, other studies that include forest floor carbon have demonstrated a net increase in total SOC stocks under conifer plantations compared to adjacent non-forested areas (Hüblová and Frouz 2021; Jandl et al. 2021; Shah et al. 2021). While forest floor carbon may be less stable than mineral SOC, this distinction is not yet well understood, highlighting the need for further research into long-term carbon stability across soil layers and forest types.

Positive impacts of conifer reforestation on soil biology have also been documented (Mehrnoosh et al. 2022). Measures of biological activity, such as soil respiration, microbial biomass carbon and enzyme activity, are often highest under conifers when compared to agricultural, pasture or even broadleaf forest systems. For example, soils under pines inoculated with ectomycorrhizae have shown significantly improved biological function relative to uninoculated plots or standard conifer plantations (Zhang et al. 2022). These results suggest that conifer stands can create biologically active and functionally diverse soil environments, though responses vary based on management practices and site conditions.

Substrate type plays a critical role in determining reforestation outcomes, yet this variable remains underexplored. In calcareous soils, for instance, the acidifying capacity of conifers, typically seen as a drawback, could actually enhance fertility by weathering minerals, increasing exchangeable cation stocks and expanding root-exploitable soil volume. Such site-specific responses demand more tailored research, particularly in Mediterranean and montane contexts.

Further evidence from Andinopatagonia, Argentina, showed that *Pinus ponderosa* reforestation significantly increased A-horizon depth compared to steppe vegetation, where substantial soil losses were documented (Broquen et al. 2003). Similarly, research in silvopastoral systems suggests that conifer cover contributes to landslide erosion reduction, and reduces avalanche risks (Cianfaglione 2014a), reinforcing the value of tree cover in stabilizing vulnerable slopes (Liguori 1928; Pavari and Susmel 1961; de Philippis 1961; Vetturini 1989; Conti 1995; Scimia 2012; Sfregola 2017).

In high mountain environments, where soil formation is an extremely slow process (estimated at 100–1000 years per centimeter of topsoil) (Carnicelli and Costantini 2013; Stockmann et al. 2014; Chiti et al. 2024), even modest improvements in soil stability and structure under conifer plantations should be viewed as ecologically valuable. Many of the Italian conifer reforestations over a century were established on bare or near-bare substrates. In such contexts, the very presence of developed soil horizons and organic layers today points to a net gain in soil quality.

The concern that conifer litter resists decomposition and leads to problematic soil development is often exaggerated. While it is true that coniferous litter decomposes more slowly and can lead to acidic or more humus layers in other contexts, this process may actually support the re-establishment of saprotrophic and mycotrophic organisms, or plant communities adapted to nutrient-poor, carbon rich or acidic conditions, such as calcifuge and oligotrophic taxa including many Ericaceae. In Italy, for example, recolonization by acidophilic species could restore or even expand primary habitats and plant communities that have retreated due to land-use change or climate pressure. This hypothesis aligns with observations elsewhere in Europe and opens avenues for further investigation into the ecological restoration potential of coniferous soils (e.g. Mallik 2003; Pipins 2024).

Another overlooked aspect is the role of conifer reforestation in enhancing biodiversity through habitat diversification. Even if conifers stand in favor of certain species over others, this is no different from any natural or semi-natural formation, such as chestnut groves, secondary grasslands or cork oak forests, which each support distinct but complementary biotic communities. There is also an excessive tendency to contrast, in an exaggeratedly alarmistic manner, non-natural habitats such as secondary grasslands, which are often described as being threatened by conifer species (i.e. Torregiani 2025), despite the fact that they are themselves environments of anthropogenic origin, like many pine plantations. These grasslands are therefore frequently kept artificially open, or even reopened and expanded, to the detriment of both conifers and broadleaved species, through projects that are highly costly both economically and environmentally, which may consequently give rise to significant debates, disputes, or formal objections. When managed thoughtfully, conifer reforestation can contribute meaningfully to the landscape-ecological mosaic, fostering diversity in fungi, flora, micro-organisms and soil fauna.

3.3.5. *Susceptibility to fires, climate change and pathogens*

Wildfires in Italy are increasing both in frequency and intensity. While this trend can be partially attributed to climate change, a significant role is also played by human activities and the artificial nature of many vegetation systems (e.g. Frate et al. 2018). Although wildfire occurrence is often correlated with vegetation cover, it should also be examined in relation to economic interests associated with fires, as well as forest management practices, vegetation dynamics and ecosystem processes. This highlights how much remains to be investigated and understood regarding these systems, particularly concerning post-disturbance regeneration following fires, storms, pathogen outbreaks and other stressors.

Not all trees are destroyed following catastrophic events such as wildfires (Fernandes et al. 2008; Niccoli et al. 2019). In fact, individual trees or entire patches of vegetation may remain unscathed, and even fire-damaged individuals can survive for several years, continuing to produce seeds, as observed in various conifer species. Some fire-adapted (pyrophytic) species such as *Pinus halepensis* and *Cupressus arizonica* (= *Hesperocyparis arizonica*) are capable of generating dense and extensive post-fire regeneration. However, these regenerating cohorts are often threatened by grazing pressure from both domestic livestock and wild herbivores, and frequent wildfires. This cancels out, or at the very least drastically reduces, the concerns that have at times been raised among people involved in managing these formations, regarding an overgrowth of certain conifer species following wildfires. It is enough to let them evolve freely, allowing natural selection to take its course, possibly protecting the areas from herbivores.

Recurrent fire events can progressively deplete the seed bank, even for conifers that are more prone to fires, ultimately driving substantial vegetation regressions. This lends support to the hypothesis that anthropogenic fire regimes, particularly those historically implemented to improve grazing conditions may have played a significant role in the long-term decline of conifer forest ecosystems in fire-prone regions. After catastrophic events such as fires, pests or storms, even when unforeseen, it is not uncommon for the final blow to be delivered by human actions in the forest. Alternatively, we can see healthy forests completely destroyed or heavily impacted by human activities aimed at preventing the risks of forest catastrophes. This ideological approach does not solve the problem; it merely eliminates it, warranting more thoughtful reflection, reduced impulsiveness, with greater ethical and deontological considerations. Conifers, although vulnerable to fires, should not be eliminated for this reason. Fire risk management must focus on prevention and protection strategies rather than actions that could damage the ecological structure of forests. Furthermore, it is crucial to avoid interventions that reduce biodiversity or damage the self-healing capacity of forests, such as the removal of dead wood, litter, undercover or indiscriminate cutting (Landi et al. 2023). Beyond the clichés on conifer formations, everything burns when set on fire. Destroying them merely because they are threatened (e.g. by fire, pathogens or climate change) is illogical. It would be akin to demolishing works of art simply because there is a risk that vandals may damage them. Clear-cutting operations, thinning or the opening of reforested areas, including those carried out for wildfire prevention, may become potentially hazardous over the years, as they can inadvertently promote fire spread by increasing the abundance of flammable grasses, shrubs and subshrubs. In contradiction, deadwood and litter are often considered hazardous fuel loads; however, even if this were partially true, their removal is difficult to justify, as it can cause significant environmental degradation and contribute to the artificial simplification of these ecosystems. Moreover, coarse woody debris and forest litter generally retain high moisture levels, and undecomposed wood remains dense and compact, making it less susceptible to ignition during wildfires. In some instances, images have circulated of trees allegedly consumed by fire, when in fact they were hollow individuals merely blackened by flame exposure.

More attention needs to be paid to the targeted suppression of residual embers smoldering under the ashes, which have been increasingly recognized in recent years as a major cause of subsequent fire re-ignition. Furthermore, there has been a recent and concerning increase in attributing wildfires to natural causes such as lightning, particularly in cases where the actual origin remains uncertain. More effective fire suppression strategies are urgently needed, along with a critical re-evaluation of current policies and protocols, particularly those that limit night-time firefighting and the control of slow-burning fires (embers and under-ash embers).

The experience gained in Italy shows us that everything can burn, and everything burns during fires, even coppices, beech tall forests, and vineyards, olive groves and other cultivated fields. Thus, conifer formations cannot be blamed for their susceptibility to fires or to other threats, nor can prevention efforts be justified by eliminating them or destroying everything.

If conifers and related formations are deemed vulnerable to fires and other dangers, their potential fragility should prompt us to protect them even more, rather than eliminate them, as is often the case with other species and formations considered fragile or threatened. A contradiction lies in the treatment of other culturally significant artificial plantations, such as chestnut groves, in Italy. These plantations are often considered valuable forests worth preserving and conserving, despite the fact that chestnut trees are extremely fragile, highly susceptible to various problems and diseases, and often non-native or of uncertain origin. Meanwhile, conifer formations are frequently viewed negatively, with growing efforts focused on their removal. How can we apply such contrasting criteria to these vegetation formations that both hold high cultural and environmental values?

Both preventive and post-fire activities can unintentionally encourage those with an interest in forest burning. Policies and strategies must take this into greater consideration than they do today. These strategies should focus on continually improving the efficiency of detection, prevention and firefighting techniques, with minimal invasiveness, while discouraging those who stand to benefit from forest fires or from the extinction of wildfires. After all, the Italian national registry of areas affected by forest fires is unreliable, relying too heavily on local authorities that are often inefficient (Mandara 2022). The Italian situation unfortunately today groups together various circumstances that can end up favoring "arsonists".

Promoting the natural growth and regeneration of forests should be recognized as an important fire risk management strategy (Taylor et al. 2020).

Regarding the susceptibility to fires, pathogens and climate change, the idea of eliminating these formations and the conifer species should be reviewed, as interventions often end up eliminating plants that have survived or could survive such adversities, effectively removing both the regeneration and mature trees that could continue to produce seeds or have developed, or could develop, resistance or immunity.

Similarly, thinning interventions carried out due to the presence of processionary caterpillars (e.g. *Thaumetopea pityocampa*) often risk opening up the ecosystems, creating better conditions for their proliferation or for the introduction of new problems, potentially exacerbating existing issues. Rather than exacerbating these issues or resorting to alarmism and sensationalism, it would be more constructive to explore alternative strategies, including approaches aimed at fostering acceptance when unavoidable (Cianfaglione 2022).

Alternative approaches to pests are often feasible, requiring neither biomass removal, nor strong human impact, nor detrimental modifications to forest structure, but instead focusing on the (re)introduction or population reinforcement of strategic species, as advocated for by Pavan (1981) and Panzavolta and Tiberi (2001).

Therefore, we should act and think differently, adopting more sustainable and thoughtful approaches that not only consider immediate emergencies, but also the long-term effects on the ecosystem and its ability to self-regenerate.

One of the main concerns regarding the implementation of drastic preventive measures is that such actions, although driven by the urgency to counter a potential threat, may ultimately cause damage equal to or even greater than that caused by the threat itself, particularly when the threat has not yet fully materialized. In any case, a significant risk is the premature removal of specimens that could be resistant, immune or capable of adapting over time. This could lead to the loss of genetic diversity and reduce the adaptive potential of populations, both of which are critical for responding to future environmental changes and emerging ecological or phytosanitary challenges.

Avoiding exaggerated alarmism is in any case essential. This situation highlights a striking contradiction: for some, these formations are considered so fragile that they should be eliminated, while for others, they are seen as too invasive and resistant and should therefore also be eliminated. This clearly points to a fundamental inconsistency in the way these ecosystems are perceived and managed.

3.3.6. *Renewal and biodiversity issues in conifer reforestation*

In our experience, generally, we observed very good regeneration of conifers, with even better results in broadleaf species, as well as in understory. The more undisturbed these reforestations are, the more remarkable the regeneration of advanced successional species becomes. Trees tend to grow mono-stem, straighter and taller in these areas, whereas in open vegetation successional stages, they often develop in a more shrubby and multiple-stem manner.

In the literature, we observed contrasting examples and positions on this matter, following the reforestations age (Tesei et al. 2021) and human impact (i.e. in more anthropized environments and in those that are less impacted by human intervention), as well as following fire events and other catastrophes.

The elevation does not seem to be very significant (Allegrezza et al. 2020), and this can be explained by the pioneering nature of this vegetation or by the type of soil or by the soil history that is being restored. All this demonstrates how much depends on the context and conditions of the reforestation, as noted earlier, and how we need to study these formations.

We observed that, in general, under conditions of minimal or no human disturbance, regeneration could be considered a constant, particularly in terms of broadleaf trees (e.g. Cianfaglione 2014a), even after fires. However, trees and regeneration are often threatened by pressure on grazing animals from both wild and domestic species.

In these cases, the removal or lack of maintenance of the planned shelter fences has not always proved to be the right choice. Based on the conditions and the cases in question, the use of fencing (implemented in the past) to protect reforested areas from large herbivores (mainly domestic as well as wild ones), while still allowing access for people and other animals, should be reconsidered.

Following forestry interventions, a mixed regeneration of conifers and broadleaves was also observed. It remains unclear how and to what extent these interventions are positive and effective in the restoration and natural regeneration of artificial conifer plantations (Mercurio et al. 2009). This confirms the need to study conifer reforestations, even in this regard. Many actions may be necessary, rather than focusing on the elimination or manipulation of these reforestation efforts.

After environmental catastrophes in forests, particularly after wildfires, the situation should become more complex (e.g. van Gils et al. 2010) and variable, depending on the specific case. It is also evident that large openings can create areas of risk for the colonization of alien species (Ricci et al. 2024); the same risk is possible after thinning, clearing and cutting activities. Furthermore, the effects on the soil are variable, although they tend to be more significant in the understory and litter layer (Marfella et al. 2023).

A contradiction can be observed when, in some cases, regeneration is described as scarce, while in other cases it is considered abundant, to the point that skeptics may view it as invasive or even dangerous. This underscores how, too often, regardless of the state of regeneration, it can always be interpreted negatively, to the detriment of these reforestation formations.

Regeneration, as well as functionality and structure, can also be threatened by the excessive use of forestry and environmental interventions, and other types of human impact, which can compromise natural regeneration and the succession that has taken place, simplifying and impoverishing the vegetation structure, promoting the desertification (increase in aridity and temperature) or the entry of dangerous species, and water pollution. These phenomena have been observed in similar contexts, in various types of ecosystems (see Sender et al. 2017). A more prudent approach, based on monitoring, on the precautionary principle, and on protection, would be more effective in ensuring the long-term conservation of these forest formations.

In addition to the natural regeneration of conifer species, it is evident that the success of reforestations that have been left to evolve spontaneously lies in their ability to promote secondary succession, making a “forced succession”. This process enables ecological facilitation phenomena thanks to the formation of bio-groups, in geobotanical terms. These reforestations have also played a crucial role for the same reasons in soil formation and development. Moreover, such ecosystems have proven to be valuable for wildlife: they support species such as squirrels, even in highly urbanized or anthropized environments.

The critique often directed at conifer reforestations, namely that they should be removed to promote spontaneous vegetation, seed bank activation and support for pollinator populations, lacks a strong ecological foundation. Conifers can in fact provide valuable resources for some plants and pollinators; for example, even bees can use conifers to make honeydew and propolis. Furthermore, not all spontaneous, non-coniferous plant species rely on animal pollination. Each plant species can facilitate one or more species instead of others; it is the existence and the diversity of plants that can support a broader and more resilient biodiversity network directly or through cascading ecological interactions. This highlights the need for a more nuanced and comprehensive approach to forest and ecosystem management.

Conifer formations can be defined as environments that have enabled many other forest-related species, previously rare before artificial reforestation, to become more common. In less disturbed habitats, bears are often drawn to these forests’ because of food availability and their preference for anthills. Predators are attracted to conifer formations in search of prey (e.g. micromammals and insects), which flourish in these forests, particularly when left enough undisturbed. Ungulates prefer rubbing against conifer trunks rather than those of broadleaved species, and even bears have been observed using them in the same way. Conifer resin can be beneficial for the health and well-being of these animals. It is not uncommon to find wild animals, even the rarest and the most remarkable species, in areas where there are conifer reforestations that have not been excessively disturbed by humans (e.g. Riva 2023).

Reforestations have demonstrated ecological success even where non-native species, such as *Pseudotsuga menziesii*, *Cedrus* spp. *Cupressus* spp., and others, have been used, consistently performing according to the same ecological mechanisms (Pavari and de Philippis 1941; Ciancio et al. 1984, 1997; Cianfaglione 2014a), even when established under the most rural or synanthropic conditions. Remarkably, such formations are increasingly disregarded or even no longer considered forests, following recent Italian policy trends supported by a sector of the national scientific community (Abrami 2020), thus contradicting both the very notion of reforestation and the original reasons behind their establishment. These trends are deeply concerning, as they risk facilitating speculative developments and the appropriation of natural resources, leading to reductions in forest quality, woody cover, and the associated processes of land grabbing and soil consumption, in line with recent Italian policy orientations in this field (Abrami 2020).

According to our experience, conifer reforestations are generally rich in plant, animal and fungal diversity, with a site-specific variability. The presence of those species largely depends on site characteristics, the environmental quality of the surrounding context, the age of the trees, the reforestation site history in relation to human activity, and the current level of disturbance affecting them. Various studies, even when preliminary (e.g. Tiberi and Panzavolta 2000; Bosa and Portoghesi 2002; Cianfaglione 2023; Châline et al. 2024; Kletty et al. 2024; Mendicino et al. 2024; Pica et al. 2025), undoubtedly highlight that conifer reforestations, including those that are not very old, established after the 1950s, can develop high levels of biodiversity, especially when they have been left relatively undisturbed over the past four decades. This includes not only a remarkable fungal component (Leonardi et al. 2017), but also a rich assemblage of plant and animal species, with the occurrence of taxa typically associated with mature forests and environments rich in deadwood. Such findings highlight the potential of these formations to serve as ecologically valuable systems, warranting greater recognition and more conservative and nuanced management strategies (Cianfaglione 2023, 2024). Even in this context, however, there are papers that could offer different, if not contrasting perspectives about biodiversity in conifer formations (e.g. Selvi et al. 2016). This suggests that context, human disturbance, necromass and structural quality may exert a stronger influence than the mere presence or absence of conifers. In any case, this underscores how much remains to be studied and analyzed in broader aspects, emphasizing the need for a more contextualized and case-by-case understanding before making exaggerated, or overly simplistic and too generalized decisions regarding the management of conifer vegetation.

A similar variability can be observed with regard to fungal diversity. Although an extraordinary presence of fungi, often of edible interest, is easily observable in conifer stands (including artificial reforestations), an analysis of the literature reveals that alongside various authors who confirm this pattern (e.g. Leonardi et al. 2017), a study suggests that forest thinning and other invasive interventions may enhance mushroom occurrence (Cantiani et al. 2021). Conversely, another study has found no significant effect of such activities (e.g. Maccherini et al. 2021). These contrasting findings highlight the need to avoid exaggerated generalizations and underscore the importance of further research with a deeper and wider regard for the ecological dynamics and biodiversity of these environments. A broader range of case studies need to be investigated through long-term research in order to deepen our understanding in case-by-case contexts. To achieve those goals, it is essential not only to halt the widespread elimination, but also to limit the disturbance of such formations, thereby enabling accurate and long-term ecological monitoring.

It is also important to study the variation of biodiversity from simplified structures to increasingly complex structures (see Zhu et al. 2024). It is not solely the age of those formations that influences the expression of biodiversity and functionality in conifer reforestations; rather, the key factor is the low level of anthropic disturbance (both in terms of frequency and intensity) over time. Therefore, we cannot use cases where these reforestations show poor undergrowth or limited biodiversity as a reference. Such cases should be analyzed individually, allowing them to evolve naturally in the least disturbed way possible, until they reach maturity and functionality, such as those that have already succeeded. This line of reasoning also alleviates concerns about the presence of non-native species in artificial reforestations carried out in the past.

The study of the ecological dynamics of conifer plant formations is indispensable for understanding ecosystem functionality, and imagining and predicting future trends in forests, given their faster development and wood decomposition patterns compared to hardwood forests, similarly to what can be observed in woody formations in wetlands (see Bănăduc et al. 2022).

3.3.7. *Criticisms on the aesthetics of reforestation projects*

A further criticism of reforestation projects, particularly those involving conifers, is the regular geometric shapes often found in these forests, with clearly defined rows, steps, straight lines and right angles, and trees planted too closely together. This criticism is rooted in valid concerns; as such, artificial arrangements often disrupt the natural harmony of the landscape, appearing both visually jarring and incongruent with the surrounding environment. From this, we can learn valuable lessons for future reforestation projects, especially when the goal is forest restoration. By emphasizing less geometric regularity and allowing for a better spatial arrangement, we could enhance both the ecological function and the visual integration of reforestations into their natural surroundings.

However, above all, we must urgently learn from this to prevent the widespread practice of forest harvesting in Italy, such as in high forests or coppices, which disrupts the structure of the forest cover and creates geometric shapes, circles, straight lines and right angles. These cutting patterns remain visible for decades, leaving a negative aesthetic impact, such as scars that disfigure the landscape far more than even the most unsightly artificial reforestation projects. For the same reasons, we should also avoid the construction of roads, ski slopes and other infrastructures that create large, geometric, regular and linear openings in the landscape. Similarly, we should be cautious about firebreaks, often advocated by some, despite the fact that they are difficult and excessively costly to manage efficiently, and their unattractive and unnatural aesthetic impact on the landscape. Additionally, the expansion of wind turbines or photovoltaics should also be carefully considered for the same primary reasons.

Let us not forget, however, that sometimes the unnatural shapes of artificial conifer reforestations have created aesthetically interesting scenarios at a landscape level, with contrasting shapes and colors that vary throughout the seasons, and which can now be defined as typical and historicized in the local landscape.

It is also important to note that some reforestation projects have been intentionally designed with unnatural shapes to create specific scenarios and patterns, much like a painting or other works of art. One of the most notable examples is the famous reforestation shaped like Italy, located in the surrounding mountains of the “Piana di Castelluccio di Norcia”, in the Sibillini Mountains, central Italy. This unique reforestation has become a major attraction in the area, drawing both local and international visitors.

Criticism is sometimes directed at conifer reforestations, based on the idea that their evergreen nature makes them aesthetically unsuitable in areas dominated by broadleaf species. This argument is not particularly convincing, considering the presence of numerous evergreen broadleaf trees, as well as formations of grasslands, junipers and other spontaneous conifers. The Mediterranean Macchia, such as the entire Mediterranean forest Biome, naturally includes both broadleaf and conifers. Reforestation with evergreen species contributes to the aesthetic and chromatic richness of the landscape, especially during the foliage seasons. In some cases, particularly at higher altitudes and in the Apennines, *Larix decidua*, a deciduous conifer, has also been used in reforestation programs, adding further nuances to the chromatic variety and creating distinctive landscapes in these mountainous areas. Unfortunately, such reforestations have often been removed in recent years, even though larch is one of the species that has the greatest difficulty regenerating spontaneously in these Apennine environments.

With snow, conifers and all types of conifer forests offer picturesque views that are unmatched by broadleaf forests, contributing to landscapes that are often more vibrant, colorful and beautiful; showing that each forest, however, possesses its own unique charm, dignity and importance.

3.3.8. *From one extreme to another; considerations on the weak and strong points, between past and present*

Generally, these reforestations were not intended to become timber plantations, nor were they meant to be subject to constant intervention (Schirone 1998). They were only managed during the early years, with attention given to protecting the young plants and ensuring their regeneration and undergrowth through fences specifically designed to protect them from the predation of domestic and wild herbivores, while still allowing free movement for humans and other animals.

Over time, these reforestations have become very important landmarks and subjects of local landscape and cultural identity. For a long time, they have been the only forests or woodlots growing in certain areas, even today. For the Italian Territory, they are often the closest thing to a true forest, with thick trunks, tall trees and abundant dead wood (Cianfaglione 2014a). This is far from negligible in a country such as Italy, where most forests are impoverished in both structure and productivity due to overexploitation or youthful growth, to the point that even today the majority of Italian high forests are largely still made up of conifer forests, plantations and reforestations.

When, in other cases, these conifer forests have been surrounded by returning broadleaf woodlands, they remain taller forest parcels, standing out with a better structure if compared to the surrounding broadleaf forests, which are either too young, more shrub-like than tree like shape (i.e. clump-forming, low-branching), or overexploited. What benefit is there in eliminating these conifer forests to make room for more broadleaf forests, especially if these new broadleaf forests will eventually be reduced to the same condition as the others? In that case, would it not be better to keep the conifer forests for environment purposes?

These reforestations have played a significant socio-economic and cultural role, especially for local communities, providing employment opportunities for local workers and serving as an important avenue for social advancement, including for women, and creating forests of high historical and cultural value, contributing to our human heritage. These efforts also supported the national economy and helped families navigate the challenging post-war and reconstruction periods when the economy had been devastated.

Unfortunately, artificial reforestation efforts have been marked by missteps and exaggerations. Even though plantations have increasingly consisted of broadleaf trees, these reforestation projects were often carried out where there was no real need, driven solely by available funding policies.

This mirrors what happens today: driven by the pursuit of funding, projects and incentives linked to current policies, natural forests are often altered or harvested, even when there is no real necessity; and also, those ancient conifer reforestation efforts are often destroyed, again, for financial gain, despite a lack of genuine need for the forests or the environment. These activities are frequently framed as conservation, protection or forest management, while in reality they destroy local heritage and are often justified under the guise of benefiting local populations or promoting the well-being and development of disadvantaged areas.

In the latest wave of reforestation, it became increasingly common to find plantations with poor outcomes, which was masked behind exceptional events (e.g. pathogen attacks, fungi, droughts) or by the incompetence of the workforce.

During the last conifer reforestation period, the greatest failures have been linked to excesses and exaggerations. Furthermore, there has been an increasing tendency to entrust reforestation work to local authorities and private companies, which has led to increasing conflicts of interest. These types of interventions are still recognizable today because we can observe a few, if any, well-developed trees, with the only remnants consisting of steps or holes made, that are now more or less colonized by secondary meadows or garriga vegetation. While this secondary vegetation has often regenerated successfully, sometimes even richer in Orchidaceae than before, there is no trace left of the once-abundant mushroom beds, which were highly productive and a key resource for locals. In the worst cases, planting has become more about exploiting available funds than achieving genuine ecological or environmental objectives.

Before any reforestation work, a careful floristic and vegetation analysis of the intervention site should be foreseen, and a record should be kept. This was foreseen in the intervention protocols of the former State Forestry Agency (abolished in 1977), whose archives have since been lost. This has not always been successful. An interesting example in this context can be found in the reforestation efforts in the L'Aquila area (Province of L'Aquila), where, alongside the conifer reforestation, *Goniolimon tataricum* subsp. *italicum* was later discovered in open conditions. In this case, we may consider whether the reforestation preparation activities could have damaged areas inhabited by this endemic and unique Italian subspecies. On the other hand, we could also consider the possibility that this species was not present at that time. Certainly, all of these are possibilities; unfortunately, there is no historical evidence to clarify this matter. It is also possible to consider that this species could be a relic and a remnant plant from past eras; or alternatively, it could be considered as a species that arrived thanks to human activities. Can we imagine that this species will be able to find spaces within the forest? How would it be better to act to conserve that territory? Could we instead think of eliminating conifers, broadleaf trees and any shrubs and other competing herbaceous plants, creating a completely oriented, gardened and artificial environment? Today, we can only hypothesize and seek further evidence, speculating whether the population is threatened by the spread of the conifer forest, broadleaf forests, environmental changes, or human impact.

In any case, each hypothesis must acknowledge the current state of the environment, avoiding one-size-fits-all solutions and instead promoting nuanced, balanced actions. This means applying the precautionary principle, managing some areas in a highly targeted way, others left to natural evolution and others with mixed approaches. This strategy increases the chances of preserving at least part of the population, and at the same time, it avoids creating a strong impact on other species. At the same time, this allows us to try to understand how this species may or may not integrate into the forest in the future. In any case, it could provide reference conditions for future scientific analysis.

Looking ahead, we must avoid deforestation, as well as be cautious with new plantations, especially in open areas where this herbaceous species still survives. Likewise, land clearing, urbanization, or livestock and wildlife overgrazing must be carefully limited to prevent the destruction of these habitats. This highlights the importance of avoiding excessive planting schemes and ensuring a thorough understanding of local ecosystems before undertaking artificial reforestation, conservation plans or other activities. Unnecessary reforestations and tree plantations in the past have unfortunately often overlooked not only pre-existing herbaceous vegetation, but also existing woody vegetation. In many cases, reforestation was carried out in areas where trees and shrubs were already present, or where clear signs of a more advanced stage of natural succession were evident (e.g. Biondi and Ballelli 1973). To make room for new plantations and take advantage of available funding, existing shrubs and young trees were frequently removed or damaged, and in some cases, mature trees were felled or extensively pruned.

In other instances, we have seen highly thermophilic species such as *Quercus ilex* or *Pinus halepensis* planted in sites that are too high, cold or humid for them, while less thermophilic species such as *Pinus nigra* have been placed too low, where more xerophilic or heat-tolerant species would be more suitable. These mismatches have led to poorly established plantations, or trees that are more vulnerable to pests such as the pine processionary moth and other environmental stressors.

Today, perhaps influenced by current policies, incentives and economic interests focused more on agriculture and forest exploitation and biomass production, rather than planting. The trend is the same, but has in many ways reversed the direction, resulting in a different set of excesses and distortions. Similar concerns persist today, as financial incentives continue to drive forest "management" or exploitation, frequently under the guise of environmental care, often resulting in harmful interventions for the environment. The same is true when we can see woody formation destroyed to obtain pastures or arable fields; or when secondary meadows rich in orchids or other valuable species are destroyed to make way for agricultural expansion, driven often by incentives and subsidies, following current policies, raising a major serious ecological concern.

Paradoxically, today, some non-native broadleaf species are still permitted for reforestation purposes, while native or regionally characteristic conifers remain prohibited. The bans on planting conifers should certainly be reconsidered, even in mountain reforestation, especially when the species in question are native or typical to the local area. These regulations, which have been increasingly enforced by various local authorities and protected areas in recent years, deserve thorough re-evaluation. At the same time, we should rethink our current approach to environmental and forest management, embracing greater caution and more ecological awareness, adopting a more balanced perspective on conifer reforestation. Once regarded as widely beneficial (Clauser 2005) and as ideal places for walks, picnics and therapeutic outings, conifer reforestations were often recommended, especially by local doctors, for people with respiratory and skin conditions, due to the perceived benefits of their balsamic vapors. These traditional uses have gained renewed significance in light of recent studies (Meneguzzo et al. 2019; Bach et al. 2020; El Omari 2021; Donelli et al. 2023), urging a more thoughtful approach to conifer reforestations. Beyond timber, these forests offer opportunities not only through their secondary metabolites, such as essential oils (Benelli et al. 2017; Rosato et al. 2018), for wellness, wellbeing and human health, but also for gastronomy (Kolosova 2017), resination (Craveri 1915), tourism, recreation, and the harvesting of mushrooms, truffles (Baldacci 1948; Granetti et al. 2005; Leonardi et al. 2017) and other non-wood forest products. These uses align closely with the goals of hydrogeological protection, carbon storage, environmental education, biodiversity preservation, and the continuity or revival of traditional practices, while potentially fostering new ones.

Such reforestations hold not just ecological and economic values, but a deep human significance, forming part of the cultural landscape and collective memory of many communities. They are linked to personal and local histories, often reflected in official toponyms or local unofficial place names. Many view them not only as natural heritage, but also as a form of "family heritage", planted by grandparents or great-grandparents. In Italy, conifer reforestation is sometimes called "Bosco dell'Impero", a term recalling the extensive reforestation initiatives undertaken during the era of the Italian Empire, both on the mainland and in its colonies, to enhance the environmental quality of the territories. In former Austro-Hungarian territories, this designation could reflect the historical association first with the Austro-Hungarian Empire, which initiated some plantations there, and later with the Italian Empire, which continued or established new ones after their unification into Italy. Almost every city and village in Italy has its own reference forest originating from a conifer reforestation, and this is even more evident throughout the Apennines.

The "Bosco della Città" of Rovereto is one famous example of how, even in urban or peri-urban areas, conifer reforestations can be often removed, even though clear cutting, in the name of renaturation (Marsili 2013). Is all of this truly necessary, even in such contexts, without risking a senselessness or certain degree of contradiction?

3.3.9. *Research needs*

There is an urgent need for more site-specific, long-term research to fully understand the multifaceted effects of conifer reforestation on soil systems and biodiversity. Areas for further investigation include:

– the stability and long-term turnover of carbon in the forest floor versus mineral horizons;

– the interaction between soil parent material, forest species and nutrient dynamics;

– the interaction between conifer formations and other ecosystems;

– long-term monitoring of soil, forest structure and ecosystem spontaneous dynamics;

– the balance between soil acidification and mineral weathering, particularly in calcareous or erosion-prone environments;

– biodiversity outcomes at the soil and understory levels across different conifer systems;

– the potential role of conifers in re-establishing lost or declining ecological niches, especially in montane and subalpine landscapes;

– in parallel, research should assess how conifer plantations interact with climate adaptation strategies, legacy land uses and socio-ecological needs in multifunctional landscapes.

Importantly, we must move away from a simplistic "positive or negative" framing of conifer reforestation. Many conifer plantations, particularly historical ones, were established under different objectives and knowledge regimes. Rather than viewing them as ecological mistakes, they should be seen as opportunities: mature or maturing forests that can be integrated into modern, diversity-oriented restoration strategies.

In this light, hybrid models that combine existing conifer stands with structural enrichment, species diversification, understory restoration and selective thinning can serve both conservation and carbon goals. Such approaches would enhance ecological function while retaining the positive soil impacts of conifers, such as erosion control and organic matter input, already in place. In sensitive landscapes, especially mountains, re-thinking our actual approach about reforestation into more autonomous dynamic, mosaic-based frameworks could help restore species and ecosystem types lost to historical land degradation or overexploitation.

Ultimately, acknowledging the context-dependent value of coniferous reforestation allows us to more effectively design integrated strategies for soil health, biodiversity conservation and climate resilience. Rather than discarding conifer landscapes, we should consider how they could contribute meaningfully to long-term conservation of human heritage environments, nature restoration and sustainable landscape mosaics.

3.4. Final considerations

In addition to the testimonies of communities and organizations, there are also stories of remarkable humanity and far-sighted vision, where ordinary yet extraordinary individuals have taken it upon themselves to engage in reforestation using conifers. One such example is Girardelli (2017) and his extensive work with *Pinus cembra* in the Italian Alps, healing the wounds of the mountains caused by both natural elements and human activity.

We must consider all the values that conifers and related formations can represent in a broader and deeper way. Those who promote their elimination or denigration cannot base their stance solely on subjective ideas, personal emotions or opportunism. Not only does it erase cultural ecosystems, human values, stories and traditions, but it also severs an umbilical cord between humans, nature and the land, while simultaneously undermining those individuals and organizations who created these forests, as well as those who regard them as a family legacy or an emblematic place to which they feel a deep connection. We cannot act by destroying these formations in the name of renaturalization, nor in the name of the local population's welfare, or the nation's benefit, without causing damage to both the memory of local communities and the environment, in any case. Actions must be taken with an understanding of the deep values and the profound bond formed between conifer reforestation and local communities, a connection that has endured across generations and should not be severed through impositions, nor by attempting to persuade local populations to sever it. Attempting to sever this temporal link, even under the guise of environmental or economic reasoning, should be viewed as an act of cultural violence against local people, their collective memory and national history. Only time and the gentleness of spontaneous dynamics can allow many of these conifer formations to gradually give way to more natural habitats in the most appropriate manner, while other reforestations may remain preserved by human decisions in more suitable locations. In Italy, we often hear statements such as "local people are the guardians who preserve biodiversity", while at the same time hearing claims that "local populations are responsible for human impact and the destruction of nature". These reforestations demonstrate that human communities can embody both realities, first through deforestation and later through reforestation. Today, reforestation and forest studies demonstrate that, in some cases, the main environmental and forest threats may come more from certain experts, businesses, environmentalists, forest or protected area managers, or public administrators than from local populations.

The attention towards reforestation has often suddenly changed almost from one opposite to another, from one drift to another, from one exaggeration to another, following the change in economic policies related to mountains, forests, development, energy, agriculture and the environment. Policies are full of criticism and should be increasingly revised (e.g. Gomiero et al. 2010; Bottacci 2018; Chiarucci and Piovesan 2018; Pedrotti 2018; Searchinger et al. 2018 and subsequent; Cazzolla Gatti et al. 2021; Clauser 2021; Manni 2021; Ciervo 2022). This highlights the need for greater caution and thoroughness when advocating for specific laws and addressing these issues, to avoid negatively impacting scientific research, historical, cultural and environmental heritage, as well as field activities and, most importantly, land, forestry and environmental management decisions (Brancalion et al. 2025).

Today, we can observe a growing and concerning convergence between certain agronomic-forestry and environmentalist perspectives in their opposition to conifers and historical artificial reforestation efforts. This may be explained by opportunism linked to the economic interests supported by current political agendas. The push to eliminate some conifer reforestations and woodlands in favor of broadleaf species also serves, in this context, to secure funding for removal operations and replace them with forests better suited to profitable harvesting under market-oriented logic.

It may also arise from a tendency towards less informed, or too superficial thinking, or from unrealistic and extremist ideologies concerning some environmental issues that can be conveniently exploited for an economic gain. Ideological extremism, even when driven by good intentions, is particularly evident among those who view rewilding as inherently "anti-human", aimed at erasing traces of human presence, history and heritage as much as possible. At the same time, extreme views can also emerge in the opposite direction, through exaggerated fears directed towards trees and forests, forest expansion and rewilding processes. Since conifers are exceptionally effective pioneer species, both they and the ecosystems they help to establish often become convenient scapegoats in these debates.

These attitudes become especially evident when we observe clear contradictions, such as a preference for human-shaped open landscapes or artificially created woodlands such as chestnut groves or cork oak plantations, while rejecting conifer formations. Such contradictions appear to reflect prejudiced stances rather than any rational or evidence-based evaluation.

The discussion remains open as to whether potential vegetation should always be defined by idealizing a reference state, typically understood as a past condition preceding human influence, particularly in territories where humans have been present and active for centuries or even millennia, thus representing a past that no longer exists and will never exist again. Consequently, potential vegetation should not necessarily be reconstructed as if humans had never existed, or as if continuous changes in biotic, climate and other environmental conditions had not occurred over this period.

The debate on reforestation and agricultural conifer plantations has been increasingly polarized during the last decades, with some advocating for their elimination and others for new plantations to address issues such as plant diseases and climate change. A more balanced approach is necessary. Well-planned plantations can benefit biodiversity-poor or erosion-prone agricultural areas (e.g. Chirici et al. 2000; Cunningham et al. 2015), but should be avoided in areas undergoing natural renaturalization. Depending on the context, it can be legitimately considered that some of these areas may be used for timber production, provided they are managed ensuring integrity in human and ecological ethical terms. In other cases, future plantations for experimental, scientific or conservation purposes can be considered legitimate and should not be prohibited, as long as they are implemented with discernment and within sound ecological and ethical boundaries.

There is also an urgent need for targeted legislation to halt the destruction of conifer forests, reforestation efforts and abandoned plantations, free from political, ideological or economic bias. Policies and actions that harm these cultural ecosystems are ultimately hindering research on biodiversity and ecosystem functionality.

We should support mainly non-invasive research on forests. Allowing artificial reforestation, with and without native species, to evolve naturally will foster more natural vegetation in Italy, aligning with the most advanced principles of silviculture, based on ecological forestry (i.e. Ciancio 2009). This, causing less human impact and less economic wastefulness. Alternatively, when possible, in an appropriate context, another viable approach is to actively conserve these formations, while enhancing their specificities in biodiversity and cultural values through increasingly sustainable and ethical practices.

Regarding the conifer reforestations that are the subject of our study, although the areas involved are significant in terms of the effort and work invested, they still represent only a very small portion of both the national forested area and the total national territory. Therefore, eliminating these formations will never be sufficient to positively affect the current environmental situation. Their removal will not provide significantly more space to enhance existing nature conservation strategies, nor to allow the regeneration of natural forests that are still too young or overly exploited, nor to establish more economically viable forests for future use. To effectively improve forest cover and quality, it would be necessary to identify such opportunities in areas where environmental pressures and competing land uses are more extensive or intense. At the same time, forest management and nature conservation strategies should be revised to better support spontaneous ecological dynamics. Such a shift could increase the overall effectiveness, durability and sustainability of conservation measures, while also reducing human impact and management costs.

The environmental context in Italy has undergone significant changes since the period of large-scale reforestation efforts, which took place from the late 19th to the late 20th century. Due to widespread land-use abandonment and the decline in human pressure across many areas in Italy, natural processes of woody vegetation recovery are now occurring spontaneously. In this context, traditional reforestation is no longer a general priority. However, in areas that are still heavily or extensively affected by human activities, such as agricultural landscapes, urban and peri-urban zones, or regions dominated by grasslands or garriga where natural secondary succession is limited, reforestation or tree-based cultivation may still represent valuable opportunities. In these cases, such practices can serve productive purposes, support local value chains and indirectly contribute to the conservation of more natural forest ecosystems.

It is nevertheless essential to avoid repeating the mistakes of the past by learning from previous excesses and misguided interventions, whether involving broadleaf or coniferous, native or non-native species.

Reforestation should not be excluded a priori, even in more natural and specific contexts, particularly those affected by environmental degradation. In such settings, reforestation can serve additional strategic purposes, such as mitigating soil erosion, improving hydrogeological stability, contributing to ecological restoration efforts, providing ecological corridors, providing sites for experimental research or supporting ex situ conservation initiatives, among other objectives.

3.5. Conclusions and outlook

This work does not aim to be a comprehensive treatise on reforestation projects and conifer formations, nor does it seek to take sides. Rather, it offers an overview of the current state of knowledge on conifer reforestations, often referred to in Italy as "Pinete" or simply "Rimboschimenti". The goal is to identify possible contradictions, strengths, unfounded beliefs and weaknesses in the current situation, learn from past experiences, and explore whether alternative interpretations are possible. The aim is to encourage a deeper and broader reflection to improve current perspectives and strategies, looking ahead to the future in the context of the environmental changes that humanity will inevitably face.

Conifer reforestation and conifer species can be defined as having significant environmental and human heritage values. Their characteristics depend on their conditions, and their impact depends mainly on the perspective from which we could consider them.

They have been successful in promoting vegetation succession, soil restoration and hydrogeological balance, preventing erosion and providing habitats for various species of animals, plants and fungi, also favoring various moss species (e.g. Biondi and Ballelli 1973).

It is not fair to say that these environments are better or worse than others are. They have their own characteristics, and they can facilitate some species over others, which happens with any other type of vegetation or ecosystem.

These formations, whether with native or non-native species, can be useful to improve the resilience facing climate change, extreme events and the arrival of new pests. In any case, they can be a resource, not only for cultural and environmental purposes, but also from the economic point of view.

In any case, these reforestation efforts and conifer species can undoubtedly contribute to biodiversity and enriching the mosaic of ecosystems, and the landscape. Instead of trying to cut down them or accelerate their elimination, the efforts should be oriented to study conifer species and related formations. The tampering and removal of conifer forests and artificial reforestations is unlikely to significantly improve Italy's environmental conditions, but it does risk eliminating ecosystems that, although artificial and with non-native species, still contribute to the diversity of the landscape and habitats across the territory.

In recent years, we observed increasing pressure to eliminate conifer plantations and reforestations, often justified by interventions aimed at renaturalization or fire prevention purposes. If in the past we witnessed certain excesses and misguided approaches in the establishment of these reforestations, today we should avoid making the opposite mistake, exaggerating in the other direction by eliminating or negatively affecting them. Another concern is related to an increasingly common trend: the granting of public and private areas, including natural forests, secondary succession and reforested zones, for domestic livestock grazing. This practice ultimately hinders the regeneration of both coniferous and broadleaved species and can easily damage already established trees, including very old individuals. It further aggravates the already significant and increasing impact caused, especially in some areas, by browsing and bark-stripping by wild animals.

The tendency to manipulate ecosystems through land cover modifications, vegetation control measures, forest exploitation and tree cutting is becoming increasingly widespread, even affecting broadleaf forests and threatening the most natural forest formations. This trend reflects, on the one hand, a misguided and overly ideological interpretation of rewilding, as if it consisted of erasing all traces of human presence. On the other hand, it stems from a certain extremist disdain and disregard for forests and trees, favoring open or other artificial environments.

Additionally, it follows current economic interests, driven by trends in energy, development, forestry, agriculture and environmental policies, as extensively discussed in the cited literature. Unfortunately, reforestation efforts often manage to find common ground among these three perspectives. A change of direction on this front is clearly necessary.

Conifer reforestations merit attention and should not be discarded, not least because of the presence of important species or provenances; not only because they could be defined as important ecosystems, as well as priority habitats according to European Community habitat interests; but also, for all of the other reasons analyzed in this chapter, and the values these formations may represent. However, as highlighted in our study, regardless of the perspectives we may hold about conifers and reforestation efforts, it is crucial to stop the widespread destruction and manipulation of these ecosystems, allowing them to evolve freely. Alternatively, we should act with greater knowledge, care and attention to manage them in a less impactful and more sustainable way, even to preserve and use them, especially within more anthropized contexts.

This approach is essential to safeguard the biodiversity they support, prevent ecological shocks and reduce the economic burden on society, while also minimizing environmental impacts and risks for local communities.

3.6. Acknowledgments

The authors express their deep gratitude to Dr. Fabio Clauser, the almost 106-year-old veteran of Italian forestry. He is the oldest living Italian forestry engineer ("forestry doctor" in Italy). His illustrious career spans various roles, including forester and officer with the "Corpo Forestale dello Stato" – Italy's national forest and environmental police, where he rose to the rank of Deputy Director. He also served as director of the "Parco Nazionale dello Stelvio", superintendent of the "Foreste Casentinesi", and played a key role in the last major reforestation initiative in the Casentino region. He was a driving force behind the creation of the "Riserva Naturale Integrale del Sassofratino" forest and held a senior position at the Italian Ministry of Agriculture and Forestry. We are deeply thankful for his invaluable advice on this chapter and for generously sharing his extensive knowledge, insights and lifelong testimony on forests, conifers and reforestation.

3.7. References

AA.VV. (1962). *Atti del Congresso nazionale sui rimboschimenti e sulla ricostituzione dei boschi degradati : tenutosi a Firenze dal 12 al 15 aprile 1961/Accademia Italiana di Scienze Forestali*. Vol I and II. Tipografia Coppini, Florence.

AA.VV. (1975). Tavola rotonda su "Rimboschimenti e conservazione dell'ambiente" (Firenze, 2 marzo 1974). Società Botanica Italiana, Gruppo di lavoro per la conservazione della natura. *Informatore Botanico Italiano*, 7, 35–105.

AA.VV. (2018). *Rinaturalizzazione dei rimboschimenti di pino nero : aspetti storici e gestione odierna*. Edizioni Polistampa, Florence.

Abrami, A. (2020). Il decreto forestale n. 34 del 2018: una legge sul bosco o sulla produzione di legname? *Rivista Giuridica dell'Ambiente*, 13, 6 [Online]. Available at: https://rgaonline.it/articoli/il-decreto-forestale-n-34-del-2018-una-legge-sul-bosco-o-sulla-produzione-di-legname/.

Allegrezza, M., Pesaresi, S., Ballelli, S., Tesei, G., Ottaviani, C. (2020). Influences of mature *Pinus nigra* plantations on the floristic-vegetational composition along an altitudinal gradient in the central Apennines, Italy. *iForest – Biogeosciences and Forestry*, 13(4), 279–285. doi: 10.3832/ifor3215-013.

Bach, A., Yáñez-Serrano, A.M., Llusià, J., Filella, I., Maneja, R., Penuelas, J. (2020). Human breathable air in a Mediterranean forest: Characterization of monoterpene concentrations under the canopy. *International Journal of Environmental Research and Public Health*, 17(12), 4391. doi: 10.3390/ijerph17124391.

Bagnato, S., Mercurio, R., Scarfò, F. (2012). Conifer afforestations in Italy: An opportunity for wood energy and forest restoration. *L'Italia Forestale e Montana*, 67(2), 103–114. doi: 10.4129/ifm.2012.2.03.

Baldacci, E. (1948) Alcuni aspetti del rimboschimento in relazione al fenomeno della micorrizia. *L'Italia Forestale e Montana*, 3(2), 67–72.

Bănăduc, D., Simić, V., Cianfaglione, K., Barinova, S., Afanasyev, S., Öktener, A., McCall, G., Simić, S., Curtean-Bănăduc, A. (2022). Freshwater as a sustainable resource and generator of secondary resources in the 21st century: Stressors, threats, risks, management and protection strategies, and conservation approaches. *International Journal of Environmental Research and Public Health*, 19(24), 16570. doi: 10.3390/ijerph192416570.

Bartolucci, F., Peruzzi, L., Galasso, G., Albano, A., Alessandrini, A., Ardenghi, N.M.G., Astuti, G., Bacchetta, G., Ballelli, S., Banfi, E. et al. (2018). An updated checklist of the vascular flora native to Italy. *Plant Biosystems*, 152(2), 179–303. doi: 10.1080/11263504.2017.1419996.

Benelli, G., Pavela, R., Canale, A., Cianfaglione, K., Ciaschetti, G., Conti, F., Nicoletti, M., Senthil-Nathan, S., Mehlhorn, H., Maggi, F. (2017). Acute larvicidal toxicity of five essential oils (*Pinus nigra*, *Hyssopus officinalis*, *Satureja montana*, *Aloysia citrodora*, and *Pelargonium graveolens*) against the filariasis vector *Culex quinquefasciatus*: Synergistic and antagonistic effects. *Parasitology International*, 66(2). doi: 10.1016/j.parint.2017.01.012.

Biagioni, A., Corsi, F., Pezzo, F., Tassi, F. (2014). Pinete costiere e necessità di conservazione forestale, faunistica e paesaggistica. Il Tombolo di Grosseto. In *Proceedings of the Second International Congress of Silviculture*, Florence, 26th–29th November, 329–336.

Biondi, E. and Ballelli, S., (1973). Osservazioni su due rimboschimenti a *Pinus nigra* ssp. *nigra* nella zona di Fabriano (Marche). *Arch. Bot. Biogeogr. Ital.*, LI, 163–171.

Biondi, E. and Vagge, I. (2015). The forests of *Pinus pinaster* Aiton subsp. *pinaster* of the NW-Italian Tyrrhenian sector. *Acta Botanica Gallica*, 162(3), 239–250. doi: 10.1080/12538078.2015.1063451.

Bisbiglia, V. (2023). Contro gli incendi (che non ci sono), il Parco Nazionale d'Abruzzo vuole tagliare i pini protetti di Villetta Barrea. Maraini: "Illogico". *Il Fatto Quotidiano* [Online]. Available at: https://www.ilfattoquotidiano.it/2023/11/01/contro-gli-incendi-che-non-ci-sono-il-parco-nazionale-dabruzzo-vuole-tagliare-i-pini-protetti-di-villetta-barrea-maraini-illogico/7339608/ [Accessed 4 April 2025].

Bonari, G., Acosta, A.T.R., Angiolini, C. (2017). Mediterranean coastal pine forest stands: Understorey distinctiveness or not? *Forest Ecology and Management*, 391, 19–28. doi: 10.1016/j.foreco.2017.02.002.

Bonari, G., Acosta, A.T.R., Angiolini, C. (2018). EU priority habitats: Rethinking Mediterranean coastal pine forests. *Rend. Fis. Acc. Lincei*, 29, 295–307. doi: 10.1007/s12210-018-0684-9.

Borghetti, M., Ferrara, A., Moretti, N., Nolè, A., Pierangeli, D., Ripullone, F., Todaro, L. (2024). Prendersi cura dei boschi di un'area interna nell'era del cambiamento climatico: il caso della Basilicata. *Forest@*, 21, 10–36. doi: 10.3832/efor0042-021.

Bosa, B. and Portoghesi, L. (2002). I giganti della Sardegna : la particella Pavari della foresta demaniale di Fiorentini. *L'Italia Forestale e Montana*, 57(6), 554–560.

Bottacci, A. (2018). Il TUFF, la gestione attiva dei boschi e le generazioni future. *L'Italia Forestale e Montana/Italian Journal of Forest and Mountain Environments*, 73(4–5), 207–214.

Bottacci A. (2020). Lo spazio e il tempo per le foreste resilienti. *L'Italia Forestale e Montana/Italian Journal of Forest and Mountain Environments*, 75(2), 1–12.

Brancalion, P.H.S., Hua, F., Joyce, F.H., Antonelli A., Holl K.D. (2025). Moving biodiversity from an afterthought to a key outcome of forest restoration. *Nat. Rev. Biodivers*. doi: 10.1038/s44358-025-00032-1.

Broquen, P., Falbo, G., Apcarián, A., Candan, F., Girardin, J., Pellegrini, V. (2003). Afforestations and its relationships with soil erosion and productivity in forest-steppe transition (Andinopatagonia, Argentina). *Forest Systems*, 12(2), 99–110. doi: 10.5424/804.

Bruschi, P., Di Santo, D., Grossoni, P., Tani, C. (2006). Indagini tassonomiche sul pino nero, *Pinus nigra* Arn. (Pinaceae), della Majella. *Informatore Botanico Italiano*, 38(2), 357–362.

Cantiani, P., Becagli, C., Bertini, G., Cazau, C., Venturi, E. (2013). Influenza del diradamento in pinete di pino nero sulla biodiversità micologica e floristica. In *Proceedings of the 9th SISEF National Congress "Multifunzionalità degli Ecosistemi Forestali : Sfide e Opportunità per la Ricerca e lo Sviluppo"*, Bolzano [Online]. Available at: http://www.sisef.it/sisef/congresso-ix/ [Accessed 4 April 2025].

Cantiani, P., Di Salvatore, U., Romano, R. (2018). La selvicoltura delle pinete artificiali di pino nero : analisi delle legislazioni regionali italiane. *Forest@*, 15, 99–111. doi: 10.3832/efor2985-015.

Carnicelli, S. and Costantini, E.A. (2013). Time as a soil forming factor and age of Italian soils. In *The Soils of Italy*, Costantini, E. and Dazzi, C. (eds). Springer, Dordrecht. doi: 10.1007/978-94-007-5642-7_5.

Cazzolla Gatti, R.C., Piovesan, G., Chiarucci, A. (2021). Italy: Forest harvesting is the opposite of green growth. *Nature*, 595, 353. doi: 10.1038/d41586-021-01923-x.

C.E.O. (2023). A retrospective and interview with Dr. Kevin Cianfaglione – Editorial Board Member of Conservation. *Conservation Editorial Office*, 3(2), 319–333. doi: 10.3390/conservation3020022.

Châline, A., Castellucci, F., Boutaud, E., Marrec, R., Cianfaglione, K. (2024). Biodiversity of carabid beetles in different types of forest in the region of Abruzzo, Italy. In *Proceedings of the 3rd International Electronic Conference on Diversity*. MDPI, Basel, 79 [Online]. Available at: https://sciforum.net/events_files/1067/customs/d924f86fe06da0e32bf13ecfdbaa7e2f.pdf [Accessed 4 April 2025].

Chiarucci, A. and Piovesan, G. (2018). Sustainable forest management can not disregard the current knowledge on ecology and conservation. *Forest@ – Journal of Silviculture and Forest Ecology*, 15, 51–55. doi: 10.3832/efor2782-015.

Chirici, G., Corona, P., Marchetti, M., Vannuccini, M. (2000). Procedura di valutazione degli interventi di rimboschimento come strumenti di composizione dell'ecotessuto paesistico. *L'Italia Forestale e Montana*, 55(4), 253–267.

Chiti, T., Benilli, N., Mastrolonardo, G., Certini, G. (2024). The potential for an old-growth forest to store carbon in the topsoil: A case study at Sasso Fratino, Italy. *Journal of Forestry Research*, 35, 10. doi: 10.1007/s11676-023-01660-z.

Ciancio, O. (2000). Metodi innovativi di rinaturalizzazione e gestione sostenibile dei sistemi forestali. *L'Italia Forestale e Montana*, 55(4), 209–210.

Ciancio O. (2009). Quale selvicoltura nel XXI secolo? In *Atti del Terzo Congresso Nazionale di Selvicoltura per il miglioramento e la conservazione dei boschi italiani*. Taormina di Scienze Forestali, Florence, 16–19 October, 3–39.

Ciancio, O., Mercurio, R., Nocentini, S. (1984). La sperimentazione di specie forestali esotiche in Italia risultati dopo un sessantennio. *Annali dell'Istituto Sperimentale per la Selvicoltura*, XIII(1982), 107–730.

Ciancio, O., Mercurio, R., Nocentini, S. (1997). Introduction and cultivation of exotic forest trees in Italy. In *Proceedings of a Discussion Meeting: Native and Non-Native in British Forestry*, University of Warwick, 193–200.

Cianfaglione, K. (2014a). *L'importanza degli alberi e del bosco : scienza e coscienza del territorio*. Vol I and II. Temi, Trento.

Cianfaglione, K. (2014b). The yew (*Taxus baccata*) between ethnography and ecology. *Der Eibenfreund*, 20, 59–64.

Cianfaglione, K. (2022). La processionaria: tra necessità di lotta e di convivenza. *Simbiosi*, 7, 74–78.

Cianfaglione, K. (2023). Pine reforestation in Mediterranean mountains. An ecological problem or an opportunity? In *XVII OPTIMA Meeting*, Palermo and Erice, 20–23 September [Online]. Available at: https://univ-catholille.hal.science/hal-04259095/file/pini%20e%20pinete%202.pdf [Accessed 4 April 2025].

Cianfaglione, K. (2024a). An editorial to introduce the new journal wild: Issues, approaches, ideas and proposals. *Wild*, 1, 30–38. doi: 10.3390/wild1010003.

Cianfaglione, K. (2024b). Conifers reforestation: An environmental problem or a possible resource? In *ECOPOSS 2024 International Conference: Rethinking Networks in Times of Transition: Ecological, Energy and Social Issues.*

Cianfaglione, K., Damiani, G., Schirone, B., Pirone, G., Ciaschetti, G., Manzi, A., Di Felice, P.L., Colazilli, A., Marras, T. (2014). Relevant aspects of the Abruzzo coast transformation during last centuries (Central Adriatic Italy). *Plant Sociology*, 51(2), S1, 73–80. doi: 10.7338/pls2014512S1/10.

Cianfaglione, K., Damiani, G., Desalme, D., Schirone, B. (2021). Historical vegetation survey on *Pinus* presence in Abruzzo (Central Adriatic Italy). A study case from Pescara. *Historical Ecology for the Future, International Conference on Historical Ecology*, Metz. Université de Lorraine, AgroParisTech, CNRS, INRAE [Online]. Available at: https://www.youtube.com/watch?v=XdjaMQER_2o [Accessed 4 April 2025].

Ciervo, M. (2022). *La Strategia di bioeconomia è sostenibile? Territori, impatti, scenari.* SdT Edizioni, Florence.

Clauser, F. (2005). Meditazione sui boschi di origine artificiale. *L'Italia Forestale e Montana*, 6, 729–731.

Clauser, F. (2021). Produzione biologica, produzione industriale del bosco e Recovery Plan. *L'Italia Forestale e Montana/Italian Journal of Forest and Mountain Environments*, 76(3), 137–140.

Conti, G. (1995). Omaggio al lavoro dei Forestali. Scritti sul rimboschimento nell'Abruzzo sulmonese. *Club Alpino d'Italia*, Sulmona, 105.

Corona, P., Bernardini, V., Muzzalupo, I., Turco, R., Garofalo, K., Pignatti, G. (2024). Experimental reforestation with Bosnian pine under Mediterranean mountain conditions. *iForest*, 17, 331–337. doi: 10.3832/ifor4635-017.

Craveri, C. (1915). *Le Conifere da rimboschimento e l'industria resinifera.* Hoepli Ed., Milan.

Csilléry, K., Ponta, N., Pistone, A., Cianfaglione, K. (2023). MyGardenOfTrees : un progetto internazionale sull'abete bianco per l'Abetina di Rosello. In *Conference: Alla scuola dell'abete bianco. Nuove iniziative tra ricerca, formazione e conservazione*, Rosello, 20 October.

Cunningham, S.C., Mac Nally, R., Baker, P.J., Cavagnaro, T.R., Beringer, J., Thomson, J.R., Thompson, R.M. (2015). Balancing the environmental benefits of reforestation in agricultural regions. *Perspectives in Plant Ecology, Evolution and Systematics*, 17(4), 301–317. doi: 10.1016/j.ppees.2015.06.001.

Di Bérenger, A. (1871–1872). Memoria I. Dell'assoluta influenza delle foreste sulla temperatura atmosferica, giuntovi un compendio di storia antica e moderna della coltura, economia ed amministrazione dei boschi. *Giornale di Economia Forestale*, 1, 3–102.

Donelli, D., Meneguzzo, F., Antonelli, M., Ardissino, D., Niccoli, G., Gronchi, G., Baraldi, R., Neri, L., Zabini, F. (2023). Effects of plant-emitted monoterpenes on anxiety symptoms: A propensity-matched observational cohort study. *International Journal of Environmental Research and Public Health*, 20, 2773. doi: 10.3390/ijerph20042773.

Eccher, A., Fusaro, E., Gemignani, G., Mughini, G., de Philippis, A. (1986). Interventi organici di forestazione a scopi produttivi nel Mezzogiorno (Progetto speciale n. 24). Divisione Forestazione, Ripartizione Progetti Promozionali. Centro di Sperimentazione Agricola e Forestale (S.A.F. S.p.A.), Gruppo E.N.C.C. 110. Edito a cura dell'Ufficio Documentazione, Divisione Stampa e Pubbliche Relazioni dell'Intervento Straordinario nel Mezzogiorno, Commissario del Governo, Rome.

El Omari, N., Guaouguaou, F.-E., El Menyiy, N., Benali, T., Aanniz, T., Chamkhi, I., Balahbib, A., Taha, D., Shariati, M. A., Zengin, G. et al. (2021). Phytochemical and biological activities of *Pinus halepensis* Mill., and their ethnomedicinal use. *Journal of Ethnopharmacology*, 268, 113661. doi: 10.1016/j.jep.2020.113661.

Fermanelli, A., Baiocco, M., Gabrielli T., (2008). Il recupero della biodiversità vegetale del parco. Voci dal Parco, periodico di informazione del Parco Nazionale dei Monti Sibillini, 2 [Online]. Available at: https://www.sibillini.net/comunicazione/Voci%20dal%20parco/vdp_2_2008/_20_13.html. [Accessed 4 April 2025].

Fernandes, P.M., Vega, J.A., Jiménez, E., Rigolot, E. (2008). Fire resistance of European pines. *Forest Ecology and Management*, 256(3), 246–255. doi: 10.1016/j.foreco.2008.04.032.

Filibeck, G., Cancellieri, L., Rosati, L. (2022). Carta dei Tipi di Habitat di Interesse Comunitario (Direttiva 92/43/CEE) del Parco Nazionale d'Abruzzo, Lazio e Molise, e dei siti Natura 2000: ZSC IT7110205, ZPS IT7120132, ZSC IT6050018, ZSC IT6050020, ZSC IT7212121. Illustrative Note, Università degli Studi della Tuscia, Viterbo [Online]. Available at: https://www.parcoabruzzo.it/pdf/Note_Illustrative-GENNAIO2022.pdf. [Accessed 4 April 2025].

Frate, L., Fabrizio, M., Ciaschetti, G., Spera, M. (2018). Analisi spaziale dell'incendio del Morrone nel Parco Nazionale della Majella mediante l'utilizzo di immagini satellitari. *Forest@*, 15, 59–64. doi: 10.3832/efor2775-015.

Gellini, R. (1968). Posizione sistematica del pino nero di Villetta Barrea in base ai caratteri anatomici degli aghi. *Annals of the Italian Academy of Forest Sciences*, 17, 101–122.

Giacobbe, A. (1973). A proposito della var. *apennina* Giac. dell'*Abies alba*. *L'Italia Forestale e Montana*, 28(1), 30–32.

van Gils, H., Odoi, J.O., Andrisano, T. (2010) From monospecific to mixed forest after fire?: An early forecast for the montane belt of Majella, Italy. *Forest Ecology and Management*, 259(3), 433–439. doi: 10.1016/j.foreco.2009.10.040.

Girardelli, A. (2017). *Una Vita Con Entusiasmo*. La Grafica, Mori.

Gomiero, T., Paoletti, M.G., Pimentel, D. (2010). Biofuels: Efficiency, ethics, and limits to human appropriation of ecosystem services. *Journal of Agricultural and Environmental Ethics*, 23, 403–434. doi: 10.1007/s10806-009-9218-x.

Granetti, B., De Angelis, A., Materozzi, G. (2005). *Umbria terra di tartufi*. Regione Umbria, Gruppo Micologico Ternano, Terni.

Guo, Y., Abdalla, M., Espenberg, M., Hastings, A., Hallett, P., Smith, P. (2021). A systematic analysis and review of the impacts of afforestation on soil quality indicators as modified by climate zone, forest type and age. *The Science of the Total Environment*, 757, 143824. doi: 10.1016/j.scitotenv.2020.143824.

Hüblová, L. and Frouz, J. (2021). Contrasting effect of coniferous and broadleaf trees on soil carbon storage during reforestation of forest soils and afforestation of agricultural and post-mining soils. *Journal of Environmental Management*, 290, 112567. doi: 10.1016/j.jenvman.2021.112567.

Iovino, F., Nicolaci, A., Menguzzato, G., Marziliano, P., Bernardini, V., Castaldi, C., Quatrini, V., Cutini, A. (2017). Approcci selvicolturali innovativi a sostegno della gestione forestale sostenibile in Calabria. *Forest@*, 14, 285–313. doi: 10.3832/efor2544-014.

Jandl, R., Ledermann, T., Kindermann, G., Weiss, P. (2021). Soil organic carbon stocks in mixed-deciduous and coniferous forests in Austria. *Frontiers in Forests and Global Change*, 4, 688851. doi: 10.3389/ffgc.2021.688851.

Kletty, F., Mudoy-Deverlanges, M., Trei J.-N., Malengreau, A., Wellstein, C., Cipollone, M., Cianfaglione, K. (2024). Comparative study of conifers and beech forests ecology. Preliminary evidences for forest structure and bird populations from Central Italy. In *SFE2 – International Congress of Ecology and Evolution*, Lyon.

Kolosova, V., Svanberg, I., Kalle, R., Strecker, L., Gençler Özkan, A.M., Pieroni, A., Cianfaglione, K., Molnár, Z., Papp, N., Łuczaj, Ł. et al. (2017). The bear in Eurasian plant names: Motivations and models. *Journal of Ethnobiology and Ethnomedicine*, 13, 14. doi: 10.1186/s13002-016-0132-9.

La Marca, O., Mercurio, R., Nocetini, L. (2016). Management and restoration of Italian afforestations under global change. *Reforesta*, 2, 50–59. doi: 10.21750/REFOR.2.06.21.

Landi, S., Schirone, B., Cianfaglione, K., Wellstein, C., Oliver, T., Bottacci, A., Damiani, G. (2023). Forest management for fire prevention. Are we always on the right path? In *9th International Wildland Fire Conference*, Porto [Online]. Available at: https://univ-catholille.hal.science/hal-04259018/file/Poster_Fire_Italy%20%281%29.pdf [Accessed 4 April 2025].

Leonardi, P., Graziosi, S., Zambonelli, A., Perini, C., Salerni, E. (2017). Biodiversità in un rimboschimento di *Pinus nigra*: il potenziale ecologico ed economico dei macromiceti. *Sherwood – Foreste ed Alberi Oggi*, 231, 35–39.

Liguori, A. (1928). Il rimboschimento della "Duna Feniglia" presso Orbetello. *L'Alpe*, 8, 251–264.

Maccherini, S., Salerni, E., Mocali, S., Bianchetto, E., Landi, S., De Meo, I., Di Salvatore, U., Marchi, M., Bacaro, G., Tordoni, E. et al. (2021). Silvicultural management does not affect biotic communities in conifer plantations in the short-term: A multi-taxon assessment using a BACI approach. *Forest Ecology and Management*, 493, 119257. doi: 10.1016/j.foreco.2021.119257.

Magini E. (1973). Esiste sull'Appennino una varietà dell'Abete bianco? *L'Italia Forestale e Montana*, 23(5), 173–176.

Mallik, A.U. (2003). Conifer regeneration problems in boreal and temperate forests with ericaceous understory: Role of disturbance, seedbed limitation, and keystone species change. *Critical Reviews in Plant Sciences*, 22(3–4), 341–366. doi: 10.1080/713610860.

Mandara L. (2022). Il Catasto degli incendi "dimenticato". Conoscere per prevenire, ma in troppi fanno finta di niente. Italia libera. *Tiscali notizie* [Online]. Available at: https://notizie.tiscali.it/italialibera/articoli/il-catasto-incendi-dimenticatoa-conoscere-prevenire-ma-in-troppi-fanno-finta-niente-00001/ [Accessed 4 April 2025].

Manni, C. (2021). Considerazioni sulla tutela del paesaggio forestale italiano prima e dopo il TUFF. *L'Italia Forestale e Montana*, 76(3), 127–135.

Manzi A. (2017). Roghi sul Morrone. Manzi, botanico: "Occorre rimboschire in maniera naturale" [Online]. Available at: https://www.abruzzolive.tv/roghi-sul-morrone-manzi-botanico-occorre-rimboschire-in-maniera-naturale.html [Accessed 4 April 2025].

Marchesoni, V. (1957). Storia climatico-forestale dell'Appennino umbro-marchigiano. *Annals of Botany*, 25(3), 459–497.

Marchesoni, V. (1958). Importanza del Pino nero, dell'Abete, del Tasso e dell'Agrifoglio nella storia climatico forestale dell'Appennino Umbro-marchigiano. *Monti e Boschi*, 10, 535–541.

Marchesoni, V. (1959a). Importanza del fattore storico-climatico e dell'azione antropica nell'evoluzione della vegetazione forestale dell'Appennino Umbro-marchigiano. *Ann. Acc. It. Scien. Forestali – Firenze*, VIII, 326–343.

Marchesoni, V. (1959b). Storia climatico forestale dell'Appennino Umbro-marchigiano. Analisi pollinica dei depositi torbo-lacustri di Colfiorito. *Studi sull'ambiente naturale degli altipiani di Colfiorito*, Camerino, 541–554.

Marfella, L., Marzaioli, R., Pazienza, G., Mairota, P., Glanville, H.C., Rutigliano, F.A. (2023). Medium-term effects of wildfire severity on soil physical, chemical and biological properties in *Pinus halepensis* Mill. woodland (Southern Italy): An opportunity for invasive *Acacia saligna* colonization? *Forest Ecology and Management*, 542, 121010. doi: 10.1016/j.foreco.2023.121010.

Marsilli, L. (2013). Rovereto, c'era una volta il Bosco della città. *Trentino* [Online]. Available at: https://www.giornaletrentino.it/cronaca/trento/rovereto-c-era-una-volta-il-bosco-della-città-0-1.1232867 [Accessed 4 April 2025].

Masseti, M. (2023). Sulla recente gestione del patrimonio naturalistico delle isole italiane. In *Atti della conferenza "La gestione del patrimonio naturalistico italiano*". Federazione Nazionale degli Ordini dei Biologi, Rome, 42–53 [Online]. Available at: https://www.fnob.it/wp-content/uploads/2023/10/speciale_conferenza_ambiente.pdf [Accessed 4 April 2025].

Mehrnoosh, S., Beheshti, A., Agha, A., Rakhsh, F., Pourreza, M., Safari Sinegani, A. (2022). Comparison of soil biological activity in forest with coniferous and broadleaf trees in Biston Region of Kermanshah. *Water and Soil*, 36(5), 593–610. doi: 10.22067/jsw.2022.78164.1189.

Mendicino, F., Carlomagno, F., Bonelli, D., Fumo, F., Di Biase, E., Treccosti, I., Flori, S., Bonacci, T. (2024). Saproxylic beetles in Sila National Park (Calabria, Italy) and their role as bioindicators of unmanaged forests. In *Proceedings of the 3rd International Electronic Conference on Diversity*. MDPI, Basel, 107 [Online]. Available at: https://sciforum.net/events_files/1067/customs/d924f86fe06da0e32bf13ecfdbaa7e2f.pdf [Accessed 4 April 2025].

Meneguzzo, F., Albanese, L., Bartolini, G., Zabini, F. (2019). Temporal and spatial variability of volatile organic compounds in the forest atmosphere. *International Journal of Environmental Research and Public Health*, 16, 4915. doi: 10.3390/ijerph16244915.

Mercurio, R. (2000). Esperienze e prospettive sull'applicazione del taglio a buche nelle abetine del Parco Nazionale delle Foreste Casentinesi. *L'Italia Forestale e Montana*, 55(4), 219–230.

Mercurio, R. (2013). La rinaturalizzazione dei rimboschimenti di conifere [Online]. Available at: https://robertomercurio.wordpress.com/2013/01/28/la-rinaturalizzazione-dei-rimboschimenti-di-conifere-una-politica-di-gestione-delle-risorse-forestali-che-stenta-ad-affermarsi/ [Accessed 4 April 2025].

Mercurio, R. (2015). Progetto MORINABIO per la rinaturalizzazione dei rimboschimenti di pino nero in Abruzzo [Online]. Available at: https://robertomercurio.wordpress.com/2015/09/24/il-progetto-morinabio-per-la-rinaturalizzazione-dei-rimboschimenti-di-pino-nero-in-abruzzo/ [Accessed 4 April 2025].

Mercurio, R. (2024). I nuovi sistemi forestali: come gestirli? [Online]. Available at: https://robertomercurio.wordpress.com/2024/08/03/i-nuovi-sistemi-forestali-come-gestirli/ [Accessed 4 April 2025].

Mercurio, R., Mallamaci, C., Muscolo, A., Sidari, M. (2009). Effetti della dimensione delle buche sulla rinnovazione naturale in rimboschimenti di pino nero. *Forest@*, 6, 312–319. doi: 10.3832/efor0591-006.

Niccoli, F., Esposito, A., Altieri, S., Battipaglia, G. (2019) Fire severity influences ecophysiological responses of *Pinus pinaster* Ait. *Frontiers in Plant Science*, 10, 539. doi: 10.3389/fpls.2019.00539.

Nocentini, S. (2006). La rinaturalizzazione dei sistemi forestali: è necessario un modello di riferimento? *Forest@*, 3, 376–379. doi: 10.3832/efor0394-0030376.

Nocentini, S. (2010). Specie forestali esotiche: sperimentazione di pavari e prospettive attuali. *L'Italia Forestale e Montana/Italian Journal of Forest and Mountain Environments*, 65(4), 449–457. doi: 10.4129/ifm.2010.4.09.

Ottaviani, C. (2017). *Pinus nigra* subsp. *nigra* in Appennino centrale: rimboschimenti e rinaturalizzazione. PhD Thesis, Università Politecnica delle Marche, Ancona [Online]. Available at: https://iris.univpm.it/retrieve/e18b8790-65cd-d302-e053-1705fe0a27c8/tesi_ottaviani.pdf [Accessed 4 April 2025].

Ottaviani, C., Tesei, G., Ballelli, S., Iorio, G., Montecchiari, S., Allegrezza, M. (2015). Vegetation dynamics in *Pinus nigra* Arnold subsp. *nigra* 100 years after reforestation: Two case studies in the central Apennines. *Plant Sociology*, 52(2), 79–94. doi: 10.7338/pls2015522/04.

Padula, M. (1972) Attività forestali e conservazione della natura con particolare riferimento all'Appennino Romagnolo. *Ann. Acc. It. Scien. Forestali – Firenze*, XXI, 213–241.

Paganelli, A. (1956) Analisi pollinica dei depositi torbosi e lacustri del Piano di Colfiorito (Appennino Umbro-Marchigiano). *Boll. Soc. Eustacchiana-Camerino*, XLIX, 71–91.

Paganelli, A. (1982). Histoire paléobotanique. *Guide itinéraire excursion internationale de Phytosociologie en Italie centrale*, Camerino, 39–74.

Pallotta, F. (1922). *Le abetine del Molise. Cattedra di agricoltura della provincia di Campobasso, sezione di Alpicoltura.* Tipografia Federazione Consorzi Agrari, Piacenza.

Panzavolta, T. and Tiberi, R. (2001). Test sul controllo della processionaria del pino con *Bacillus thuringiensis*. *L'Italia Forestale e Montana*, 56(6), 498–509.

Pavan, M. (1981). *Utilità delle formiche del gruppo Formica rufa (collana verde 57).* Ministero dell'Agricoltura e Foreste, Rome.

Pavari, A. (1927). La tecnica dei rimboschimenti secondo le più recenti vedute ed esperienze. *L'Alpe*, 11, 321–334; 12, 359–364.

Pavari, A. (1959). Basi ecologiche e tecniche dei rimboschimenti in Italia. In *Scritti di Ecologia, Selvicoltura e Botanica Forestale.* Pubblicazioni dell'Accademia Italiana di Scienze Forestali, Florence.

Pavari, A. and de Philippis, A. (1941). *La sperimentazione di specie forestali esotiche in Italia. Risultati del primo ventennio.* Failli, Rome.

Pavari A. and Susmel L. (1961). I rimboschimenti della catena appenninica. In *Atti del Congresso nazionale sui rimboschimenti e sulla ricostituzione dei boschi degradati*, Florence, 12–15 April, 185–223.

Pedrotti, F. (1987). I rimboschimenti di pino cembro (*Pinus cembra* L.) nella zona di Malga Spora, nel Parco naturale del Brenta-Adamello. *Informatore Botanico Italiano*, 19(2), 204–206.

Pedrotti, F. (2014). Stato attuale dei boschi dell'Appenino centrale e possibilità di recupero. In *L'importanza degli alberi e del bosco: scienza e coscienza del territorio*, Cianfaglione, K. (ed.). Temi, Trento, 29–48.

Pedrotti, F. (2018). Il patrimonio forestale italiano è gravemente minacciato. *Natura & Montagna, Unione Bolognese Naturalisti*, 65(1), 48–50.

Pedrotti, F. (2024). Opposizione ai tagli nella Pineta di Villetta Barrea, di Franco Pedrotti. *Simbiosi*, 9, 108.

Pellegrini, M., Manzi, A., Di Martino, L. (2016). Il pino nero di Fara San Martino (*Pinus nigra* J.F. Arnold subsp. *nigra* Hoss var. *italica* Hochstetter): dalla tradizione alla ricerca scientifica. Quali azioni per la conservazione? In *I granai della biodiversità. Strumenti, mezzi ed azioni per la conservazione ex situ della biodiversità vegetale. Documenti tecnico-scientifici del Parco Nazionale della Majella*, Di Martino, L., Di Santo, M., Di Cecco, V., Di Nino, O. (eds). Majambiente Ed., Caramanico/Terme, 118–128.

Pesaresi, S., Allegrezza, M., Ballelli, S., Mentoni, M., Olivieri, M., Ottaviani, C., Tesei, S. (2013). Biodiversity in the Sibillini Mountain range (Sibillini National Park, central Apennines): The example of Piè Vettore. *Plant Sociology*, 50, 57–89. doi: 10.7338/pls2013501/06.

de Philippis A., (1961). Problemi e tecnica del rimboschimento nella zona del Lauretum. In *Atti del Congresso nazionale sui rimboschimenti e sulla ricostituzione dei boschi degradati*, Florence, 12–15 April, 163–184.

Pica, A., Schirone, B., Magrini, S., Laghi, P., Cianfaglione, K., Di Filippo, A. (2025). Naturalness and tree composition determine the abundance of rare and threatened orchids in mature and old-growth *Abies alba* forests in the Northern Apennines (Italy). *Land*, 14(3), 579. doi: 10.3390/land14030579.

Piermattei, A., Renzaglia, F. Urbinati, C. (2012). Recent expansion of *Pinus nigra* Arn. above the timberline in the central Apennines, Italy. *Annals of Forest Science*, 69(4), 509–517. doi: 10.1007/s13595-012-0207-2.

Piermattei, A., Garbarino, M., Renzaglia, F. Urbinati, C. (2013). Climate influence on the expansion and tree-ring growth of *Pinus nigra* L. at high altitude in the Central Apennines. *The Open Forest Science Journal*, 6, 54–56. doi: 10.2174/1874398601306010054.

Pipins, S., Baillie, J.E.M., Bowmer, A., Pollock, L.J., Owen, N., Gumbs, R. (2024). Spatial decoupling of taxon richness, phylogenetic diversity and threat status in the megagenus *Erica* (Ericaceae). *PhytoKeys*, 244, 127–150. doi: 10.3897/phytokeys.244.124565.

Piussi P. (2004). Ricordo di Alessandro de Philippis. *Forest@*, 1, 3–8. doi: 10.3832/efor0186-0001.

Ricci, L., Farda, B., Ferrara, A., Cerasoli, F., De Simone, W., Frattaroli, A.R., Pirone, G., Di Musciano, M. (2024). Short-term functional response to post-fire vegetation dynamic: A case study in a Mediterranean *Pinus halepensis* forest. *Flora*, 310, 152415. doi: 10.1016/j.flora.2023.152415.

Riva, F. (2023). Una lunga lista di avvelenamenti avvenuti nell'Appennino Centrale nel corso del 2023. *Terre dell'orso. Rivista di Salviamo l'Orso*, 17, 110–118.

Rosato, A., Maggi, F., Cianfaglione, K., Conti, F., Ciaschetti, G., Rakotosaona, R., Fracchiolla, G., Clodoveo, M.L., Franchini, C., Corbo, F. (2018). Chemical composition and antibacterial activity of seven uncommon essential oils. *Journal of Essential Oil Research*, 30(4). doi: 10.1080/10412905.2018.1442753.

Sangüesa-Barreda, G., Esper, J., Büntgen, U., Camarero, J.J., Di Filippo, A., Baliva, M., Piovesan, G. (2020). Climate–human interactions contributed to historical forest recruitment dynamics in Mediterranean subalpine ecosystems. *Global Change Biology*, 26(9), 4988–4997. doi: 10.1111/gcb.15246.

Schirone, B. (1998). Le attività di Rimboschimento in Italia: situazione attuale e prospettive. In *Secondo Congresso di Selvicoltura. Per il miglioramento e la conservazione dei boschi italiani*, Venice, 24–27 June, 347–351.

Scimia A. (2012). *Il bosco e l'Abruzzo. Le vicende e le voci nel II millennio.* REA Ed., L'Aquila.

Searchinger, T.D., Beringer, T., Holtsmark, B., Kammen, D.M., Lambin, E.F., Lucht, W., Raven, P., van Ypersele, J.-P. (2018). Europe's renewable energy directive poised to harm global forests. *Nature Communications.* doi: 10.1038/s41467-018-06175-4.

Selvi, F., Carrari, E., Coppi, A. (2016). Impact of pine invasion on the taxonomic and phylogenetic diversity of a relict Mediterranean forest ecosystem. *Forest Ecology and Management*, 367, 1–11. doi: 10.1016/j.foreco.2016.02.013.

Sender, J., Maślanko, W., Różańska-Boczula, M., Cianfaglione, K. (2017). A new multi-criteria method for the ecological assessment of lakes: A case study from the transboundary biosphere reserve 'West Polesie' (Poland). *Journal of Limnology*, 76(S1). doi: 10.4081/jlimnol.2017.1639.

Sfregola, P. (2017). *Il bosco Bazzoni. Storia dell'imboschimento del Carso.* Luglio Ed., Trieste.

Shah, N.W., Nisbet, T.R., Broadmeadow, S.B. (2021). The impacts of conifer afforestation and climate on water quality and freshwater ecology in a sensitive peaty catchment: A 25 year study in the upper River Halladale in North Scotland. *Forest Ecology and Management*, 502, 119616. doi: 10.1016/j.foreco.2021.119616.

Sheffer, E. (2012). A review of the development of Mediterranean pine–oak ecosystems after land abandonment and afforestation: Are they novel ecosystems? *Annals of Forest Science*, 69, 429–443. doi: 10.1007/s13595-011-0181-0.

Stockmann, U., Minasny, B., McBratney, A.B. (2014). How fast does soil grow? *Geoderma*, 216, 48–61. doi: 10.1016/j.geoderma.2013.10.007.

Tassi, F. (2024). Opposizione ai tagli nella Pineta di Villetta Barrea, di Franco Tassi. *Simbiosi*, 9, 105–107.

Taylor, C., Blanchard, W., Lindenmayer, D.B. (2021). Does forest thinning reduce fire severity in Australian eucalypt forests? *Conservation Letters*, 14, e12766. doi: 10.1111/conl.12766.

Tesei, G., Allegrezza, M., Ballelli, S., Ciaschetti, G., Console, C., Montecchiari, S., Ottaviani, C., Pirone, G., Frattaroli, A.R. (2021). The oldest *Pinus nigra* plantations in the central Apennines: Distribution and floristic-vegetational characterisation. *Plant Sociology*, 58(2), 15–28. doi: 10.3897/pls.2021.58202.

Thomas, P.A. and Polwart, A. (2003). *Taxus baccata* L. *Journal of Ecology*, 91(3), 489–524. doi: 10.1046/j.1365-2745.2003.00783.x.

Tiberi, R. and Roversi, P.F. (2005). Gli insetti e la foresta di Vallombrosa: una storia secolare di difesa fitosanitaria e tutela della biodiversità. *L'Italia Forestale e Montana*, 60(2), 207–211.

Tiberi R., Niccoli A., Panzavolta T. (2000). Plant–animal interactions in forest plantations on abandoned agricultural land: Preliminary results. *L'Italia Forestale e Montana*, 55(4), 268–276.

Torreggiani, L. (2025). "Tagliare alberi per salvare la biodiversità". Ed è subito polemica attorno ad un intervento (dove è coinvolto il Wwf) per la tutela della 'landa carsica'. *L'AltraMontagna* [Online]. Available at: https://www.ildolomiti.it/altra-montagna/idee/2025/tagliare-alberi-per-salvare-la-biodiversita-ed-e-subito-polemica-attorno-ad-un-intervento-dove-e-coinvolto-il-wwf-per-la-tutela-della-landa-carsica?fbclid=IwY2xjawMsEy5leHRuA2FlbQIxMABicmlkETBnM1BrMEpXZERKODJzMUtkAR6xrWCPNSRNRNPvuwVAvS8gcHoYF6I7WkEpmadgJwdvPd7GpdODmkCzAQRB6w_aem_ul_0axZq4dTD9acb5bCVZw [Accessed 30 August 2025].

Ursitti, A. (2022). Un centenario a colpi di motosega per il PNALM? *Simbiosi*, 7, 18–19.

Vetturini, E. (1989). Settanta anni di rimboschimenti sul Subasio. Con una introduzione sull'aspetto fisico e sulle vicissitudini ambientali del monte attraverso i secoli. *Atti dell'Accademia Properziana del Subasio*, Assisi, VII(17), 73–203.

Zhang Y., Wang Q., Xu L., Ma S., Cui D., Zhu K., Feng W. (2022). Mixed conifer-broadleaf trees on arbuscular mycorrhizal and ectomycorrhizal communities in rhizosphere soil of different plantation stands in the temperate zone, Northeast China. *Frontiers in Microbiology*, 13. doi: 10.3389/fmicb.2022.986515.

Zhu, Z., Chelli, S., Tsakalos, J.L., Bricca, A., Canullo, R., Cervellini, M., Pennesi, R., Ludovico, L., Cesaroni, V., Bottacci, A. et al. (2024). How effective are different protection strategies in promoting the plant diversity of temperate forests in national parks? *Forest Ecology and Management*, 584, 1–9. doi: 10.1016/j.foreco.2025.122602.

4

Urban Interstices: Small Spaces with Big Issues

4.1. Introduction

Human activity impacts considerably on the environment. Greenhouse gas emissions that result from our lifestyles, and patterns of production and consumption, have led to an average global temperature increase of approximately 1°C compared to the pre-industrial period (1850–1900). If this trajectory continues, global warming could reach 1.5°C between 2030 and 2052 (IPCC 2023).

Although urban areas occupy only 2% of the Earth's surface, they are home to nearly 56% of the global population. These urban areas are responsible for 60% of greenhouse gas emissions (United Nations, Department of Economic and Social Affairs, Population Division 2018). Addressing the challenges posed by climate change thus requires sustainable strategies that account for both the habitability of cities and the biodiversity they support.

In this context, urban policies are increasingly focused on renaturation and the demineralization of built environments, aiming to reimagine cities as more sustainable and livable spaces. The greening of urban areas has emerged as the most important measures (Bush 2020). The development of urban green spaces and support for urban biodiversity are recognized as crucial for enhancing the quality of life and mitigating environmental impacts (Egea-Cariñanos et al. 2024; Wu et al. 2024). The recent review by Robinson et al. (2024) provides evidence that biodiversity (e.g. microbiota, green spaces, trees) directly or indirectly support the well-being by reducing stress and anxiety, fostering social cohesion and physical activity, buffering the heat island effect and lowering air pollution levels.

Chapter written by Agathe DOUCHET and Sadia BENAMROUZ-VANNESTE.

Focusing more specifically on informal urban green spaces, or urban interstices, several studies explored these spaces from either sociological or ecological perspectives (Rupprecht and Byrne 2014; Miao et al. 2023). However, few have integrated both dimensions and considered the temporal dynamics of citizens' perceptions. These spaces, despite their often marginal and unregulated status, may play a significant role in enhancing the connectivity of urban green networks. In this study, we adopt an integrative approach that combines ecological variables using metrics such as taxonomy, diversity/richness in vegetation, macro- and meso-fauna, and soil microbiota, with sociological variables related to public acceptability, perceptions and their evolution in response to increased knowledge about local biodiversity.

This research forms part of the IUPI project (Urban Interstices: small spaces with big issues), funded by ADEME in collaboration with the city of Lille (France). We investigated urban interstices of less than 3,000 m², defined as small green areas located between two or more urban elements (Erdi 2021) (e.g. buildings, parks or roads), independently of their usage or management status.

This chapter focuses only on the sociological component of the IUPI project. The primary objective was to provide urban planners and green space managers in Lille with data regarding how citizens and municipal employees perceive and use these small informal green spaces. A secondary objective was to evaluate how users' perceptions and biodiversity awareness changed after receiving scientific information about the ecological features of these sites.

4.2. Materials and methods

4.2.1. *Site mapping, selection and context*

The sites were selected depending on a typology defined by the same spatial configurations. This typology includes the following criteria: built environment and landscape conformation. A shortlist of 40 sites of less than 3,000 m^2 was drawn up using a map of the areas managed by the city of Lille. Based on a sociological observation grid, these sites were visited. The sociological observation grid considered the following criteria: frequency of use, diversity of uses and description of the immediate environment. Finally, eight sites were selected for the sociological study because they were relevant with regard to the criteria analysis grid: high frequency of use was observed, making it easier to recruit interviewees, high diversity of uses to collect interesting sociological material, and ease of access to the site (Figure 4.1). Each site was contextualized by (i) the geolocation of the site via Google Maps; and (ii) the qualification of the level of attractiveness as defined by Chaze (2017) of the urban environment of the sites. Each site was analyzed in terms of several variables: (i) urban amenities (food shops, schools, restaurants, presence of subway station and bus stops, cycle rental stations, cycle garages) as indicated by Sahraoui et al. (2014); and (ii) environmental amenities (squares, parks and public gardens). These amenities are assessed in relation to a distance scale classified into three categories: immediate (between 0 and 50 m around the site), near (beyond 50 m to less than 200 m), and distant (from 200 m to 500 m).

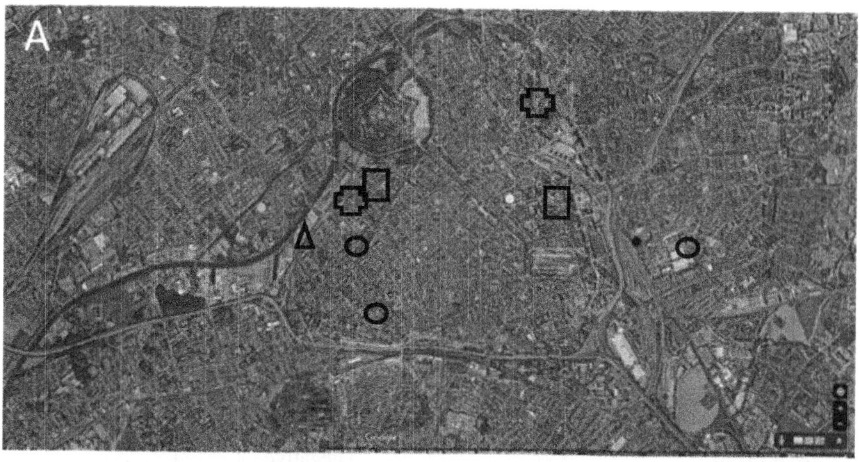

B

	Site	Typology
1	Médiathèque Jean Lévy	□ Surrounded by 3-4 buildings
2	• Square Katherine Johnson	□ Surrounded by 3-4 buildings
3	• Espace Fourchon/Bayart	Δ Surrounded by 2 buildings
4	• Espace Béranger/Fulton	○ Surrounded by 1 building
5	• Square de la Houblonnière	○ Surrounded by 1 building
6	• Rue Francis de Pressencé	○ Surrounded by 1 building
7	• Avenue de Dunkerque	✚ Surrounded by 0 building
8	• Verger conservatoire du jardin Vauban	✚ Surrounded by 0 building

Figure 4.1. *Selected sites for sociological analysis: (A) map of the city of Lille with selected sites (Google in 2025). (B) Description of selected sites. For a color version of this figure, see www.iste.co.uk/dobigny/rethinking.zip*

4.2.2. *Strategy and management methods for green urban interstices*

The aims of the study were (i) to understand the city of Lille's strategy regarding urban interstices and, more generally, nature in the city; and (ii) to study the maintenance practices of the green spaces and more specifically the urban interstices selected. The methodology used was based on several sources: (i) a literature review on the city of Lille's strategy for urban interstices, greening and, more generally, nature in the city (Lille City Council website, internal reports submitted by the local authority); and (ii) 30 interviews (individual or by group) with members of the Lille City Council staff directly involved in the management and upkeep of the studied sites.

4.2.3. *Study of perceptions and uses of selected urban interstices*

	Category 1: only positive	**Category 2: mostly positive**	**Category 3: both positive and negative**	**Category 4: mainly neutral and negative**
1st criteria: free attributes	Positive attributes only (100%) (related to well-being, site history, plants, site maintenance)	Mostly positive attributes (88%) (related to well-being, social relationships, plants, environmental quality and confidentiality)	Positive attributes (48.5%) and negatives (51.5%) (related to site potential, environmental quality) and negative attributes (related to maintenance, urban morphology) in similar proportion	Mainly neutral (25.9%) and negatives (63.2 %) attributes (related to uses, morphology)
2nd criteria: unequivocal attributes	Totally (100%) positive, unequivocal attributes	Mostly (87.5%) positive, clear-cut attributes	Positive (46,4%) and negative (53.6%) unequivocal attributes	Positive (38.9%) and negative (61.2 %) unequivocal attributes
3rd criteria: influence of the senses on satisfaction, pleasure and curiosity	Above average satisfaction, enjoyment and curiosity levels (at least +20 points compared with the overall average for the sites)	Above average satisfaction and enjoyment (at least +5 points more than the overall average for the sites). Level of curiosity within the average for sites	Below-average levels of satisfaction, pleasure and curiosity (at least -10 points below the average)	Below-average level of satisfaction and enjoyment (at least -10 points below the average)

Table 4.1. *Spontaneous site characterization by users*

A total of 92 semi-structured interviews were conducted with users of the selected sites (interviews corresponding to Time 1, T1). Each interview was transcribed. The aim of these interviews was to explore the perceptions of the users, how they use the sites and what their expectations were of these spaces. Users' profiles were also collected. A thematic analysis of the interviews was carried out by analyzing a series of questions designed to characterize the sites, illustrated by verbatims. The questions were: "What words come to mind to characterize or describe this space?"; "In your opinion, the site is rather…" (beautiful/ugly, artificial/natural, interesting/interesting, pleasant/unpleasant, attractive/repellent, useful/useless, maintained/neglected, rich/poor in biodiversity, favorable/disadvantageous for social interaction); "Do the sights, smells and sounds of this room give you satisfaction/curiosity/pleasure on a scale of 1 to 5?". The analysis of the responses allowed us to classify the level of perception into four categories (see Table 4.1).

In addition, the study of uses of these sites was conducted through the analysis of semi-directive interviews, identifying uses related to rest and contemplation, recreational and/or social activities, educational purposes, ecological functions and as a transit or parking area.

4.2.4. *Temporal study of perception, uses and knowledge of biodiversity in selected urban interstices*

Based on the interviews performed in T1, the users were invited to take part in a site visit in the presence of scientists who worked on the biodiversity of the selected sites. The aim of these visits was to provide users with data about the biodiversity present in the sites to raise residents' awareness of the diversity of visible and invisible flora and fauna on the ground. On each visit, between three and six scientists from different disciplines were present to give a comprehensive overview of the issues raised during the project. Educational materials were produced to illustrate the scientific explanation. A total of 30 of the 92 users interviewed in T1 agreed to take part in the site visits and interviews in T2, which were conducted in the same way as in T1. The aim of these interviews was to analyze the impact of the site visits on perceptions and uses of the sites, and on the level of users' knowledge in terms of biodiversity.

4.3. Results

4.3.1. *Site mapping, selection and contextualization*

Eight sites were selected for the sociological study (Figure 4.1). Of these eight sites, at least seven were close to urban amenities "immediate or near", while only four out of eight were close to environmental amenities (see Table 4.2). For example, "Square Katherine Johnson" and "Verger conservatoire du jardin Vauban" are close "Immediate or near" to both urban and environmental amenities, while "Espace Fourchon/Bayart" is close to urban amenities but distant from environmental amenities.

	Immediate	Near	Distant	Total
Access to urban amenities				
Close to food shops, schools, restaurants and other services	3	4	1	**8**
Proximity to public and soft transport networks: railway station and metro stations, bus stops, V'Lille terminals, etc.	3	5		**8**
Access to environmental amenities				
Availability of squares, gardens and parks	1	3	4	**8**

Table 4.2. *Distance of selected sites from urban and environmental amenities*

4.3.2. *Strategy and management methods for urban interstices and nature in the city*

Since the 2000s, the city of Lille has been concerned about nature in the city and, more specifically, in the urban interstices. This concern has been formalized in the territory's strategic documents (Strategy for the Ecological Management of Green Areas adopted in 2001, Lille Biodiversity Plan 2010, Metropolitan Local Town Planning Plan since 2020, Lille Low Carbon Pact 2021) and through actions related to nature in the city. The first eco-district was created in 2004, and an ecologist position was created in 2005, and implementation of biodiversity protection measures, such as the phasing out of pesticides for the maintenance of green spaces in 2008. Involving residents in greening projects is a priority. This can be seen in various programs such as participatory budgets and the greening license. The city of Lille promotes its involvement in preserving biodiversity, including through its participation in various competitions for local authorities (e.g. the French Capital of Biodiversity).

At the level of the city of Lille, public green spaces, including urban interstices, are managed by three areas of activity: design, maintenance and cleanliness. The city's green spaces cover 335 hectares. The type of management varies according to the size of the area and the level of technical expertise required. The dominant trend is as follows: small areas are mainly managed by the public service (approximately 150 hectares) and are subject to horticultural management; on the other hand, large areas (approximately 185 hectares) are subject to two types of management: mixed management by the public service and contractors, and management by contractors alone because of cost and availability of equipment (e.g. extensive areas requiring mowing). The frequency of intervention (mowing frequency: the only information we have) varies from site to site (Table 4.3).

Interstices	Site management	Intervention frequency	Categorization of spontaneous site characterization
Médiathèque Jean Lévy	Public	14 mowings a year	Mostly positive characterization
Verger conservatoire du jardin Vauban	Public	26 mowings a year	Only positive characterization
Espace Béranger/Fulton	Public	10–13 mowings a year	Positive as well as negative characterization
Rue Francis de Pressencé	Private	Few to none mowings	Positive as well as negative characterization
Espace Fourchon/Bayart	Public	13 mowings a year	Mostly neutral and negative characterization
Avenue de Dunkerque	Public	13 mowings a year	Mostly neutral and negative characterization
Square Katherine Johnson	Public	14 mowings a year	Mostly positive characterization
Square de la Houblonnière	Public	13–15 mowings a year	Mostly positive characterization

Table 4.3. *Management and characterization of the urban interstices selected*

The transition to "zero phyto" is now almost unanimously supported by city staff. However, they continue to face challenges in managing and maintaining urban spaces, including paradoxical practices, the proliferation of urban areas, a lack of management plans, pronounced compartmentalization between design and management, and increasing issues related to local urban management.

4.3.3. *User profiles, uses and perceptions of selected urban interstices*

The 92 interviews conducted with local residents revealed a relatively homogeneous group of respondents: more than half the respondents were couples (57.6%); almost half were aged 55 or over (46.7%); the level of education was high (65.2% of respondents had at least three years of higher education); the majority were owners of single-family homes, with strong local attachment and had lived in their homes for a long time: 68.5% of respondents were homeowners or first-time buyers, 53.3% of respondents lived in single-family homes, 56.5% had lived in their current district for at least 15 years and 65.3% had lived in their current home for at least ten years.

The users' characterization of the sites is summarized in Table 4.4. We noted that the "Verger conservatoire du jardin Vauban" and "Square Katherine Johnson" are the sites with the most positive characterization according to the users interviewed. On the other hand, "Espace Fourchon Bayart" and "Avenue de Dunkerque" are the two sites with the most neutral and negative characterization.

The analysis of the interviews reveals a variety of uses depending on the site. There are uses for contemplation and rest, uses for recreation and/or socializing and/or creating social links, educational uses, ecological uses and uses related to passage or parking area (Table 4.4).

The "Médiathèque Jean Levy", "Square de la Houblonnière", "Verger conservatoire du jardin Vauban" and "Square Katherine Johnson" are the sites where the greatest diversity of use has been identified. Otherwise, the "Fourchon/Bayart" and "Avenue de Dunkerque" sites are only used as a passage and/or car park.

	Use related to rest and contemplation	**Recreational and/or social use and/or creating social links**	**Educational use**	**Ecological use**	**Passage and/or parking area**
Médiathèque Jean Lévy	x	x	x	x	
Verger conservatoire du jardin Vauban	x	x	x	x	x
Espace Béranger/Fulton		x			x
Rue Francis de Pressencé		x			x
Espace Fourchon/Bayart					x
Avenue de Dunkerque					x
Square Katherine Johnson	x	x	x		x
Square de la Houblonnière	x	x	x	x	x

Table 4.4. *Inventory of the uses for each selected site*

Regardless of its appearance, each green space studied was appreciated by the respondents for its mere presence. The respondents' feedback underscores the numerous benefits associated with visiting or simply viewing a green space within an urban environment dominated by concrete: health benefits (both physical and mental well-being), enhanced feelings of safety, strengthened social connections, environmental advantages, educational opportunities and a positive impact on perceptions of noise and visual pollution. An analysis of the respondents' comments reveals a strong attachment to the sites, and more broadly to the green spaces of the city of Lille. Construction projects that have led to the disappearance of green spaces are viewed very negatively by the respondents, who express regret and incomprehension regarding this practice, especially considering current environmental concerns. The respondents expressed a strong desire to understand the intricacies of construction projects. Additionally, they envisioned each green space as having potential for development in terms of landscaping, activities, access arrangements and street furniture. Users find it easy to project themselves into the future of these sites. The respondents recognize the heritage value offered by the site from both mineral and plant elements. The historic character of certain sites, such as the "Verger conservatoire du Jardin Vauban", is frequently highlighted. Respondents also recognize the environmental value of these sites. The services provided by each site, including their role as urban cool islands and their ability to capture CO_2 through vegetation, were emphasized. These sites are perceived as offering significant environmental benefits in contrast to the built-up area of Lille. The "Verger conservatoire du Jardin Vauban" and the "Médiathèque Jean Lévy" sites received the most positive comments.

Weak points were also highlighted and can be divided into five types: 1) lack of physical and functional visibility: lack of signage, absence of rules of use; 2) low-quality urban and landscaping development: urban furniture absent, in poor condition or unsuitable, lack of security at sites; 3) lack of accessibility: access is unique and not very visible from the outside; this deficit is perceived specifically for sites with a significant historical heritage; 4) lack of connection is a weak point identified for more than half of the sites; and 5) problems of incivility are mentioned for the squares.

4.3.4. *Temporal evolution of perception, uses and level of knowledge of biodiversity*

An analysis of changes in the profile, perceptions, level of knowledge and practices regarding biodiversity reveals three key findings. The socio-demographic, economic, geographical and housing-related profile of the respondents between T1 and T2 (n = 29) showed little variation. Between T1 and T2, environmental benefits and well-being benefits were cited more frequently: environmental benefits were mentioned 23 times in T1 and 31 times in T2, while well-being benefits were cited nine times in T1 and 19 times in T2. Conversely, the benefits related to social connections were mentioned less frequently between T1 and T2: ten times in T1 and four times in T2. Between T1 and T2, the weak points concerning the poor quality of urban and landscaping development were more recurrent: cited 34 times in T1 and 38 times in T2. However, the lack of physical visibility and rules of use were mentioned less frequently: nine times in T1 and six times in T2.

The 30 users surveyed were very positive about the site visits and identified several benefits: the discovery of an invisible biodiversity essential to life, the value of these visits in driving change and raising awareness of environmental issues among the population, increased interest in biodiversity and the associated knowledge, and changes in practices regarding biodiversity and the management and maintenance of green spaces. The suggested points for improvement from users include enhancing the accessibility of the site visit content, clarifying the purpose of the visits and highlighting the connections between humans and biodiversity.

The level of biodiversity knowledge of the residents/users interviewed changed between T1 and T2 (Table 4.5) In T2, there were some changes: no users were classified in category 0, category 1 had fewer users (17.8%), the proportion of residents with an intermediate level of knowledge of the concept of biodiversity became the majority in T2 (rising from 44.8% in T1 to 64.3% in T2) and the proportion of category 4 users did not change. Overall, users have a better understanding of the concept, and the issues involved (Table 4.5).

	T1	T2
Category 0 "Lack of knowledge"	3	0
Category 1 "Basic level of knowledge"	8	5
Category 2 "Intermediate of knowledge"	13	18
Category 3 "Initiated knowledge level"	3	5
Category 4 "Expert level of knowledge"	0	0

Table 4.5. *Improvement of knowledge regarding the concept of biodiversity*

4.4. Discussion

4.4.1. *Social acceptability of changes in management and upkeeping practices of urban interstices*

Most small green spaces are managed by the public service. This observation is confirmed for the urban interstices selected for this sociological study: seven out of eight interstices are managed by the Lille city council. The case of "Rue Francis de Pressencé" is specific in that it belongs to a consultancy firm operating in the environmental sector.

The way in which the city of Lille upkeeps and manages its urban interstices has evolved toward ecological management since the early 2000s, but not without difficulties, obstacles and reluctance. Today, more than 20 years after its introduction, the level of social acceptance among staff and residents is more favorable. This has also been observed in studies carried out in other cities. This is the case in the city of Rennes, also in France, where a survey of the city's managers showed that the differentiated management implemented in 1981 also took approximately 20 years before these changes in practices were accepted (Martin et al. 2015). Gathering the perceptions of green space managers shows that the introduction of new activities for technicians (manual weeding, integrated biological pest management, etc.) has generated fears, particularly that of losing "the creative character" as described by Robert et al. (2015). According to municipal employees, user acceptance of the changes would have been easier if they had been accompanied by information and awareness-raising campaigns on the issues involved in differentiated management.

For several years now, phenomena linked to climate change have been emerging, such as heatwaves leading to periods of drought, which have resulted in restricted watering times and the appearance of pests (e.g. the box tree borer, *Cydalima perspectalis*). In response to these challenges, the municipal employees are adapting their maintenance or upkeeping practices by implementing adaptive solutions. These include the use of mulching techniques to help shade the soil, conserve moisture and reduce the need for weeding and watering. Additionally, they select plant species that are tolerant of dry spells (e.g. mulberry, *Morus* sp. and hackberry, *Celtis australis*). These changes in practices are part of a broader concept of territorial resilience, which is defined by Villar and Guézo (2018).

4.4.2. *Interconnections between site utilization, perception, attractiveness and management practices*

The results obtained in terms of uses and perceptions should be put into perspective, as the sample of respondents is not entirely representative of the city of Lille: more than half of the respondents were couples (57.6%), compared with 32.6% for the city of Lille as a whole (Insee 2020); almost half of those questioned were aged 55 and over (46.7%), compared with 19% for the city of Lille (Insee 2020). The sample is not entirely representative, but our aim was to carry out a qualitative study that would enable us to interview a wide range of profilers who frequently use these sites.

The uses identified in the eight urban interstices are diverse and complementary. The proposed typology can be compared with that proposed by *ResearchGate* (2005) which identifies cultural services (recreational, educational, aesthetic, spiritual) as part of ecosystem services. The identification of uses is linked to the profile of the population surveyed, as previously shown by Adinolfi et al. (2014): for example, recreational use was recorded on sites where the population surveyed consisted mainly of couples with children: this was the case in particular for "Square de la Houblonnière" (60% of couples with children surveyed) and "Square Katherine Johnson" (46% of couples with children surveyed).

In addition, our study shows that users' perceptions of small green spaces are more positive when these spaces offer a wide variety of uses. Sites with a high diversity of uses correspond both to sites with the most positive spontaneous characterization and to sites with a high level of attractiveness (distance from amenities).

The study also shows a link between positive perception and the frequency of intervention in urban interstices (see Table 4.3). This is in line with what was noted by Lizée et al. (2014). Nevertheless, there are exceptions such as the Fourchon/Bayart space (unsignalized parking space) and Avenue de Dunkerque (grassed space along an avenue) which have a high frequency of intervention, but which do not benefit from a positive perception. This finding shows that management alone is not enough and that the quality of urban planning and the location of sites also influence perception, as demonstrated by Van Herzele and Wiedemann (2003) and Bjerke et al. (2006).

Among the strengths of green spaces identified by users, it is first and foremost the health benefits (physical and mental well-being) that are highlighted as unanimously emphasized by other studies (Long and Tonini 2012; Konijnendijk van den Bosch et al. 2013).

Users find it very easy to envision the future of these sites. This can be partly attributed to the expertise and knowledge of the interviewees, who have strong local ties at various levels of the city (from neighborhoods to housing). These deep-rooted connections also explain the high emotional value expressed by users of these green spaces.

4.4.3. *Impact of the site's visits*

The aim of the site's visits was to initiate dialogue between the team of researchers and the users, to gather their perceptions of the urban interstices and to raise awareness among residents of the diversity of the flora, fauna and micro-organisms in the soil. Measuring changes in the level of knowledge about biodiversity shows that the users who took part in the site's visit have generally improved their knowledge of biodiversity. This clearly shows the positive impact of communication and awareness-raising initiatives on participants' perception of the biodiversity as shown previously by other studies (Dejean et al. 2019).

Based on the analysis of interviews with municipal employees, several recommendations are proposed to enhance the visibility and quality of use of urban interstices. These include developing an urban interstices strategy for the city of Lille, creating an inventory of urban interstices, fostering closer collaboration between the land registry department and the "Nature en Ville" department, incorporating staff expertise into design criteria and maintenance methods for the sites, developing management plans for each site, promoting the expertise of city of Lille staff and raising awareness of biodiversity issues among residents

4.5. Conclusion

The study revealed that the city of Lille has had a long-standing commitment to the renaturation of the city, dating back to the early 2000s. This strategy is reflected in the ecological management of its green spaces and the involvement of residents in greening projects. Nonetheless, the Lille City Council staff still face daily challenges in managing and maintaining these urban interstices.

In response to the issues encountered by the staff, we proposed recommendations to enhance the visibility and quality of use of urban interstices, as well as to improve management methods.

The study also demonstrated that, despite their small size, these spaces hold significant sociological interest. Firstly, we observed a strong emotional attachment to these spaces, regardless of the level of appreciation. Respondents highlighted the multiple benefits and perceived the heritage value from both a mineral and plant perspective. The results of the sociological analysis indicate that positive appreciation is linked to a diversity of uses and a high level of site attractiveness (proximity to amenities).

To improve user satisfaction at the studied sites, it appears necessary to implement (i) more physical facilities to enhance the user experience; (ii) more facilities to promote biodiversity; (iii) a wider range of activities; (iv) better accessibility to these sites; and (v) improved quality and signage for landscaping and urban facilities.

Developing green urban strategies with the participation of the population: upstream integration of residents in site design and redevelopment projects, and throughout the evolution of site development.

4.6. References

Adinolfi, C., Suárez-Cáceres, G.P., Cariñanos, P. (2014). Relation between visitors' behaviour and characteristics of green spaces in the city of Granada, south-eastern Spain. *Urban Forestry & Urban Greening*, 13(3), 534–542. doi: 10.1016/j.ufug.2014.03.007.

Bjerke, T. Østdahl, T., Thrane, C., Strumse, E. (2006). Vegetation density of urban parks and perceived appropriateness for recreation. *Urban Forestry & Urban Greening*, 5(1), 35–44. doi: 10.1016/j.ufug.2006.01.006.

Bush, J. (2020). The role of local government greening policies in the transition towards nature-based cities. *Environmental Innovation and Societal Transitions*, 35, 35–44. doi: 10.1016/j.eist.2020.01.015.

Chaze, M. (2017). L'attractivité territoriale : éléments de définition. Research Report, UMR Territoires [Online]. Available at: https://shs.hal.science/halshs-02520675v1.

Dejean, A., Hild, A., Rotaru, R., Sasso, M., Vuilleumier-Papaloïzos, D., Clergeau, D. (2019). Des leviers d'action pour favoriser la biodiversité urbaine dans le cadre du processus de métropolisation. *Cybergeo*. doi: 10.4000/cybergeo.32578.

Egea-Cariñanos, P., Calaza-Martínez, P., López Roche, D., Cariñanos, P. (2024). Uses, attitudes and perceptions of urban green spaces according to the sociodemographic profile: An exploratory analysis in Spain. *Cities*, 150, 104996. doi: 10.1016/j.cities.2024.104996.

Erdi (2021). Urban ınterstices between appropriations and resistances. *User Experience and Urban Creativity* [Online]. Available at: https://shs.hal.science/halshs-03891512v1.

Insee (2020). Recensements de la population. Données infra-communales-Couples-Familles-Ménages.

IPCC (2023). Climate change 2023: Synthesis report. Contribution of working groups I, II and III to the sixth assessment report of the intergovernmental panel on climate change. Report, Geneva. doi: 10.59327/IPCC/AR6-9789291691647.

Konijnendijk van den Bosch, C., Annerstedt, M., Nielsen, A.B., Maruthaveeran, S. (2013). Benefits of urban parks – Ifpra review identifies strength of the evidence. *IFPRA World*, 2013(3) [Online]. Available at: https://res.slu.se/id/publ/54134 [Accessed 20 February 2025].

Lizée, M.-H., Bourdil, C., Barthélémy, C., Deschamps-Cottin, M. (2014). Approche socio-écologique des parcs publics marseillais : relations entre gestion et biodiversité urbaine. *Méditerranée*, 123, 123–132. doi: 10.4000/mediterranee.7470.

Long, N. and Tonini, B. (2012). Les espaces verts urbains : étude exploratoire des pratiques et du ressenti des usagers. *Vertigo*, 12(2). doi: 10.4000/vertigo.12931.

Martin, B., Gutleben, C., Colson, F. (2015). Aménagement et gestion des espaces verts pour produire, sous contraintes réglementaires, des bouquets de services écosystémiques : témoignage de la ville de Rennes. *Innovations Agronomiques*, 45, 95–99. doi: 10.15454/1.4622658170018E12.

Miao, X., Pan, Y., Chen, H., Zhang, M-J., Hu, W., Li, Y., Wu, R., Wang, P., Fang, S., Niu, S. et al. (2023). Understanding spontaneous biodiversity in informal urban green spaces: A local-landscape filtering framework with a test on wall plants. *Urban Forestry & Urban Greening*, 86, 127996. doi: 10.1016/j.ufug.2023.127996.

ResearchGate (2005). Millennium Ecosystem Assessment. Ecosystems and human well-being: Synthesis [Online]. Available at: https://www.researchgate.net/publication/297563785_Millennium_Ecosystem_Assessment_Ecosystems_and_human_well-being_synthesis [Accessed 20 February 2025].

Robert, A., Yengué, J.L., Lamare, J. (2015). Espaces verts urbains et transition écologique, perception par les gestionnaires municipaux. In *5e colloque de l'ARPEnv (association pour la recherche en psychologie de l'environnement) Transition écologique : de la perception à l'action*, Nanterre [Online]. Available at: https://hal.science/hal-01165321v1.

Robinson, J.M., Breed, A.C., Camargo, A., Redvers, N., Breed, M.F. (2024). Biodiversity and human health: A scoping review and examples of underrepresented linkages. *Environmental Research*, 246, 118115. doi: 10.1016/j.envres.2024.118115.

Rupprecht, C.D.D. and Byrne, J.A. (2014). Informal urban greenspace: A typology and trilingual systematic review of its role for urban residents and trends in the literature. *Urban Forestry & Urban Greening*, 13(4), 597–611. doi: 10.1016/j.ufug.2014.09.002.

Sahraoui, Y., Lunardi, N., Antoni, J.-P. (2014). Essai de typologie des espaces résidentiels à partir d'indicateurs désagrégés à l'échelle du bâtiment. Application à Besançon et à Metz'. *Cybergeo*. doi: 10.4000/cybergeo.26431.

United Nations, Department of Economic and Social Affairs, Population Division (2018). World Urbanization Prospects: The 2018 Revision.

Van Herzele, A. and Wiedemann, T. (2003). A monitoring tool for the provision of accessible and attractive urban green spaces. *Landscape and Urban Planning*, 63(2), 109–126. doi: 10.1016/S0169-2046(02)00192-5.

Villar, C. and Guézo, B. (2018). La résilience des territoires aux catastrophes. Report [Online]. Available at: https://www.cerema.fr/fr/actualites/resilience-territoires-aux-catastrophes-document-du-cgdd [Accessed 28 March 2025].

Wu, C.-C., O'Keefe, J., Ding, Y., Sullivan, W.C. (2024). Biodiversity of urban green spaces and human health: A systematic review of recent research. *Frontiers in Ecology and Evolution*, 12. doi: 10.3389/fevo.2024.1467568.

5

From a Climate-Centered Vision to a *Vivo-Centered* Vision: A Paradigm Shift for a Genuine Socio-Ecological Transition?

5.1. Introduction

In this chapter, we will explore the extent to which current socio-ecological transition strategies perpetuate an extractivist and anthropocentric model, prioritizing techno-solutionist approaches and focusing on a single symptom: CO_2 emissions. We will examine how this "pseudo-transition", by overlooking the root causes of the socio-ecological crisis, ultimately limits our capacity to move towards a truly sustainable system. In light of this observation, we will investigate the potential of a transition model based on strong sustainability, one that places the flourishing of the living at the core of its concerns.

We will analyze how such a *vivo-centered*[1] approach could not only strengthen the resilience of our societies in the face of environmental, social and economic challenges, but also make the transition project more desirable and engaging. Finally, we will translate these reflections into a few concrete proposals for implementation within a university ecosystem, illustrating possible levers of action to embed this vision into practice.

Chapter written by Alice FOUILLOUZE.

For a color version of all figures in this chapter, see www.iste.co.uk/dobigny/rethinking.zip.

1 From the French word "vivant": the living world including human and nonhuman.

5.2. A meta-crisis reduced to a molecule, CO_2

Although the efforts currently being made for the transition are multiplying, they remain insufficient or, paradoxically, even reinforce unsustainable trajectories.

The ecological crisis is frequently reduced, in Western societies, to the sole issue of climate change (the "carbon tunnel vision"). Carbon neutrality and decarbonization are now placed at the heart of our transition policies, reducing ecological transition to an energy transition aimed at replacing fossil fuels with clean energy. This focus on the carbon footprint raises several issues (Savasta-Kennedy 2014).

Focusing on reducing carbon dioxide (CO_2) emissions tends to obscure other equally crucial issues, such as biodiversity loss, resource depletion and the associated economic, social and geopolitical challenges. For instance, the Reducing Emission from Deforestation and Forest Degradation program (REDD+), launched by the United Nations to address climate change issues through carbon offsetting, resulted in the planting eucalyptus monocultures that did not absorb as much CO_2 as expected, made the soil sterile and did not benefit local communities (they even deprived them of their land) (Nuesiri 2017)[2]. Furthermore, this CO_2-focused approach encourages the use of essentially technical responses (such as electric cars or geo-engineering) and fuels a techno-solutionist mindset (the belief that technology can solve all problems, including socio-ecological ones). Yet, the widespread use of technology intensifies mining[3], exacerbates resource depletion, generates new forms of pollution and contributes to the collapse of biodiversity, leading to an increase in the ecological footprint[4].

As a result, transition strategies, primarily conceived through the lens of carbon, fail to achieve long-term sustainability because, in attempting to solve one problem, they negatively impact others, ultimately risking worsening the overall socio-ecological balance, rather than improving it.

---

2 See also Tordjman, H. (2021). *La croissance verte contre la nature.* Éditions La Découverte, Paris.

3 See, for example, Izoard, C. (2024). *La Ruée minière au XXIe siècle. Enquête sur les métaux à l'ère de la transition.* Éditions du Seuil, Paris.

4 While technology has contributed to social progress, it has also amplified our capacity to destroy nature, fueling the ecological crisis. Seeking purely technical solutions to a problem created by technology risks generating an endless spiral of new challenges. Moreover, relying on energy- and resource-intensive technology seems unwise in a context of resource depletion. For all of these reasons, it seems essential to adopt not a technophobic stance, but a techno-critical one, in order to make technology a tool in service of the transition, rather than an end in itself perpetuating an unsustainable model.

This focus on carbon perpetuates a compartmentalized view of problems that reflects our inability to approach challenges in a systemic way, i.e. to understand problems in their globality and complexity, considering their interactions and common root causes[5].

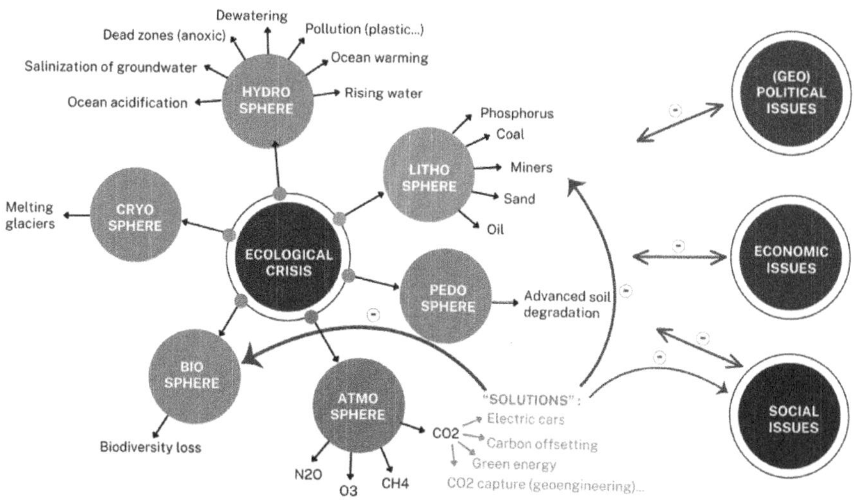

Figure 5.1. *Examples of negative impacts of climate change mitigation measures on other socio-ecological issues. The impacts are represented by arrows with minus symbols (source: Alice Fouillouze, inspired by lectures by Keller)*

This siloed vision of socio-ecological issues also reveals a fundamental confusion between the causes and the consequences of the crisis. By focusing on *symptoms* such as climate change, biodiversity loss or pollution, transition strategies remain superficial and miss the root causes of the problem. Understanding the crisis as a systemic crisis means recognizing that it is a crisis of a global system, shaped by destructive economic, social and cultural logics. A true transformation, therefore, requires a radical approach – in the etymological sense of the word *radix*, which "takes at the root" – addressing not only the manifestations of the crisis but, above all, its deep causes. Only under these conditions can we halt the destruction of the living and move towards a truly sustainable system.

5 Adopting a systemic approach is crucial to ensure coherence in transition strategies. Otherwise, we risk waging war on CO_2… and on the living itself. For Arthur Keller, "without a coherent systemic approach, we relieve, at best, some symptoms while aggravating others. We do not solve the problem, we shift it".

5.3. Toward an inevitable energy and material descent

Arthur Keller, a specialist in systemic risks and resilience strategies, offers a systemic and radical perspective on the socio-ecological crisis. He describes our civilization as a gigantic machine that converts natural elements into waste (Keller 2013, 2020). Upstream, we exploit resources, transform them using energy and use them in the form of goods and services. Downstream, these processes generate waste and pollution that accumulate in the environment. This linear model leads to a continuous degradation of ecosystems, producing solid, liquid and gaseous emissions, some of which, such as greenhouse gases, disrupt the climate balance.

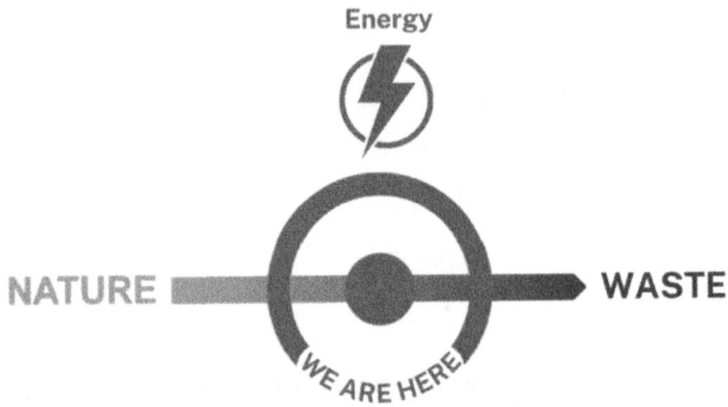

Figure 5.2. *Socio-ecological crisis: our civilization, a mega-machine that turns nature into waste (inspired by Keller's lectures)*

The central problem lies in the fact that this flow far exceeds the planet's resilience capacities: resources are extracted faster than they can regenerate; pollution is produced faster than ecosystems can absorb it; the environment is altered at a rate higher than its adaptive capacity; and natural habitats are destroyed more rapidly than they can regenerate. As a result, since the 1970s, the ecological footprint (see Figure 5.3, red curve) has exceeded the Earth's biocapacity (see Figure 5.3, green curve).

Although we cannot predict the future, one thing is now certain: the Earth system is literally exhausted. At least six of the nine planetary boundaries have been crossed (Richardson et al. 2023), compromising the planet's capacity to maintain ecosystem functions and meet our growing needs. As a result, the biocapacity curve is likely to continue to decline, dragging with it, in the coming years, the ecological footprint curve. In other words, there will, *no matter what we do*, be an energy and material descent in the coming decades, which will concretely result in a forced reduction in production and consumption within our society (Herrington 2021). This statement is not based on collapse theory, but rather on biogeophysical laws and the dynamics of complex systems: "there will always be limits to growth. If we do not impose them on ourselves, the system will do so" (Meadows 2009).

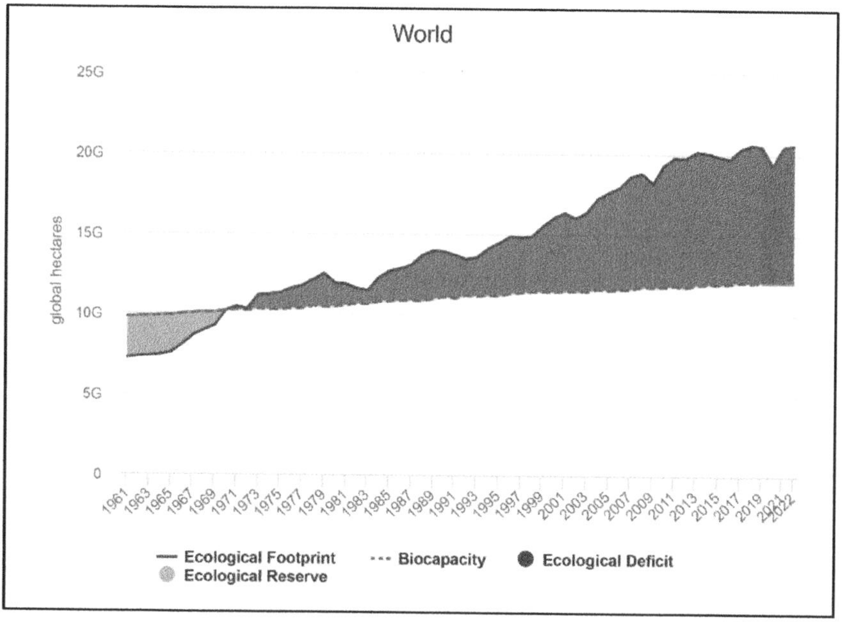

Figure 5.3. *Evolution of global ecological footprint*[6] *and biocapacity*[7] *(in global hectares, gha)*[8] *between 1961 and 2022 worldwide: an increasing ecological deficit. Note: last three years are estimates (source: York University, FoDaFo, Global Footprint Network, 2023 National Footprint and Biocapacity Accounts)*

5.4. Weak sustainability versus strong sustainability: two ways of conceiving sustainability

In the face of the upcoming energy and material descent, two main trends are available to us.

6 The ecological footprint measures the pressure exerted by humanity on ecosystems.

7 The biocapacity measures the capacity of ecosystems to provide us with resources and absorb our waste.

8 A global hectare is a hectare on the Earth's surface with the capacity to produce resources and absorb waste corresponding to the global average.

5.4.1. *We maintain the current trajectory ("business as usual" approach)*

Maintaining the current trajectory – often described as a "business as usual" approach – means responding to the socio-ecological crisis through the lens of green growth. This paradigm assumes that economic growth can be decoupled from environmental degradation, an increasingly contested assumption (Fletcher and Rammelt 2017; Hickel 2019). It promotes technical solutions (e.g. green energy) and regulatory tools (e.g. market mechanisms to internalize environmental costs) to mitigate environmental impacts without altering existing socio-economic structures.

This approach reflects the logic of *weak sustainability* (Costanza and Daly 1992; Vivien 2005). It assumes that different forms of capital – natural, human technological – are substitutable, so that replacing pollinators with drones could be considered a sustainable solution. It treats the human and natural spheres as largely separate, and often overlooks biogeophysical constraints (Keller 2013).

This paradigm, which prioritizes technological innovation and tends to subordinate living systems to economic objectives, remains prevalent within dominant political and economic frameworks, including in global agendas such as the United Nations Sustainable Development Goals (Eisenmenger et al. 2020). However, the changes it promotes – often referred to as type-1 changes, characterized by superficial or marginal adjustments – appear insufficient to reverse current ecological trends. They particularly fail to ensure a significant reduction of humanity's ecological footprint to within the limits of planetary biocapacity. What emerges instead is a form of "pseudo-transition" that maintains the trajectory of the prevailing ultra-liberal, extractivist economic model.

Under such conditions, socio-ecological systems may be exposed to a forced reduction in energy and material throughput, with potentially severe consequences, including economic recession, rising social tensions and the intensification of environmental crises. This trajectory entails a growing risk of undermining the resilience of societies and increasing their vulnerability to future systemic disruptions (see Figure 5.4, orange curve).

5.4.2. *We anticipate and prepare for this upcoming energy and material descent*

The underlying rationale here is that only a substantial reduction in production and consumption can lead to a meaningful decrease in ecological footprint and a restoration of biocapacity, ultimately moving towards socio-ecological sustainability.

This reasoning is emblematic of the *strong sustainability* paradigm (Costanza and Daly 1992; Vivien 2005), which holds that a system can only be considered sustainable if its ecological footprint remains below the level of biocapacity. It acknowledges the fundamental dependence of the anthropic sphere on the natural sphere and affirms that technical capital (e.g. pollinating drones) cannot substitute for natural capital (e.g. pollinating insects). Rather than relying on incremental technical adjustments, this paradigm calls for a profound reconfiguration of institutional frameworks, cultural values and modes of living. Strong sustainability explicitly integrates the necessity of respecting planetary boundaries while promoting long-term social justice. It thus supports a more radical – in the etymological sense of addressing problems at their root – and systemic response to socio-ecological challenges, recognizing the deep entanglement of environmental and socio-economic issues and the need for comprehensive transformation. Although marginal within current policy frameworks, such type-2 changes – structural and transformative – arguably offer a more robust potential to mitigate the consequences of the socio-ecological crisis and to strengthen territorial resilience, notably by promoting the relocalization of flows, production and consumption, thereby reducing dependence on increasingly fragile global systems (see Figure 5.4, gray curve).

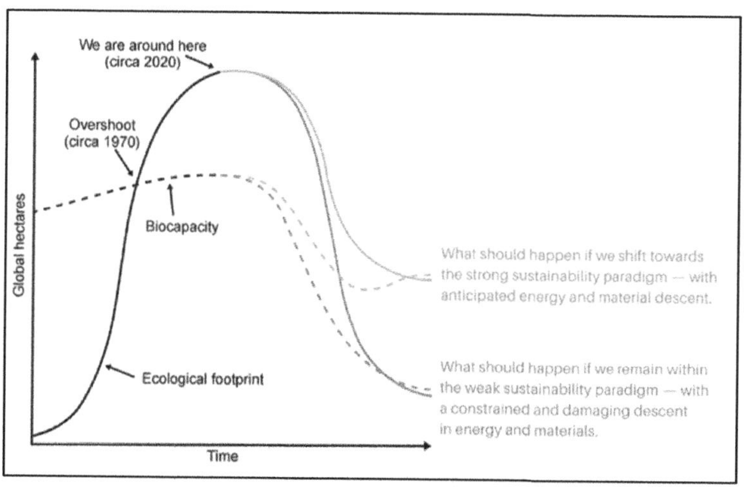

Figure 5.4. *Projected trends of ecological footprint and biocapacity based on weak or strong sustainability paradigms*

5.5. The vivo-centered transition model

Current technical and economic solutions have shown their limitations. There is a growing urgency to adopt a transition model grounded in the strong sustainability paradigm; one that rethinks strategies in a radical and systemic manner, addressing the root causes of socio-ecological issues while integrating both environmental and social dimensions (McPhearson et al. 2021).

In this perspective, we propose a socio-ecological transition model that is no longer climate-centered, as is currently the case, but *vivo-centered* (i.e. a framework that redefines how we think and act for the transition, with the primary goal of fostering the flourishing of life, both human and nonhuman). Here, human flourishing is not about material comfort or immediate gratification but about living in harmony with other beings.

This view echoes Arne Naess's ecosophy, which holds that true human fulfillment arises from recognizing our interdependence with all life and evolving in symbiosis with the natural world (Naess 1989). The aim is not only to limit harm to the biosphere, but to engage in an inner shift; to "experience the flourishing of other life forms as part of our own" (Pelluchon 2020). By "non-human flourishing", we refer to a regenerated natural world, where ecosystems function fully and regain their integrity: a nature restored to its rightful place, whether as ordinary biodiversity within human systems, or as wild spaces free from harmful human influence.

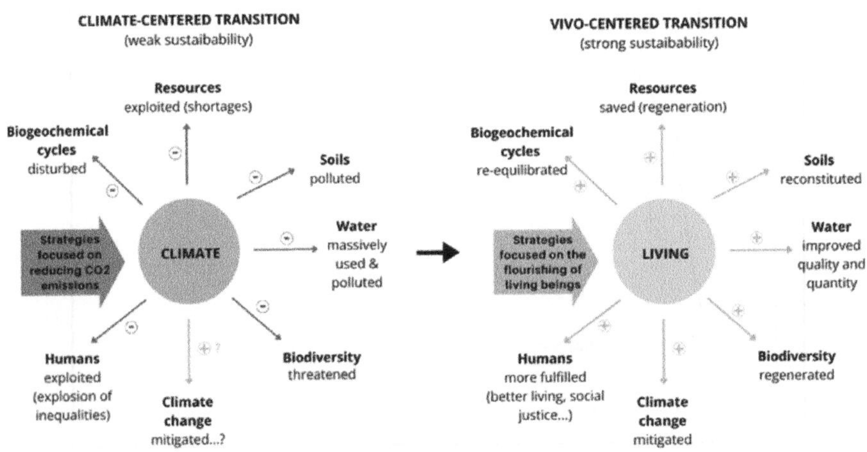

Figure 5.5. *Climate-centered versus vivo-centered transition models and their socio-ecological impacts. On the left: current strategies primarily focused on decarbonization, often implemented at the expense of other critical issues such as biodiversity loss, resource depletion and global pollution. On the right: strategies oriented towards the flourishing and regeneration of human and nonhuman lives, offering a systemic response to multiple interconnected challenges (source: Alice Fouillouze)*

5.5.1. *The principles of the vivo-centered transition*

The flourishing of life should not be regarded as a distant or utopian ideal, but as a concrete compass guiding transition strategies. We propose two foundational principles for implementing a *vivo-centered* transition, calling for a redefinition of our societal project based on:

1) The degrowth of nonessential and nonvirtuous activities, to reduce the global ecological footprint in a just and democratic way, while safeguarding collective well-being and social protections.

2) Reconnection with the living world, to transform our relationship with nature and foster deep cultural and societal change.

These principles aim to reorient not only individual behaviors but also social, political and economic systems towards a shared objective: reducing our ecological footprint while restoring a respectful and symbiotic relationship with the living world.

5.5.1.1. *Degrowth, for a structural transformation in favor of the living*

Sustainability – from a systemic perspective – will only be achieved if the ecological footprint falls below the biocapacity. Reducing our harmful impact on the biosphere and contributing to its regeneration must therefore be the fundamental objectives of any *vivo-centered* socio-ecological transition. To achieve this, it is necessary to rethink current socio-economic models in order to anticipate and prepare for the energy and material descent (and not merely endure it). The sooner a country begins to degrow, the lighter (and potentially calmer) that degrowth will be. On the other hand, the longer countries continue to grow beyond their ecological budget, the more brutal and uncontrolled the degrowth will be, making any planning difficult and increasing the risks of inequalities and social tensions. In this perspective, degrowth, understood as a planned and democratic reduction of production and consumption, would provide a relevant framework for carrying out this transition (Parrique 2023).

More than just a new economic model, degrowth would be a societal project, a holistic and systemic vision aimed at transforming economic structures to achieve environmental and social sustainability. Far from being a step backward or a forced austerity, degrowth would be an intentional choice for transformation, allowing society to move away from the extractivist and consumerist logic in favor of sustainable and socially beneficial activities. This would involve, in particular, slowing down or even eliminating the most polluting and superfluous sectors (excessive highway infrastructure, production of goods with planned obsolescence, unnecessary jobs, etc.), in order to preserve and strengthen those that are essential (agroecology, solidarity, health, education, etc.). This selective degrowth would therefore reallocate resources towards activities that meet the fundamental needs of societies while respecting planetary boundaries. Degrowth would thus require democratic questioning of our real needs and what gives meaning to our lives.

By placing the economy in the service of the living, degrowth would trigger a profound transformation that, while limiting our impact on the natural world, would ensure social justice and collective well-being. By promoting the relocalization of material and energy flows, degrowth would also contribute to increasing the resilience of territories in the face of geopolitical, economic and ecological shocks. By reducing their dependence on globalized infrastructures and long supply chains, societies would become less vulnerable to future systemic crises.

Furthermore, it is important to remember that degrowth would primarily concern privileged countries and social classes that overproduce and overconsume far beyond what is necessary for a good standard of living. It would not be about depriving those who lack the essentials but about shedding excess weight to achieve full socio-ecological health, much like a beneficial diet. Degrowth would therefore be a transitional state leading to post-growth, a stable state where the economy operates without the need for infinite growth.

Finally, reducing production does not in any way mean impoverishment, as it is possible to decrease the monetary added value of an economy (GDP), while increasing the social and ecological added value, for example, through increased free time and improved ecosystem services. Such degrowth is therefore not synonymous with recession. It is a reorientation towards a model that prioritizes quality over quantity and cooperation over competition. Thus, far from being a constraint, degrowth represents an opportunity to rethink our life and production models so that they are compatible with the flourishing of all forms of living, present and future[9].

5.5.1.2. *Reconnecting humans to nature for a cultural shift in favor of the living*

The current ecological crisis partly finds its roots in a profound disconnection between humans and nature. This phenomenon results from a Western, anthropocentric and dualistic imagination that has historically placed humans in a position of domination over a natural world perceived as a mere inert resource. This view, inherited from modernity and reinforced by technoscientific development, has shaped our economic, political and cultural systems, promoting massive exploitation

9 For more information about degrowth, see Fitzpatrick et al. (2022) and Parrique (2022, 2023, 2025).

of living beings and a break in the sensitive and symbolic bonds with nature (e.g. Descola 2005; Latour 2010).

Therefore, the socio-ecological transition is not limited to a simple reduction of our ecological footprint. It involves a profound transformation of our representations and our ways of being in the world. More than an ecological challenge, it is a philosophical challenge (Maris 2010). It is therefore urgent to rethink our relationship with the living: to move beyond the utilitarian conception of the living in order to embrace a more integrated and symbiotic vision, where living beings are no longer seen as just another set of objects in the world, but as our deepest identity (Morizot 2020).

Research in conservation psychology shows that the degree of connection to nature directly influences pro-environmental behaviors. Increased disconnection leads to growing indifference towards ecological issues (e.g. Miller 2005; Soga and Gaston 2016), while a strong connection with the world fosters attitudes and practices that support its preservation (Rosa and Collado 2019; Barragan-Jason et al. 2023).

Beyond the individual benefits for mental and physical health, forging strong ties with the living world would generate a sense of communion (i.e. the feeling of being part of a larger whole that transcends us) and activate deep levers for social transformation: the stronger the connection to nature is felt internally – through emotional, cultural or spiritual experiences – the more it would induce a lasting change in values and ways of thinking, capable of steering our societies towards a systemic and enduring ecological transition (Ives et al. 2018).

Thus, the process of reconnecting to nature, understood as strengthening and enriching positive interactions between humans and nature, could influence the goals and values embedded in a system and bring about a profound and global societal transformation. By embedding humans in an expanded community of the living and fostering a new way of inhabiting the world, the process of reconnection to the living world would pave the way for a paradigm shift, constituting a key leverage point to transition to a truly sustainable system (Abson et al. 2017; Riechers et al. 2021).

5.5.2. *The vivo-centered transition, a model rooted in strong sustainability*

The *vivo-centered* transition proposes a model of transition fully embedded within the paradigm of strong sustainability. By placing the living at the heart of the issues, this model can do the following:

1) Recognize the physical limits of the biosphere and integrates the need for energy and material reduction (natural laws take precedence over human laws).

2) Tackle the socio-ecological crisis at its roots, by questioning the dominant paradigm and initiating economic degrowth aimed at reducing our ecological footprint (a radical approach).

3) Simultaneously address various socio-ecological issues: by reducing our pressure on the living (while allowing biocapacity to increase), we recreate the conditions for a sustainable system, where a regenerated nature can fully play its role in mitigating climate change, regulating biogeochemical cycles and maintaining essential ecological balances, thus ensuring dignified living conditions for humans (a systemic approach).

In this regard, we believe the *vivo-centered* transition is a promising pathway towards a truly sustainable system.

5.5.3. *The vivo-centered transition for a desirable ecology*

In addition to ensuring the sustainability of our society in the face of upcoming economic, environmental and social challenges, we argue that a *vivo-centered* transition could make the transition project desirable.

Ecology, as it is currently conceived and perceived, is characterized by a highly rational and global approach to sustainability issues (e.g. statistical reports on "climate change", "biodiversity collapse", etc.). This "external" view of sustainability issues struggles to generate popular support. Moreover, ecology – as promoted by transition strategies mainly focused on decarbonization and still confined to individual efforts – tends to resemble constraint, effort or even sacrifice, and fosters feelings of powerlessness, guilt and despair (eco-anxiety), if not denial and rejection. As a result, the socio-ecological transition is not perceived as an emancipatory, desirable and motivating project, but divides society and struggles to define itself as a political force and a social cause (Stoknes 2014; Comby 2024).

A *vivo-centered* transition, based on reconnection to the living world and a new socio-economic model prioritizing being over having, could overcome these obstacles. By leading people to experience rich and varied encounters with nature, this approach would engage affective dimensions and generate positive emotions (awe, well-being, unity, attachment, etc.). This form of "internal" ecology would touch on our deep and intrinsic (or "visceral") motivations, allowing ecology to become not just an "ecology of constraint or knowledge", but a "lived ecology", potentially far more inspiring and mobilizing (Barbé et al. 2023).

It is important to clarify that this is not about opposing internal ecology and external ecology, but rather about viewing them as complementary approaches leading to "integral ecology": a holistic vision of ecology that seeks to integrate psychological and spiritual well-being with concrete actions to protect the environment, emphasizing that sustainability requires changes both in external systems and in states of consciousness and values at both the individual and societal levels (Esbjörn-Hargens and Zimmerman 2009).

5.6. Application of the vivo-centered transition: the case of the university ecosystem

This final section presents concrete proposals for deploying this transition within the university ecosystem. As we have seen, the *vivo-centered* transition calls for a more systemic, radical approach focused on the flourishing of the living, aiming at bringing the ecological footprint below the biocapacity. From this perspective, advocating for eco-friendly behaviors or reducing the carbon footprint on a university campus, for instance, is not enough. We need to rethink and deeply transform the way the university operates, teaches, finances and interacts with society, in light of planetary challenges. The transition must unfold across all spheres (governance, education-research, campus), requiring the commitment of all system stakeholders: leadership, administrative, technical and educational staff, as well as students.

5.6.1. *A driving and committed governance*

Strong sustainability primarily calls for deep and structural changes. These are essential to create fertile ground for the deployment of all initiatives: by changing the structure, individual and collective transitions will become possible, desirable, meaningful and truly useful.

Governance therefore plays a particularly fundamental role in the deployment of the *vivo-centered* transition. It is the guarantor of the effective and efficient implementation of this dynamic and its overall coherence. Its support and leadership in the process are indispensable.

The fundamental measure here is to place the socio-ecological transition at the heart of the university's global strategy (so that it is not just one strategy among others but the strategy that will determine all the others). This requires redefining the university's purpose (the system's finality) in advance, so that all actions are conceived and directed in favor of strong sustainability and the flourishing of living beings.

5.6.2. *A resilient and regenerative campus*

A resilient university is one capable of enduring in a context where materials and energy are expected to become increasingly scarce. *Vivo-centered* transition strategies must therefore enable a significant reduction in its ecological footprint (not just its carbon footprint), while ensuring the university's functions and guaranteeing the flourishing of its students and staff. The challenge is to reconcile ecological imperatives, operational efficiency and the well-being of humans within this socio-ecosystem.

Several pathways are available to achieve this.

5.6.2.1. *Overcoming the growth paradigm while improving the university's overall prosperity*

We could rethink the university's economic model considering planetary limits and the flourishing of the living. The challenge here is to consider long-term socio-ecological issues without compromising short-term strategic, economic and technical concerns. Seeking advice and support from ecological economists[10] is recommended.

5.6.2.2. *Organizing energy and material degrowth*

We could reduce and localize the flow of energy and materials (e.g. organic, local, plant-based food; energy and digital sobriety; reclaiming low-tech skills; reducing new product purchases; establishing carbon quotas for travel, etc.). This requires redefining, with all stakeholders, the essential goods, services and activities, as well as individual and collective needs and values. It also requires strengthening territorial anchoring by developing virtuous partnerships with local authorities, associations and engaged businesses to promote local resilience (agroecology, short circuits, social justice, etc.).

5.6.2.3. *Promoting all initiatives favorable to the return and flourishing of biodiversity*

We could encourage projects aimed at "rewilding" the campus (e.g. shared gardens, abandoned spaces, urban farms, green rooftops, de-impermeabilizing soils, etc.) and promote workshops for reconnecting with the living (e.g. "The Work That Reconnects" (Macy and Brown 2008)).

10 Environmental economists primarily work within the framework of neoclassical economics and aim to "green" the current economic system. In contrast, ecological economists adopt a more radical perspective, recognizing the need to fundamentally rethink the relationships between the economy, society and the environment. They focus on new economic models such as degrowth and circular economy, among others.

5.6.2.4. *Making the campus a living laboratory for the transition degrowth*

The campus, a place for education and student life, can be both an exemplary and inspiring space. To foster this, we could create narratives to inspire collective transition, and provide spaces for exchange and action so that the campus becomes a resource hub where the vivo-centered transition is experimented with and embodied (urban agroecology, rewilding, etc.).

5.6.3. *Conscious and engaged students, equipped to face the complex challenges of our time with compassion, integrity and competence*

	Understanding the challenges of the Anthropocene	**Reconnecting to the living (to ourselves, to others and to nature)**	**Imagining other possibilities for a sustainable world**	**Acting for a sustainable and desirable world**
Objective	Understanding the socio-ecological challenges in their entirety and complexity	Understanding and experiencing our bonds with the living world	Re-enchanting the goal of reinscribing our societies into a finite world	Understanding the levers for action to embody the transition
Disciplines	Scientific ecology, ecological economics, environmental history, etc.	Environmental philosophy, anthropology of nature, ecopsychology, body practices and meditation, etc.	Foresight, eco-art, design fiction, etc.	Social and solidarity economy, governance and participatory democracy, etc.
Skills developed	Systemic thinking	Emotion regulation and openness to alternative ways of relating to the living world	Creative, ethical, critical, and reflective thinking	Collective intelligence, sense of individual and collective effectiveness
Approach	Head	Body, heart	Head, heart	Body

Table 5.1. *Proposal for multidisciplinary course modules for a mandatory common foundation in higher education (bachelor's level)*

Today, teaching remains largely disconnected from ecological realities. Often depoliticized and dominated by technoscientific rationality, it reinforces a fragmented understanding of knowledge and perpetuates the view of the living world as external to humans (Renouard 2022). Ecological knowledge can no longer be considered as mere transversal or optional supplements. They must become a mandatory foundational core, integrated into the essential skillset of every student.

Training students on the challenges of the Anthropocene involves learning to live on a planet with limited resources. This requires a profound pedagogical overhaul, based on several key principles: promoting participatory and cooperative approaches (bottom-up), strengthening an integral and transdisciplinary pedagogy, particularly through a "head–body–heart" approach, and refocusing education on the ultimate purposes of knowledge (what will students do with this knowledge?). Such training must ultimately enable students to acquire not only knowledge but also skills and attitudes that foster awareness and the ability to take appropriate action.

5.6.4. *An ethical and sustainable research service for socio-ecological challenges*

Research must become a key lever in the *vivo-centered* transition. This involves prioritizing funding for projects that have a positive impact on socio-ecological issues and dedicating a significant portion of research topics to these themes. Each project or pre-project should be evaluated based on its contribution to the transition. This dynamic requires encouraging interdisciplinary and transdisciplinary approaches, incorporating systemic, ethical and radical reflections. Furthermore, researchers' practices must be aligned with these imperatives, integrating concrete measures such as assessing their carbon footprint, setting air travel quotas and developing low-carbon alternatives for research and academic collaborations.

5.6.5. *Employees and students at the heart of transition projects*

Co-constructing with all stakeholders is essential to ensure ownership and effectiveness of transition projects. Leveraging the expertise of staff, particularly those on the ground, helps address technical and practical challenges while overcoming obstacles together. By starting from the concrete realities of teaching teams and existing initiatives, a top-down approach is avoided, ensuring an adapted and practical implementation. Involving employees and students from the outset, gathering their expectations and needs, is therefore a key condition for a rooted and sustainable transition.

To support this transformation, it will also be crucial to train and support administrative and technical staff, as well as management teams, so that they integrate the implications of these issues into their roles and missions. Additionally, the entire teaching and pedagogical staff should be trained on transition-related challenges and supported in adapting their teaching methods and practices[11].

5.6.6. *The vivo-centered transition: a strategic asset for universities*

A vivo-centered transition represents a major asset for a university, allowing it to effectively address contemporary challenges while honoring its educational mission. It would promote integral education, valuing intellectual, social, emotional and ecological dimensions, to form engaged and conscious citizens capable of addressing the complex challenges of our time with compassion, integrity and competence.

In terms of resilience and sustainability, such an approach would enable the university to reduce its reliance on nonrenewable resources and secure its infrastructure by prioritizing local models of production and consumption. This approach would contribute to mitigating risks related to disruptions in the globalized system. Adopting a vivo-centered transition would also serve as a powerful lever to enhance the university's image, positioning it as a pioneering and responsible player in ecological and social matters. By becoming a model of successful transition while maintaining academic excellence, the university could inspire other higher education institutions and play a catalytic role at the local and regional levels, collaborating with businesses, local authorities and civil society.

Finally, this transition would respond to the growing demand from students, who are seeking to be prepared for future challenges (Pour un réveil écologique 2021). By training individuals with transdisciplinary skills and a systemic understanding of the issues at hand, the university would enhance its appeal, providing graduates highly sought after to tackle the challenges faced by future institutions and businesses.

5.7. Conclusion

Current transition strategies, primarily focused on decarbonization and technological solutions, overlook the systemic causes of the ecological crisis and reinforce our vulnerability to current socio-ecological crises. In light of these limitations, the vivo-centered transition model offers pathways to break away from the current paradigm and move towards a truly sustainable system. By placing the

---

11 For more information on the higher education institutions involved in the transition, see the reports: INSA Lyon 2022; Campus de la Transition 2023; Université de Rennes 2023.

flourishing of both human and nonhuman living at the heart of its objectives, this transition model adopts radical and systemic strategies that ensure the system's sustainability.

More than just an ecological project, the *vivo-centered* transition is a true transformation of our way of being in the world. It invites our society to resolutely turn towards a desirable future that connects science and spirituality, reason and emotions. This transition, conceived in a holistic and inclusive manner, would strengthen the resilience of our socio-ecosystems in the face of tomorrow's crises. By placing the flourishing of the living at the core of its strategies, it offers an alternative to models of infinite growth that have failed to ensure the sustainability of life on Earth. It is a model that calls for action and commitment, while being both ethical, systemic and deeply humanistic. This transition project is ambitious, but we believe it is essential if we genuinely wish to address the ecological and social crises of the 21st century.

5.8. Acknowledgments

We acknowledge the support of the Université Catholique de Lille and, in particular, the guidance of Benoît Robyns during the preparation of this chapter. We also thank Arthur Keller for his valuable insights concerning the figures.

5.9. References

Abson, D.J., Fischer, J., Leventon, J., Newig, J., Schomerus, T., Vilsmaier, U., von Wehrden, H., Abernethy, P., Ives, C.D., Jager, N.W. et al. (2017). Leverage points for sustainability transformation. *Ambio*, 46(1), 30–39.

Barbé, H., Béchet, L., Lafond, M., Delvaux, B. (2023). *Incarner l'écologie. S'ouvrir aux mondes vivants.* Les Liens qui Libèrent, Paris.

Barragan-Jason, G., Martin, A., Barou-Dagues, M., Grimonprez, A., Moritz, L., Dufour, A., Schaller, E., Schradin, C. (2023). Psychological and physical connections with nature improve both human well-being and nature conservation: A systematic review of meta-analyses. *Biological Conservation*, 277, 109842. doi: 10.1016/j.biocon.2022.109842.

Campus de la Transition (2023). Transformer l'offre de formation des universités à l'aune des défis de la transition écologique et sociale. Retour d'expérience [Online]. Available at: https://www.campus-transition.org/fr/accompagnement-de-cy-cergy-paris-universite/ [Accessed 2 January 2025].

Comby, J.-B. (2024). *Écolos, mais pas trop : les classes sociales face à l'enjeu environnemental.* Éditions Raisons d'Agir, Paris.

Costanza, R. and Daly, H.E. (1992). Natural capital and sustainable development. *Conservation Biology*, 6(1), 37–46.

Dardelet, C., Béchet, L., Lafond, M., Delvaux, B. (2022). *Transformation des campus*. Les Liens qui Libèrent, Paris.

Descola, P. (2005). *Par-delà nature et culture*. Gallimard, Paris.

Eisenmenger, N., Bowyer, C., Bjelle, E., Nielsen, P.S., Meidhof, M., Kuokkanen, A., Lazarus, M. (2020). The sustainable development goals prioritize economic growth over sustainable resource use: A critical reflection on the SDGs from a socio-ecological perspective. *Sustainability Science*, 15(4), 1101–1110.

Esbjörn-Hargens, S. and Zimmerman, M.E. (2009). *Integral Ecology: Uniting Multiple Perspectives on the Natural World*, 1st edition. Integral Books, Boston.

Fitzpatrick, N., Parrique, T., Cosme, I. (2022). Exploring degrowth policy proposals: A systematic mapping with thematic synthesis. *Journal of Cleaner Production*, 365, 132764. doi: 10.1016/j.jclepro.2022.132764.

Fletcher, R. and Rammelt, C.F. (2017). Decoupling: A key fantasy of the post-2015 development agenda. *Globalizations*, 14(4), 450–467.

Herrington, G. (2021). Update to limits to growth: Comparing the World3 model with empirical data. *Journal of Industrial Ecology*, 25(3), 614–626.

Hickel, J. (2019). Is green growth possible? *New Political Economy*, 25(4), 469–486.

INSA Lyon (2022). Les cahiers d'évolution de la formation. Numéro 2. Training Progress Report, Calaméo, Paris [Online]. Available at: https://www.calameo.com/insalyon/read/001288714f2e5150fc172 [Accessed 22 April 2025].

Ives, C.D., Giusti, M., Fischer, J. (2018). Reconnecting with nature for sustainability. *Sustainability Science*, 13, 1389–1397.

Izoard, C. (2024). *La ruée minière au XXIe siècle : enquête sur les métaux à l'ère de la transition*. Éditions du Seuil, Paris.

Keller, A. (2013). De la croissance à la soutenabilité forte. Mémoire de master 'Responsable de programmes internationaux'. Course, Institut de relations internationales et stratégiques, Paris.

Keller, A. (2020). Construire un nouveau système. In *L'effondrement de l'empire humain, regards croisés*, Commaret, M. and Pantel, P. (eds). Rue de l'échiquier, Paris.

Latour, B. (2010). *Nous n'avons jamais été modernes : essai d'anthropologie symétrique*. La Découverte, Paris.

Macy, J. and Brown, M.Y. (2019). *Écopsychologie pratique et rituels pour la Terre*. Le Souffle d'Or, Paris.

Maris, V. (2010). *Philosophie de la biodiversité : petite éthique pour une nature en péril*. Buchet-Chastel, Paris.

McPhearson, T., Iwaniec, D.M., Bai, X. (2021). Radical changes are needed for transformations to a good Anthropocene. *npj Urban Sustainability*, 1(1), 5.

Meadows, D.H. (2009). *Thinking in Systems: A Primer*. Earthscan, London.

Miller, J.R. (2005). Biodiversity conservation and the extinction of experience. *Trends in Ecology & Evolution*, 20(8), 430–434.

Morizot, B. (2020). *Raviver les braises du vivant : un front commun*. Actes Sud, Arles.

Naess, A. (1989). *Ecology, Community and Lifestyle: Outline of an Ecosophy*, Rothenberg, D. (trans.). Cambridge University Press, Cambridge.

Nuesiri, E. (2017). Feigning democracy: Performing representation in the UN-REDD Funded Nigeria-REDD Programme. *Conservation and Society*, 15(4), 384.

Parrique, T. (2022). *Ralentir ou périr : l'économie de la décroissance*. Éditions du Seuil, Paris.

Parrique, T. (2023). Degrowth. In *Handbook of the Anthropocene*, Wallenhorst, N. and Wulf, C. (eds). Springer, Cham.

Parrique, T. (2025). Defining degrowth. Working Paper, 2025-1 [Online]. Available at: https://timotheeparrique.com/wp-content/uploads/2025/01/Parrique-T.-2025.-Defining-degrowth-V1-1.pdf [Accessed 10 April 2025].

Pelluchon, C. (2020). *Réparons le monde. Humains, animaux, nature*. Rivages, Paris.

Pour un réveil écologique (2021). L'écologie aux rattrapages. L'enseignement supérieur français à l'heure de la transition écologique : état des lieux et revue des pratiques. Summary, Pour un réveil écologique [Online]. Available at: https://pour-un-reveil-ecologique.org/documents/6/RESUME_Rapport_Ecologie_aux_rattrapages-Pour_un_reveil_ecologique-fev_2021.pdf [Accessed 15 February 2025].

Renouard, C. (2022). *Pédagogie de la transition*. Les Liens qui Libèrent, Paris.

Richardson, K., Steffen, W., Lucht, W., Bendtsen, J., Cornell, S.E., Donges, J.F., Drüke, M., Fetzer, I., Bala, G., von Bloh, W. et al. (2023). Earth beyond six of nine planetary boundaries. *Science Advances*, 9(37), eadh2458.A.

Riechers, M., Balázsi, Á., García-Llorente, M., Loos, J. (2021). Human-nature connectedness as leverage point. *Ecosystems and People*, 17(1), 215–221.

Rosa, C.D. and Collado, S. (2019). Experiences in nature and environmental attitudes and behaviors: Setting the ground for future research. *Frontiers in Psychology*, 10(763). doi: 10.3389/fpsyg.2019.00763.

Savasta-Kennedy, M. (2014). The dangers of carbon reduction tunnel vision. *SSRN Electronic Journal*. doi: 10.2139/ssrn.2518613.

Soga, M. and Gaston, K.J. (2016). Extinction of experience: The loss of human–nature interactions. *Frontiers in Ecology and the Environment*, 14(2), 94–101.

Stoknes, P.E. (2014). Rethinking climate communications and the 'psychological climate paradox'. *Energy Research & Social Science*, 1, 161–170.

Tordjman, H. (2021). *La croissance verte contre la nature : critique de l'écologie marchande.* La Découverte, Paris.

Université de Rennes (2023). Rapport du groupe de travail de l'université de Rennes Formation Transition Ecologique et Sociale (TES). Report, Université de Rennes, Rennes [Online]. Available at: https://www.univ-rennes.fr/sites/www.univ-rennes.fr/files/medias/files/Rapport_GT_TES_vf.pdf [Accessed 14 February 2025].

Vivien, F.D. (2005). *Le développement soutenable.* La Découverte, Paris.

PART 2

Energy Networks

6

Electricity Transmission Networks, the Keystone of Neo-electrification at the Crossroads of Multiple Issues: Nuclear Power Generation and Conflictualization of European and French Energy Policies

6.1. Introduction

While climate change is now a known and accepted reality, there are many strategies and methods to significantly reduce its effects on our society, ranging from the Cornucopian ideal[1] and the growth model, to degrowth[2] and the promotion of energy sobriety[3]. However, the main path taken remains the one of energy and ecological transition, which is supposed to enable a gradual phase-out of fossil fuels, replacing them with so-called green energies. The main lever to be activated in this strategy lies in the most extensive electrification of our uses and practices, or more accurately in a neo-electrification, since electrification as it stands has already been achieved over the course of the 20th century by the gridding of French and European

Chapter written by Lucas LOPEZ.

1 Bihouix, P. (2015). Les technosciences, ou l'utopie corrompue. Quand l'idéologie du progrès capture notre avenir. *Revue du Crieur*, 2, 112–127. doi: 10.3917/crieu.002.0112.

2 Latouche, S. (2022). *La décroissance*. Presses Universitaires de France, Paris. doi: 10.3917/puf.latou.2022.01.

3 Romary, P. (2024). De la Frugalité à la Sobriété : une histoire de la simplicité volontaire. ⟨hal-04495815v2⟩.

territories by large-scale transmission networks and electricity production facilities[4]. This implies a new deployment and expansion of the electricity transmission network and interconnections to strengthen the resilience of our supplies, by improving continental coverage. A global interconnection that covers an entire continent such as Europe is a long-standing project, dating back to the years 1990–2000 (with, in particular, the ambition of creating a "super-dispatching" system[5]), which is now finding a certain degree of resonance through ENTSO-E[6] and the extended continental coverage of the transmission network, and which should therefore be subject to further development with neo-electrification.

However, this neo-electrification cannot be considered exclusively on the scale of the transmission network, as the nature of the primary energies used to produce electricity is just as important. Indeed, what is the point of electrifying as many of our uses as possible if power stations continue to run on coal and fossil fuels, which are sources of greenhouse gas emissions? Across Europe, one of the crucial aspects of the energy issue is the place that nuclear power should or should not occupy in the energy transition. France, for example, is the leader of the European nuclear power industry, with one of the largest nuclear reactor fleets in the world and the highest proportion of nuclear power in the electricity mix (63% of the electricity produced in 2022 was generated by nuclear power stations in France, in the least productive year for the French industry since 1988[7]), while Germany, a country that denuclearized itself after the Fukushima disaster in 2011, opposes the French position, favoring renewable energies instead[8]. This opposition can be a source of conflict, with two main camps: pro- and anti-nuclear. This can be observed at many levels, whether in the establishment of the European green taxonomy, the "European green deal" or even in the planning of new networks, such as the new pipelines needed to transport hydrogen, a potentially essential element in the energy transition (we are thinking here in particular of the conflict surrounding the H2MED pipeline

4 RTE (2022). Futurs énergétiques 2050, rapport complet. [Online]. Avaialble at: https://www.rte-france.com/analyses-tendances-et-prospectives/bilan-previsionnel-2050-futurs-energetiques#Lesresultatsdeletude.

5 Lagendijk, V. (2008). Histoire de l'idée d'un système européen de l'électricité : projet, progrès, persistance. *Annales historiques de l'électricité*, 6, 57–79. doi: 10.3917/ahe.006.0057.

6 European Network of Transmission System Operators for Electricity.

7 RTE (2023). Bilan électrique 2022 [Online]. Available at: https://analysesetdonnees.rte-france.com/bilan-electrique-production.

8 Valérian, F. (2019). Transition électrique : la fin d'un consensus allemande ? *Annales des Mines – Responsabilité et environnement*, 93, 71–75. doi: 10.3917/re1.093.0071.

between Spain and Germany and crossing France[9], and the opposition surrounding so-called "pink" hydrogen, produced using nuclear power).

In other words, in the current climate emergency and energy transition context, thinking about the resizing of the national, or even continental, electricity transmission network is inextricably linked to the reflection on the nature of the electricity production fleet. The inextricable link between these two elements (transport and production) is central to the decarbonization strategy chosen by European leaders to achieve a global energy transition requiring a widespread electrification of uses. However, while nuclear power seems to offer an undeniable advantage in terms of electricity production capacity, it has a significant downside in the form of the conflict it can generate. So, how can we combine the need for rapid neo-electrification with the high level of conflict that nuclear power can generate, even though it is the most efficient form of carbon-free electricity production in terms of power per unit of production and as it is already installed in many European countries that rely on this source in their energy mix?

In this chapter, we will therefore begin by looking at the history of neo-electrification and its nervous system, in the form of national and international electricity transmission networks, which are at the crossroads of multiple issues, particularly when it comes to nuclear power generation, including a necessary contextualization of the first electrification, to have a better understanding of the issues involved in a new electrification, after one that relied heavily, in France in particular, on nuclear power generation. This same nuclear power generation, which may in some ways represent an obstacle or even a repellent, is a subject that cannot be separated from the issue of the conflicting European and French energy policies, which we will address in a second step.

6.2. From the first electrification to a neo-electrification

As a society, we are all confronted with the known and accepted reality of global warming. In the second half of the 20th century, it began to be mentioned as a real risk for our societies, and in 1972, the Meadows report, entitled "Limits to growth", seriously raised the issue of the impact of human activity on the environment and the need to implement various measures to guard against a catastrophe[10].

9 Collen, V. (2023). Hydrogène : clash au sommet entre la France, l'Allemagne et l'Espagne. *Les Échos* [Online]. Available at: https://www.lesechos.fr/monde/europe/hydrogene-clash-au-sommet-entre-la-france-lallemagne-et-lespagne-1905009.

10 Meadows, D.H., Meadows, D.L., Randers, J., Behrens III, W.W. (1972). *The Limits to Growth*. Potomac Associates, Falls Church, VA. Commissioned by the Club of Rome.

Numerous solutions have since been considered as a way out of the headlong rush represented by global warming, ranging from the Cornucopian ideal of innovation and growth as providers of solutions, to degrowth or the promotion of energy sobriety[11]. However, at the crossroads of all these possibilities lies the concept of energy transition, first mentioned in 1967 by Harrison Brown[12], an American atomic scientist, who wanted to steer the energy mix towards an all-nuclear future[13].

The transition, which is supposed to enable a gradual phase-out of fossil fuels, replacing them with so-called green energies, is gradually becoming the preferred and dominant strategy among governments, particularly in Europe, and in some ways embodies the achievement that needs to be made to deal with climate change. In the pursuit of this objective, there is one lever that seems obvious at first sight: the most extensive electrification of our uses and practices, which could be called neo-electrification. But why talk about neo-electrification and not simply electrification? Because the first electrification of France and Europe has already taken place, in the 20th century, when power lines were gradually installed across the country and electricity became one of the most widely used forms of energy in our lifestyles. Alain Beltran's *The Electric Life*, co-written with Patrice Carré[14], traces the history of the arrival of electricity in our lifestyles, from the observation by the Greek philosopher Thales (7th century BCE) of amber emitting lightning flashes when rubbed by animal fur, to the development of electronuclear power stations.

We can also describe as electrification the propaganda campaign conducted by EDF[15] from the 1960s to promote the use of electricity in French homes, with the first washing machines, electric fridges, electric cookers, etc. (there are also traces of

11 In English, it is more common to use the term "sufficiency", but this is in fact closer to the French "resilience". With French academics, sobriety (*sobriété*) has been increasingly studied since the 2020s and expresses more of a paradigmatic change in society than just a lifestyle adjustment.

12 Fressoz, J.-B. (2024). *Sans transition. Une nouvelle histoire de l'énergie.* Seuil, Paris.

13 Ibid.: At that time, atomic scientists, particularly Americans, had an important role to play in the development and the propagation of the energy transition as a concept.

14 Beltran, A. and Carré, P. (2016). *La vie électrique : histoire et imaginaire, XVIIIe-XXIe siècle.* Belin, Paris.

15 Stands for *Électricité De France.* This is the French national electric company, which managed production, transport and distribution between 1946 and 1996. Since 1997, it has no longer dealt with electricity transport and distribution, because of the European regulation on the electricity market which brought about the separation between electricity production and transport, ending EDF's monopoly on the French electric system. See: Bouneau, C., Derdevet, M., Percebois, J. (2007). *Les réseaux électriques au cœur de la civilisation industrielle.* Timée Editions, Boulogne-Billancourt.

such propaganda campaigns elsewhere in Europe[16]), especially when France, facing the 1973 oil crisis, decided to install 54 reactors in 15 years under the Messmer plan[17]. Across Europe, the first electrification took place around the same time, starting with the explosion of electrical innovation in the 1880s (the most famous invention being Edison's light bulb in 1879, which contributed to a profound urban transformation through urban lighting, or the development of electric mobility, with tramways and trains of this kind being presented at the Paris Universal Exhibition in 1881[18]) and extending its preliminary phase until the 1920s[19].

Throughout the 20th century, the energy system, and more specifically the European electricity system, diversified through the development of interconnections between countries, first for economic reasons, then for increasingly political reasons. In his article *History of the idea of a European electricity system: project, progress, persistence*[20], Vincent Lagendijk discusses the ambition in the 1930s of Oskar Oliven, a German engineer and director of Gesfürel (a Berlin-based electricity company that existed between 1894 and 1942 and was the source of capital for German electrification from the 1900s), to set up a European supergrid. This international network, conceived by Oliven, was only really approached in 1995, when the two European electricity areas (West and East) were interconnected after the Cold War. This first electrification therefore took place over a relatively long period of time, and more than a specific historical event, it refers to a change in our societies.

The neo-electrification that we mentioned earlier is also a change in our societies, but instead of observing it in a European industrial history (as we have just done for the first electrification), it must be called for and achieved as the main lever of the energy transition in a more or less distant future given the resources available. It implies a new deployment of carbon-free electricity generation facilities and an expansion of the electricity transmission network and interconnections to carry these new electricity flows and strengthen the resilience of European supplies, in particular by improving continental coverage. Indeed, while decarbonized production methods are the first condition for this neo-electrification, interconnected electricity

16 Bouvier, Y. (2010). L'industrie électrique européenne depuis la seconde guerre mondiale, des technologies nationales aux marchés européens. In *Les trajectoires de l'innovation technologique et la construction européenne, des voies de structuration durable ?* Bouneau, C., Burigana, D., Varsori, A. (eds). Euroclio, Brussels.

17 Mangeon, M. and Roger, M. (2024). Le plan Messmer : retour aux sources du parc électronucléaire français. *Revue Générale Nucléaire*, 3 [Online]. Available at: https://www.sfen.org/rgn/le-plan-messmer-retour-aux-sources-du-parc-electronucleaire-francais/.

18 Beltran, op. cit., 53–67.

19 Bouttes, J.-P. and Dassa, F. (2016). Europe de l'électricité – Une perspective historique. Insitut Français des Relations Internationales – IFRI.

20 Lagendijk, op. cit.

transmission networks are the keystone: there is no point in producing electricity in sufficient quantities to replace fossil fuels if it cannot be transported to points of consumption. While ENTSO-E, founded in 2009 on the basis of six major regional associations[21] representing a total of 42 TSOs, currently provides excellent coverage of the European territory (extending beyond the borders of the European Union, and even as far as Northwest Africa), this coverage is set to expand as electricity production increases.

So, the main consequences of this future neo-electrification should be a new deployment of electricity generation capacity and a widening of the territory covered by the electricity transmission network, implying an increase in the number of interconnections between European countries. Significant resources therefore need to be deployed, and above all, a decision needs to be taken in a relatively short time span on the strategy to be followed in this neo-electrification process. As we have just shown, the first electrification was achieved over a long period of time (almost a century) and without any consideration for the nature of the electricity production methods (carbon-based or not), which can no longer be the case for neo-electrification. It seems even more important to draw a distinction between 20th century electrification and the concept of neo-electrification that we are using here to support the need for a thorough reflection on this important aspect of the energy transition. Furthermore, this issue is a potential source of conflict and opposition, due to one of Europe's main institutional and industrial legacies: its electronuclear production fleet.

6.3. Neo-electrification, a potential source of conflict and opposition

To properly understand the deep conflicting nature of nuclear energy, it is important to first provide a brief historical context in order to grasp and explain the role it has played for nearly 70 years. Europe as a political entity was founded on the Treaties of Rome in 1957, which established the European Economic Community (EEC), on the one hand, and the European Atomic Energy Community, better

21 – ETSO: European Transmission System Operators, including four other associations (ATSOI, UKTSOA, Nordel and UCTE).

– ATSOI: Association of the Transmission System Operators of Ireland.

– UKTSOA: UK Transmission System Operators Association.

– Nordel: founded in 1963, was a body for co-operation between the transmission system operators in Denmark, Finland, Iceland, Norway and Sweden.

– UCTE: Union for the Coordination of the Transmission of Electricity, initially the UCPTE (P stands for Production) since 1951, but decided to focus on transmission in 1999 after the creation of the internal electricity market.

– BALTSO: Baltic Transmission System Operators, organization of Estonian, Latvian and Lithuanian Transmission system operators.

known as Euratom, on the other. While France is indeed the European country that made the most massive choice in favor of nuclear power, gradually over the course of the second half of the 20th century, no fewer than 16 European countries followed suit (not counting Russia and Belarus) and were still nuclear-powered in 2021, according to the IAEA[22], out of a total of 32 countries worldwide: France, Germany, Belgium, Bulgaria, Spain, Finland, Hungary, the Netherlands, the Czech Republic, Romania, Slovakia, Slovenia and Sweden for the European Union, to which we can add the United Kingdom, Switzerland and Ukraine[23] (since 2022 in particular), which are not strictly speaking part of the EU but which participate in the Europe of energy.

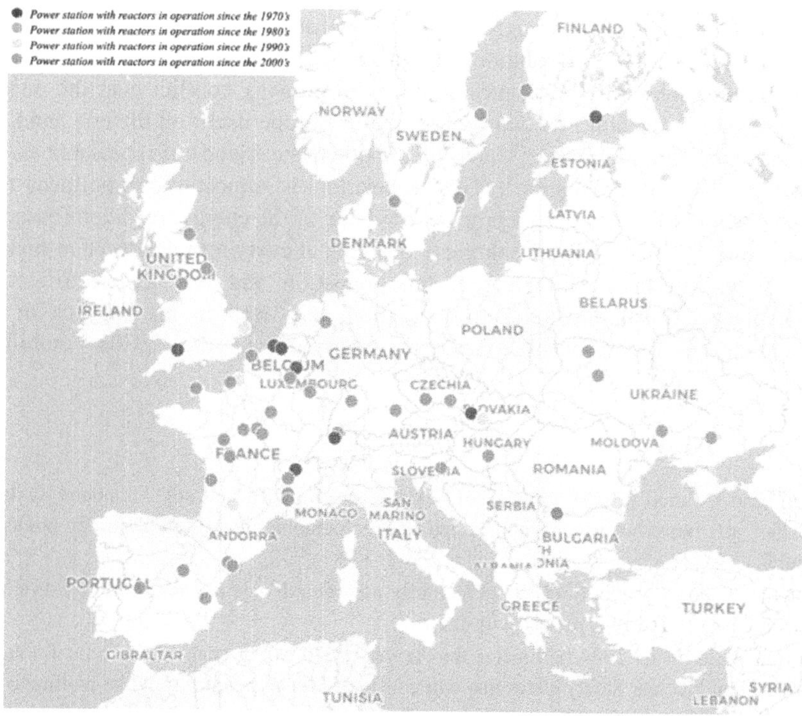

Figure 6.1. *The distribution of nuclear power plants in Europe (2022), based on the IAEA numbers[24], realized by Lucas Lopez. For a color version of this figure, see www.iste.co.uk/dobigny/rethinking.zip*

22 International Atomic Energy Agency.

23 IAEA, Power Reactor Information System (PRIS) (n.d.) [Online]. Available at: https://pris.iaea.org/PRIS/CountryStatistics/CountryStatisticsLandingPage.aspx.

24 IAEA (2022). Nuclear Technology Review [Online]. Available at: https://www.iaea.org/sites/default/files/gc/gc66-inf4.pdf.

Nuclear electricity provides approximately 25% of decarbonized electricity in Europe[25], and it should be remembered that 50% of this nuclear-generated electricity in Europe comes from France. Although these figures have stagnated since 2011 and the events at Fukushima[26], nuclear power is once again being put forward as a solution for neo-electrification, being highly valued in particular for its modularity and availability, two points that are regularly pointed out as shortcomings of renewable energies such as photovoltaics and wind power.

However, particularly since 2011, nuclear power has been the most controversial source of decarbonized energy[27], seen as too dangerous or too expensive, with recent problems at EDF's French sites (cases of stress corrosion) and at the European EPR sites (Oikiluoto, Flamanville and Hinkley Point) not helping its opponents to see nuclear power as a reliable solution. Memories of the Fukushima disaster and, to a lesser extent, Chernobyl, are contributing to the growing conflict over the nuclear issue, creating a division into two groups within Europe itself: on the one hand, the nuclear alliance around France[28], and, on the other, the friends of renewable energy around Germany and Austria[29]. This opposition is almost more political than technical, with both sides pushing their own vision of the energy transition (based on nuclear power or on renewable energies), delaying at every turn the decision-making and planning needed to implement this transition and the neo-electrification. Jean-Baptiste Fressoz illustrates this situation very well in his analysis of the exploitation of the concept of energy transition in the 1970s, mobilized

25 European Council (2025). How is EU electricity produced and sold? January [Online]. Available at: https://www.consilium.europa.eu/en/infographics/how-is-eu-electricity-produced-and-sold/.

26 Ritchie, H. and Rosado, P. (2020). Electricity mix. *Our World in Data* [Online]. Available at: https://ourworldindata.org/electricity-mix.

27 For example, in its 2024 barometer for the year 2023, the French Institute for Radiation Protection and Nuclear Safety states that while 38% of French people would be willing to live near a wind turbine, only 22% would be willing to live near a nuclear power plant (solar power is not discussed here, as it is usually installed on roofs and is therefore less visible, and does not pose the same problems): IRSN (2024). Baromètre 2024. 94–95 [Online]. Available at: https://barometre.irsn.fr/graphiques2024.

28 French Government (2024). Déclaration de l'Alliance européenne du nucléaire, March 4 [Online]. Available at: https://presse.economie.gouv.fr/declaration-de-lalliance-europeenne-du-nucleaire/.

29 Kurmayer, N.J. (2024). L'Alliance du nucléaire et le groupe pro-renouvelables prêts à se disputer les fonds de l'UE. *Euractiv* [Online]. Available at: https://www.euractiv.fr/section/all/interview/lalliance-du-nucleaire-et-le-groupe-pro-renouvelables-prets-a-se-disputer-les-fonds-de-lue/.

simultaneously by atomists and anti-nuclearists, each using this new concept to support their side[30].

Finally, it is worth mentioning an example that perfectly illustrates this conflictualization on a European scale: the case of the H2MED pipeline. While we have so far only mentioned electricity transmission networks in the energy transition, this pipeline is, in a way, directly concerned. It is a pipeline project designed to link Barcelona and Marseille via the Mediterranean Sea, and at the same time it would link Spain and France, and by extension, Germany and the rest of Europe[31]. Hydrogen is not, strictly speaking, electricity, but it is a solution that is being considered to overcome one of the main current problems: it is difficult to store electricity efficiently. The production of hydrogen by electrolysis, using peaks in renewable energy production or using the high modulation capacity of nuclear reactors, makes it possible to produce "clean" hydrogen, and this gas, in the form of a liquid at low temperature, becomes much easier to store and transport, to be converted back into electrical energy on demand. The new hydrogen networks can therefore be considered, in the same way as the electricity transmission network, as central to the energy transition and neo-electrification. However, they are not exempt from the issue of conflictualization that we mentioned earlier in relation to nuclear power. In fact, the H2MED pipeline has been strongly contested, particularly by Germany, which fears that it will enable France to inject "pink" hydrogen, produced by nuclear power stations, into the European hydrogen network, and not just "green" hydrogen produced by renewable energies, even though this would significantly increase European hydrogen flows[32]. This example, which allows us to look at the situation and understand it through a viewpoint other than that of the electricity transmission network, highlights even more the problems raised at network level by the conflicting nature of nuclear power, because if Germany or other countries in the "friends of renewables" group are opposed to nuclear power, the widespread interconnection of the electricity transmission network in Europe means that electricity produced in French, Finnish and other power stations can also be consumed in Germany or Austria. Even if these countries dismantle all their reactors, as long as their neighbors continue to operate their nuclear power plants, they will still benefit from the interconnections. This debate is therefore becoming central to the establishment of a strategy for decarbonizing energy, sometimes slowing down decision-making in an area that is heavily dependent on planning and requires a great deal of anticipation. The intrinsic conflictual nature of nuclear power thus becomes an essential component of neo-electrification and, by extension, of the energy transition.

---

30 Fressoz, op. cit.

31 See the map here: https://h2medproject.com/wp-content/uploads/2024/02/map_overview_spain_portugal.png.

32 Collen, V. op. cit.

6.4. Conclusion

In the end, while neo-electrification and energy transition are the objectives chosen by our leaders as the main guidelines for reducing the effects of climate change, the levers and paths to be taken to achieve them have not yet been fully defined, with the conflicting nature of nuclear power generation slowing down decisions. The point here is not to praise nuclear power or to denigrate any person or state that opposes it. On the contrary, what our previous reflections seem to show is that the question of the conflictualization of energy embodies a point of convergence between interconnected networks and nuclear power, the repulsion generated by the latter having a direct impact on the former and, conversely, the electronuclear production fleet having no sense in existing without the connection to an extensive interconnected network allowing the electricity produced to be transported over long distances. This can be seen very clearly in France, where the consultation and planning process surrounding the new EPR 2 reactors was partly initiated by RTE, the transmission system operator, in addition to EDF, the nuclear power plant operator, as indicated in the public debate documents available online[33], directly exposing the link between electricity transportation and production.

This controversial nature of nuclear power is currently manifesting itself in a political clash with renewable energies, although it seems reasonable to admit that the solution (if it exists) lies somewhere in the middle, within a diversified electricity mix, exploiting the various existing means of generating carbon-free electricity, and introducing new ones to intensify the electrification of our uses, as proposed in numerous prospective scenarios analyzing the energy mix[34]. Furthermore, even the scenarios least favorable to nuclear power, in the case of a highly nuclearized country such as France, do not foresee an active phase-out of nuclear power, as occurred in Germany following the Fukushima accident in 2011, but are instead based on a gradual phase-out, with existing reactors being operated until the end of their planned operating life and gradually replaced by renewable energy sources. This is, for example, what is proposed in RTE's "M" scenarios[35] or the scenario put forward by the *négaWatt* association[36]. It is difficult for countries that are already heavily reliant on nuclear power, such as France, to envisage a

---

33 Commission nationale du débat public (CNDP) (2022). Qui organise le débat ? *Nouveaux réacteurs nucléaires et projet Penly* [Online]. Available at: https://www.debatpublic.fr/nouveaux-reacteurs-nucleaires-et-projet-penly/qui-organise-le-debat-3368.

34 RTE (2022). Futurs énergétiques 2050, rapport complet [Online]. Available at: https://www.rte-france.com/analyses-tendances-et-prospectives/bilan-previsionnel-2050-futurs-energetiques#Lesresultatsdeletude.

35 Ibid., 205.

36 négaWatt (2022). Scénario négaWatt 2022–2050 [Online]. Available at: https://www.negawatt.org/Scenario-negaWatt-2022.

radical change in their electricity mix. It is therefore necessary to organize a collective reflection to reduce the conflict surrounding Europe's nuclear power plants in order to accelerate neo-electrification and the energy transition. One way to achieve this seems to lie in the concept of social acceptability, which refers to a collective judgment on a new project or its modification. This judgment can vary according to the values and beliefs of each group, and it is dynamic, as it can evolve, either positively or negatively, over the course of social debates[37]. The sociologist Rémi Barbier further specifies that "acceptability is not an intrinsic property of the object, nor is it a quality that is consubstantial with it[38]", and in fact acceptability varies according to the interactions between the object, humans and the environment, in the context of a project being rooted in a particular territory[39]. Attempting to improve the level of social acceptability of nuclear power is therefore a fully viable solution with a view to reducing the level of conflict. However, this issue is much wider and more complex, as Andrew Stirling and Philip Johnstone have shown in their article *Comparing nuclear trajectories in Germany and the United Kingdom: from regimes to democracies in sociotechnical transitions and discontinuities*[40], in which they examine the fundamental sociotechnical differences between Germany and the United Kingdom in their choices to favor or disfavor nuclear power in their energy mixes, in order to identify the parameters that determine whether or not a country becomes nuclearized, relying heavily on societal indicators inherent to each society. This reflection precisely demonstrates that the issue is entirely transversal and interdisciplinary, and above all highlights that while the question of how electricity is produced is central, it only makes sense if the European strategy for achieving the energy transition ensures that transport infrastructures have the capacity to cover a denser and increased flow of electricity[41], so as to provide fair and equitable access to electricity for all, the energy transition and neo-electrification being carried out with and for society.

---

37 Kellermann, T. (2023). *Acceptabilité et contestations face au déploiement des énergies renouvelables et à ses conséquences: focus sur l'éolien offshore de la façade Atlantique.* Chaire RESET, Pessac.

38 Barbier, R. (2021). L'épreuve d'acceptabilité sociale, ou la composition disputée du collectif. *Revue internationale de psychosociologie et de gestion des comportements organisationnels,* 27, 45–61.

39 Kellermann, T., op. cit., 9–10.

40 Johnstone, P. and Stirling, A. (2020). Comparing nuclear trajectories in Germany and the United Kingdom: From regimes to democracies in sociotechnical transitions and discontinuities. *Energy Research & Social Science*, 59, 1–27. doi: 10.1016/j.erss.2019.101245.

41 Palle, A. (2016). Les dynamiques d'évolution des réseaux de transport d'électricité européens, des réseaux techniques entre croissance et décroissance. *Flux*, 105(3), 18–32. doi: 10.3917/flux.105.0018.

7

Is it Relevant to Question the Size of Power Grids?

7.1. Introduction

Power grids are among the largest machines ever built by human, as demonstrated by the immensity of the continental European power grid. Through its lines and transformers, this grid transmits the same electromagnetic wave from Morocco to Turkey, and across Estonia. This large size is often presented as an advantage in many energy transition narratives. For instance, as part of the energy transition, the International Energy Agency (IEA) suggests adding or refurbishing a total of 80 million kilometers of grid by 2040, which is equivalent in size to the entire existing grid.

On the other hand, the narrative of energy transition, which places sufficiency at its center, seems to demand a form of proximity between users and means of production. Otherwise, energy is no longer embodied and, as such, there is no point in taking steps to reduce superfluous consumption.

This text looks at ways of reconciling the current immensity of the European electricity grid, along with its anticipated expansion under the energy transition, with the notion of sufficiency, which would require greater proximity.

This reconciliation is driven by the fact that while electrification, and therefore the grid, certainly has a role to play in decarbonization, it cannot realistically meet the needs of all applications, such as, aviation, maritime transport, cement plants, steel industry, etc. (Fressoz 2024), not even taking into account a potential "rebound" effect. In addition, the unbridled growth of power systems also has an

---

Chapter written by Jonathan COIGNARD.

impact on other planetary limits, such as the erosion of biodiversity due to mining activities. To meet the challenges of the energy transition, we therefore need to think in terms of sufficiency.

The approach proposed is interested in developing the prism of "scale effects" (i.e. in analyzing the shifts that occur when a system changes size). The aim is to highlight what changes nonlinearly (i.e. non-proportionally), as a socio-technical system such as the power grid grows. Very often, complex systems exhibit discontinuities, exponentials or other forms of nonlinearity that should prevent us from imagining the same systems as identical when they grow larger. For example, we need to understand whether the grid of tomorrow, after massive electrification, will be proportionally larger than today's grid, or whether it will encounter new difficulties that will affect its shape.

Nonlinearities related to technical constraints are historically well represented in the planning of electrical infrastructures, for example, electrical losses proportional to the square of the current. On the other hand, infrastructure planning often overlooks the social implications of changes in scale within the energy system, such as the centralization of decision-making or shifts in consumption practices (Lopez and Le Blanc 2021). However, other literature in the social field, particularly that on large technical systems (LTS), does address some of these aspects (Coutard 1999). This text is part of an interdisciplinary approach to introducing the concept of scale effects and human scale applied to the power grid.

In order to reconcile the power grid with a form of human scale that allows us to think collectively about sufficiency, this text is divided into three parts. First, it highlights the relevance and lessons to be learned from scale issues outside the strict domain of power grids. Second, the concept of scale is defined specifically for the power grid. Third, we analyze the implications of changing scale in the different dimensions of a power system. In addition, a number of research avenues are explored to imagine what a human-scale power grid might look like.

7.2. Relevance of size outside power systems

Scale problems are a source of unexpected events. They reveal what changes nonlinearly, in nonintuitive ways, when a system changes size. Haldane and Rey's (2008) work on "being the right size" illustrates this concept. For example, it is hard to imagine a warm-blooded animal (i.e. with a fixed temperature, smaller than a mouse). The smaller the animal, the greater the surface area it exposes to the environment in relation to its volume (i.e. its ability to produce heat). Beyond the size of a mouse, a tipping point occurs and organisms can no longer maintain warm blood. In the same way, the giants that populate our stories are physically unrealistic. A giant 10 times taller, wider and thicker than a human of normal height is a thousand times heavier, but the cross-section of their bones is only a hundred times larger. So, every square centimeter of those giant bones would have to support ten times the weight of a human bone, causing it to break at the first step (Rey 2014). The world of centimeters cannot be understood in the same way as the world of meters. The ant is not afraid of gravity, whereas for a human, it is lethal. On the other hand, the surface tension of water is a serious danger of asphyxiation for an ant (Haldane and Rey 2008).

The literature in cognitive psychology offers some evidence of the difficulties associated with nonlinear thinking, for example, the difficulty to interpret that "more a implies more b" may result in "k times a implies k^2 times b" (Greer 2010). Furthermore, Olivier Rey (2014) argues that the contemporary inflow of images contributes to the difficulty of understanding nonlinear scale effects. In particular, images accustom us to considering the things or beings represented independently of the scale at which these things or beings exist. On a computer screen, we can even vary the size as we like. In reality, it is a different matter. Height is not just relative to another object; it also carries an absolute dimension. A human that is 2.5 m tall is not just tall in terms of our everyday experience, but above all because a human 3 m tall is not possible.

When you grow a certain entity, there comes a time when nonlinear changes occur. In the case of the bridge, it collapses. In the field of engineering, nonlinearity leads to the development of new techniques, for example, for building ever-larger bridges (cable-stayed steel, prestressed concrete, etc.). This is also the case for power grid, where the development of alternating current and transformers provides a solution (raising the voltage) to the nonlinearity of line losses. This technique made it possible to extend Edison's DC grid, which was then limited to approximately 1 km between consumers and the production unit in the heart of Manhattan in 1882.

When faced with a change in the size of the power system, the usual planning methods do not consider the social implications, let alone the possible discrepancies with the notion of human scale. Yet, the power grid is an eminently social issue (Hughes 1983). However, in the literature on grid planning (Alvarez-Hérault et al. 2023), the social aspects are not addressed, and so the nonlinearities that would result from a change in size remain unaddressed. Planning the size of the grid implicitly assumes that it is up to society to adapt to the new technical form of the grid.

There are, however, fields related to the planning of power grid where notions of scale are considered and even put into perspective with the notion of human scale. In particular, in urban planning, where the notion of a city on a human scale suggests a city adapted to the sensory and physical capabilities of the human body (which have hardly changed in the last few thousand years). In contrast to urban planning dominated by modernist ideology, which gives priority to automobile traffic and the construction of skyscrapers isolated from their environment.

In this sense, urban planning also highlights the initial paradox between large-scale expansion of the power grid and sufficiency. While engineering is capable of resolving the mechanical nonlinearities of a 600 m high tower, it is moving away from the very function of urban space as a meeting place, and a fortiori, as a space for democratic conversation (Gehl 2013).

What the concept of the human-scale city reminds us of is that, from a human perspective, muscular, sensory, cognitive and social limitations are very real. A first threshold at 100 m determines the limit beyond which it is difficult to perceive another human being in motion (this limit also determines the maximum size of the largest stadiums, and that of old public squares in Europe). At 25 m, there is another important threshold, below which we can decode emotions and facial expressions (this limit determines the size of auditoriums). Also, the higher the action, the harder it is to watch. To do this, you have to move further and further back, so that the connection between the street and the buildings breaks down after the fourth floor. Finally, beyond a walking speed of 5 km/h, the human brain's ability to perceive details, interpret and maneuver is limited (this limit determines the size of road signs, which are disproportionately large for pedestrians) (Gehl 2013).

The relevance of thinking on the human scale is also present in other fields such as communication systems or social organization. Can we use the same word "democracy" for the city of Athens with 40,000 inhabitants and modern India with 1.4 billion citizens? Similarly to how Haldane approached the animal world, Leopold Kohr argues that different sizes of societies lead to different forms of societies.

Leopold Kohr's (1957) thesis is that, once a society reaches a certain size, its problems grow faster than the human resources needed to deal with them. As a contemporary example, in France, the number of crimes per 1,000 inhabitants is higher in large urban areas, for example, violent robberies are 75 times more frequent in large cities (Duclos-Grisier 2022). For Kohr, it is not that the inhabitants of big cities are intrinsically more vicious than those of the countryside, but rather that big cities have passed a threshold where size in itself becomes the main source of crime. Behind this threshold lie human limits that cannot be extended ad infinitum, for example, maintaining the common interest while increasing the possibilities of escaping prosecution by blending in with the masses. So, according to Kohr (1957), the unprecedented change that marks our time is not the nature of social difficulties, but their size. This vision may seem simplistic in a world that is not. Nevertheless, the point is not to deny the importance of social policies, but to propose a framework that considers size in terms of orders of magnitude, rather than pronouncing on a precise size that would depend on the environment.

For Kohr, the concept of human scale is complementary to Gehl's vision, based on sight, textures, social distances and speed of movement. It is the scale at which all members of the community have a say in the choices and policies that affect them.

This detour outside the field of power systems suggests that we should not assume that a change in a system size will have proportional consequences. Discontinuities exist, in particular for power systems, a larger size may result in a different relationship with infrastructures and energy consumption. The notion of human scale ensures that we do not lose sight of the fact that the power grid is a tool. Like any tool, it is a means to an end, not an end in itself. For Ivan Illich (1974), technology must remain commensurable with human faculties. Otherwise, it will enslave rather than benefit humankind.

7.3. Measuring the size of the power system

In Haldane's examples, the representation of size in terms of area and volume is relatively easy to think about for animals. By comparison, the size of a power grid is more complex to define. Where does the power grid stop, and therefore what size are we talking about? Depending on our point of view, the size of the grid applies to physical aspects (e.g. conductor size), spatial aspects (e.g. density, spread, etc.), temporal aspects, organizational aspects, number of consumers, functional complexity, etc.

We propose here a necessarily partial delimitation of the concept of the size of a power grid. Inspired by the French context, this vision is based on eight dimensions for understanding the size of a power grid. Each of these propositions is in itself a multidimensional object. Note that this approach also applies to sub-groups of the power grid, for example, the grid serving a city. The following paragraphs describe the eight dimensions proposed and schematized in Figure 7.1.

– Dimension 1: the number of users in the grid's service area. This first dimension links the size of the power grid to the number of people directly connected to the grid. In this sense, a grid that serves a large number of inhabitants becomes a large grid. Certain situations can be more ambiguous, for grids serving a few energy-intensive users, such as steel manufacturers. In these situations, the number of users represented by energy-intensive usage is considered.

– Dimension 2: the geographical zone of electrical synchronicity in which the grid of interest is integrated. In other words, the area where there is strong physical coupling, and where a disturbance, for example, the shutdown of a generator, is observed on the frequency of the electrical signal. This dimension is a marker for the scale of management of grid stability issues. In this respect, as England is not synchronous with the rest of the European continent, it must maintain the stability of its grid without the inertial assistance of its neighbors' rotating machines.

– Dimension 3: organizational complexity for the operation of the grid. This dimension considers the number of organizations (public, private, associations, etc.) involved in the operation of the grid. In the case of the French national grid, there is a range of organizations: a transmission system operator, distribution system operators, balance responsible entities, regulatory institutions, energy market operators, flexibility aggregators, as well as municipalities (as infrastructure owners) and many other organizations. This dimension of organizational complexity also reflects the size of one player, in particular, the distribution system operator. In France, Enedis manages 95% of the country. So, a power grid managed by Enedis necessarily has a large organizational structure.

– Dimension 4: the voltage level of overhead lines and underground cables. This dimension gives an indication of the size of the infrastructures, their underground nature and the dangers of physically intervening on equipment. A 400 kV transmission grid line requires at least 30 m high pylons with 10 m arms to hold power lines without any risk of electrical faults.

– Dimension 5: the distance between consumers and producers. This dimension takes account of the physical distance between the places where electricity is consumed and the places where it is produced. When the grid under study is part of a wider framework, the definition of a producer is extended to include transformers at the interface with other grids. One of the objectives of this dimension is to express how easy it is to imagine (and value) the infrastructure required to produce 1 kWh, i.e. the energy needed to bake a pie.

– Dimension 6: the role of "N-1" reliability devices in the event of the loss of an element on the grid. This dimension reflects the scale of the effort required to maintain the power supply and reduce outage times. It relates both to technical systems and to the acceptability of power cuts in the collective imagination. For instance, the European electricity grid is capable of guaranteeing its stability in the event of the loss of the two largest production units, i.e. 3 GW of primary reserve. Another example on a local scale is that urban consumers can usually be supplied by a second feeder in the event of a planned or unplanned fault on the first feeder.

– Dimension 7: the regulatory framework applying to the grid. This dimension considers the regulatory framework with regard to safety, as well as the standardization of grid objects and players. For example, the requirement for the distribution grid not to deviate by more than 5% from the nominal voltage. These standards also apply to equipment connected to the grid. Inverters must be disconnected in a certain number of cases, for example, if the frequency measured exceeds 52 Hz, or if the anti-islanding algorithms detect an interruption in the grid.

– Dimension 8: the complexity of devices for optimal control of the grid. This dimension covers the devices and their complexity, for example, to minimize electrical losses on the grid. This dimension looks at the presence of equipment such as reactive power compensators (e.g. FACTS), or on-load tap changers (OLTCs) to regulate voltage. In addition, this section integrates the presence of communications and digital technology in network management, for example, electricity metering in France uses power line carrier (PLC) and then telecom networks.

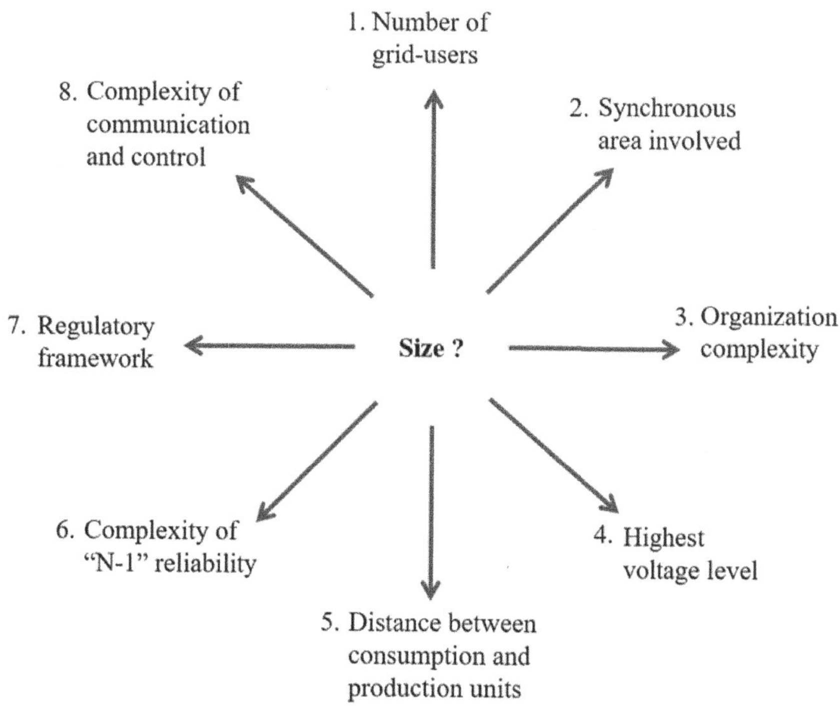

Figure 7.1. *Proposed reading prism for measuring the size of the power system*

Although partial, this breakdown of the size of a power system demonstrates the complexity of the system involved. To complete the use of such a breakdown, the next section looks at the tipping points in the power system when its size increases in one dimension. Furthermore, to what extent can we imagine a human-scale grid based on the proposed breakdown?

7.4. Nonlinearities between grid size and payoffs

From an engineering standpoint, the benefits of a very large grid are often presented as improving with the size of the power system, i.e. the larger the grid, the better. Indeed, a very large grid gives access to a greater diversity of load profiles, which reduces the total capacity of generators to be installed through a mutualization effect. In addition, the very large grid creates synergies between different intermittent production zones with different weather conditions, and access to distant hydroelectric systems. Jacobson (2021) estimates that in a scenario with 100% so-called renewable energy, Western Europe could reduce the capacity of generators to be installed by 12% by using a large, perfectly interconnected grid, rather than a sum of isolated national grids. Third, the very large electricity grid increases the security of supply by creating a neighborhood that can share electricity, as is the case during the war in Ukraine.

The question is whether these benefits scale linearly with the size of the grid, or whether, on the contrary, there are thresholds beyond which they begin to diminish. On the basis of the proposed delimitation, it is possible to identify a number of potential tipping points. An increase in the physical distance between places of production and places of consumption (dimension 5) points to the difficulties for consumers to conceive the notion of sufficiency when externalities are made invisible by long distances. Similarly, producers may find it challenging to accept externalities when they are unaware of who benefits from them. This is illustrated by the well-known "Not In My BackYard" (NIMBY) phenomenon. An increase in the number of users (dimension 1) seems to limit the ability to decide democratically on a common energy future, for example, for the construction of new nuclear power stations, or, to negotiate the distribution of resources, for example, the weight given to datacenters on the grid. An increase in the complexity of systems (dimension 8) may disrupt a stakeholder's ability to understand and act on the grid, for example, this is the case when distribution system operators have to plan a number of installations in the same area without causing any outages (Lopez and Le Blanc 2021). An increase in reliability (dimension 6) seems to involve a dependence on the electricity vector, inducing the idea that a power outage becomes unacceptable, which in turn increases the material redundancy needed in the grid. Furthermore, in an article related to grid reliability and size, Carreras et al. (2014) show that there exist a threshold for a generic power transmission system above which it is preferable to have a set of weakly interconnected grids, so as to limit the risk of large cascading failure of the entire system.

These dimensions also have implications for other dimensions, in particular on the organizational complexity (dimension 3), which is necessary given the degree of specialization of the problems addressed, for example, requiring flexibility aggregators to emerge. Thus, at a minimum, the benefits of a power grid do not grow in proportion to its size, above certain thresholds changes in form take place. Further research could be carried out to discover the different orders of magnitude at which these discontinuities and others can occur.

Unlike the "supergrid" for the energy transition, a "human-scale" grid seems more difficult to imagine. The proposed delimitation of the concept of size does, however, open up certain avenues of research in an attempt to reconcile the immensity and proximity of the grid. As Fanny Lopez reminds us, the idea is not to exclude existing parts of todays' grid, but to make it work for a different purpose.

The possibility of forming energy communities within a restricted perimeter, for example, in the sense of collective self-consumption, seems to suggest a way of reorienting the existing system. Among other things, these communities make it possible to bring the places of production and consumption closer together (dimension 5), and to tackle issues related to energy access, organization and technical complexity together in small groups (dimension 1). Another approach is to reduce the organizational complexity (dimension 3) of distribution system operators, to enable greater subsidiarity in decision-making on grid infrastructure. It should be noted that Poucineau (2023) emphasizes the importance of a minimum size of distribution system operators to mutualize resources (i.e. it is not a question of being smaller, but of being just the right size to become a lever for the energy transition). Finally, facilitating the repair and maintenance of power electronic equipment (dimension 8), which is necessary for photovoltaic production, participates in reducing the distance in our relationship with technical objects, thus increasing our autonomy.

7.5. Conclusion

The various narratives surrounding the energy transition offer visions that seem antagonistic. Some power grid narratives play on the synergies of a very large European, or even intercontinental, grid. On the other hand, narratives of sufficiency play on the proximity of citizens with energy systems. To find forms of agreement between these narratives, the proposed approach focuses on scale effects (i.e. the analysis of disruptions when the grid changes size).

In this respect, the analysis of scale effects in urban planning encourages us not to think of complex systems as being identical as they grow bigger. The concept of the city on a human scale demonstrates the importance of considering the limits of human capacity, both physical and, above all, social. As Illich reminds us, the grid as a tool must remain commensurable with human capabilities.

With the idea of achieving the right size for the power grid in mind, this chapter proposes a framework for delimiting the concept of size. This necessarily partial delimitation is broken down into eight dimensions. It includes aspects such as the physical distance between production sites and consumption sites, as well as the number of grid stakeholders, the efforts made to avoid power cuts and the technical complexity of the infrastructure, to name but a few.

This delimitation makes it possible to establish a framework which, at the very least, shows that the benefits of an electricity grid do not increase in proportion to its size, and that above certain thresholds changes in form occur. Finally, the proposed concept of size opens up new avenues of research to try and make existing systems work for a different purpose, for example, designing a grid that would make a momentary power cut understandable and acceptable.

7.6. References

Alvarez-Hérault, M.-C., Gouin, V., Chardin-Segui, T., Malot, A., Coignard, J., Raison, B., Coulet, J. (2023). *Distribution System Planning: Evolution of Methodologies and Digital Tools for Energy Transition*. ISTE Ltd, London, and John Wiley & Sons, New York.

Carreras, B.A., Newman, D.E., Dobson, I. (2014). Does size matter? *Chaos*, 24(2), 023104. doi: 10.1063/1.4868393.

Coutard, O. (1999). *The Governance of Large Technical Systems*. Routledge, London.

Duclos-Grisier, A. (2022). La géographie de la délinquance à l'échelle communale. *Vie publique* [Online]. Available at: https://www.vie-publique.fr/en-bref/284320-la-geographie-de-la-delinquance-lechelle-communale [Accessed 24 February 2025].

Fressoz, Jean-Baptiste (2024). *Sans transition. Une nouvelle histoire de l'énergie*. Éditions du Seuil, Paris.

Gehl, J. (2013). *Pour des villes à échelle humaine*. Écosociété, Montreal.

Greer, B. (2010). Overview of the papers: Why is linear thinking so dominant? *Mathematical Thinking and Learning*, 12(1), 109–115. doi: 10.1080/10986060903465996.

Haldane, J.B.S. and Rey, O. (2008). Être de la bonne taille, de J.B.S. Haldane. *Conférence*, 33, 441–454.

Hughes, T.P. (1983). *Networks of Power: Electrification in Western Society. 1880–1930*. Johns Hopkins University Press, Baltimore, MD.

Illich, I. (1974). *Energy and Equity*. Marion Boyars, London.

Jacobson, M.Z. (2021). The cost of grid stability with 100% clean, renewable energy for all purposes when countries are isolated versus interconnected. *Renewable Energy*, 179, 1065–1075. doi: 10.1016/j.renene.2021.07.115.

Kohr, L. (1957). *Breakdown of Nations*. Routledge, London.

Lopez, F. and Le Blanc, N. (2021). Utopies techniques des réseaux énergétiques. *Pour d'autres imaginaires d'interconnexion : mouvements*, 108(4), 130–144. doi: 10.3917/mouv.108.0130.

Poucineau, V. (2023). Les entreprises locales d'énergie, leviers de transition énergétique ? *Canadian Journal of Regional Science*, 46(2), 30–39. doi: 10.7202/1100212ar.

Rey, O. (2014). *Une question de taille*. Stock, Paris.

8

The Characteristics and Role of Users' Network in the Appropriation of Complex Self-built Technology: Case-study of Locally Manufactured Wind Turbines in France

8.1. Introduction

In the early 1970s, Scottish engineer Hugh Piggott left his city life for the remote peninsula of Scoraig on the west coast of the country. The peninsula located in northern Scotland is known for its harsh living conditions and its chaotic settlement dynamics that led its inhabitants unconnected to the national electricity grid. High cloud coverage in winter made photovoltaic an unfitting solution, and the locations for pico-hydroelectric generators were scarce, leaving wind turbines as the only low cost and sustainable way to produce energy locally. After a few years recharging a car battery to the closest store to get energy in his house, Piggott designed a first wind turbine model in 1978 to charge it up on the spot. Ten years and several iterations later, Piggott developed a design robust enough to face the strong winds and founded an enterprise to answer the local electricity needs. In the span of a few years, he became an international expert on the topic of locally manufactured small wind turbines, touring the world during self-construction sessions to empower local communities (Sumanik-Leary 2013).

Piggott's fame comes from its simple and effective design. The Piggott wind turbines are designed to be built anywhere with the bare minimum of tools, money,

Chapter written by Baptiste BEGUINET.

materials and technical knowledge. To maximize the diffusion of his generator, he decided not to patent the design, and wrote *A Wind Turbine Recipe Book* (Piggott 1988), detailing every step from the choice of location to the hand crafting of the whole generator and wings, and even the electrical installation. Such wind turbines are labeled as small since they measure between 1.60 m and 4.20 m in diameter, and provide a maximum power of 2,000 W compared to 2,000 kW for the smallest of big industrial wind turbines. The design can be characterized by at least three specificities: it is partly made from reused materials such as a wheel hub; the whole wind turbine can be built with nonspecialist hand tools, and almost exclusively with mechanical nonelectrical ones; the wind turbine has a mechanically triggered "flag position" when the wind is too strong, to protect it. Some parts are often customized by their builders, such as the tail or the tower. Finally, these small wind turbines can be installed linked to batteries to further increase autonomy, or connected to a conventional electricity grid, whether regional or national. Therefore, beside the construction of the wind turbine head, we must also include tower construction and electronics as the two other moments of the manufacturing process.

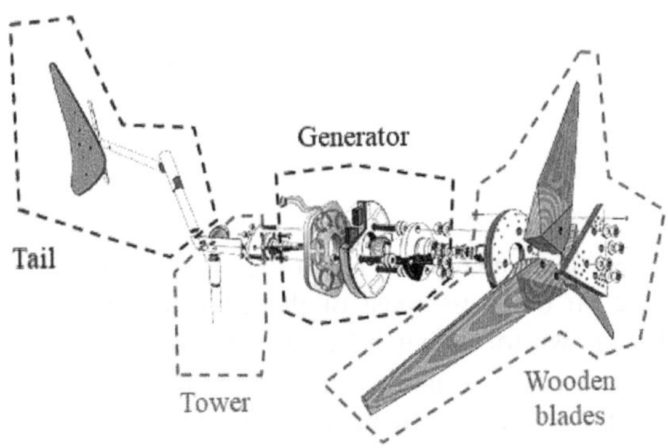

Figure 8.1. *Exploded view of a PWT (Roland Beile/Tripalium Network)*

This Piggott wind turbine soon attracted many people seeking sustainable and cheap ways to provide in energy in remote rural areas, or to become independent of the national electricity grid, or even to empower themselves in building and hand crafting techniques. In embodying these different perspectives on the same technical object, it is interesting to note the creation of at least two different NGOs, networks of pilots and dedicated Piggott Wind Turbines (PWT) researchers. In France, Tripalium was created in 2008 and maps approximately 200 Piggot wind turbines (PWT) across national territory, focusing more on empowering through hand crafting aspect. Its international counterpart, Wind Empowerment, was founded in 2011 to advocate for sustainable development through locally manufactured small wind turbines (LMSWT) (Latoufis et al. 2015).

Users can be found in remote areas with limited access to resources as well as highly urbanized countries relying on efficient national grid, such as France. This observation suggests that objectives, practices and representations among the PWT users are diverse. By focusing on the technical dimension and its appropriation, we wish to reveal the underlying motivations in favor of the PWT appropriation in the French community. We also want to understand how such a complex technology can impact practices towards sufficiency attitudes, especially in countries where energy efficiency (Shove 2018) is the dominant paradigm. As such, we distinguish two research questions: how does the users' network contribute to the PWT appropriation? Which new practices are emerging from the unique socio-technical system?

To answer these questions, we will build our thoughts upon a qualitative fieldwork with French users, based on fifteen semi-directive interviews and two participant observation of self-building sessions.

PWT can be described as complex technical objects, whose complexity must be overcome to allow its appropriation at an individual and collective level (Latzko-Toth and Proulx 2017). To paraphrase the authors' words, individual appropriation of any technique takes place through a technical understanding, its integration to our daily life and creative uses. Collective appropriation is defined as its usage, manufacture and design by a certain social group to empower itself towards other groups. Therefore, to have a better understanding of the PWT, we must consider it not only as a defined technical object, but also as a continuous process ranging from its design to its use and maintenance, through its manufacture and dissemination. This idea of a long-term relationship between a PWT and its careful users is embodied in the term "pilots" by which the users refer to themselves (Réseau Tripalium n.d.). Throughout this chapter, we will refer to the French community as a pilots' network, including but not restricted to Tripalium, despite the fact it does occupy an important part in networking. Indeed, the association provides means for the pilots to stay connected to each other through online forums, yearly meetings, group chat and interpersonal connection.

To understand the role of the users' network, we need to define who its members are and how they relate to each other. First, we wish to characterize the French users constituting the network, creating a four-profile typology. Then, we will underline the core role of the network in the appropriation, through three key moments of the PWT life cycle. Finally, we draw attention to the common representations of users and how new practices emerge, interwoven with the technology and infrastructure specific to the PWT.

8.2. A typology of French pilots

The interviewed French pilots contacted through Tripalium's member list share identity characteristics. They are all white men, aged between 30 and 60 years old, and most of them are from a technical background. Even if their wealth levels are very disparate as they were all able to afford the installation costs, from 2,000€ to 10,000€, that include materials and electronics (batteries represent the main variable here). However, to enable a nuanced understanding of their community, we can artificially create four archetypes: farmer–activist, technology enthusiast, autonomist and self-builder, indicating a range of practices and representation.

The first profile is the *farmer–activist.* What matters the most to them is the collective aspect of the project. The decision of setting up a PWT directly proceeds from their involvement in activist networks, as a means to perpetuate their professional ethics in the field of energy production. They are organic farmer, sometimes as a second career, taking pride in producing their own food, locally, while limiting their environmental impact. Energy production is not as much as a deciding factor for the installation since their needs are too important to be fully met thanks to wind energy. They reject a priori the electrical installation, perceived as a complex and time-consuming "black box". This moderate technical interest combined with their very little free time leads to a total maintenance delegation.

The second profile is the *autonomist,* who holds a pessimistic view on the future, when it is not catastrophist. We mean here the idea that "our modern societies are reaching thresholds beyond which planetary ecological dysfunctions could lead to a disruption of the material conditions required to perpetuate political communities" (Semal 2012). Their objective is to ensure a relatively comfortable life in a potentially significantly degraded environment. They reduce their electrical dependency as much as possible, but still want to be able to use electrical devices. Through the installation of a PWT, they seek a satisfying energy production and the insurance of a long-lasting installation that they can maintain themselves. However, they wish to have a functioning PWT as fast as they possibly can and chose to delegate some parts of the installation that would be too time-consuming to do on their own. In this process, they acknowledge unavoidable dependencies and arbitrates towards the most local ones.

The third profile is the *technology enthusiast,* who perceives the PWT as a complex real-sized toy that allows them to be familiar with manufacturing processes and tests all sorts of parameters. They have a technical professional background, live in suburban areas and tend to install smaller models. They have a very scientific approach of the PWT being very dedicated to the monitoring of their electricity consumption. However, they do not engage in radical sustainable lifestyles, and rather chose to focus on energy efficiency. At last, they do not really feel involved in collective dynamics, and are experimenting alone various evolutions of the models (mast, modified generator and optimized electrical installation).

Finally, the last profile is the *self-builder*, who sees the manufacturing process of the PWT as a self-teaching opportunity. They want to develop new skills that could be used in their future projects. They love learning and teaching and connecting with people sharing their interests, regardless of the geographical distance. It is a common profile among trainers of Tripalium. Their passion for self-construction proceeds from technical interests as well as economical constraints, leading them to always be looking for cost-efficient solutions to their various technical issues. At the level of their household, they tend to adopt very sufficient lifestyles.

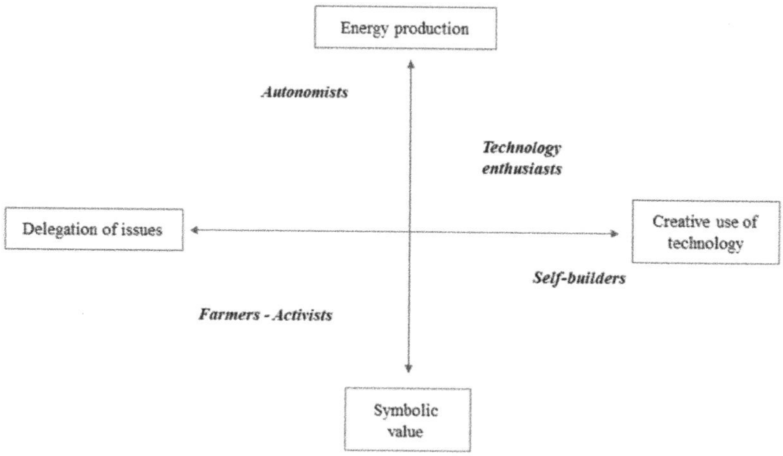

Figure 8.2. *Representation of four archetypes of French pilots*

All of these profiles show different representations and practices regarding the PWT and are summarized in Figure 8.1. The users can be roughly characterized by two dimensions. The first one is their level of technical mastery, ranging from total delegation of building parts to design innovation. The second dimension is the perception of the wind turbine as an efficient means of energy production versus a symbol of empowerment through self-learning and chosen dependencies. However, regardless of the diversity of their profiles, users overcome the apparent complexity of the PWT design to become a pilot, which can be understood as both the result and a necessary condition of appropriation.

8.3. Technical appropriation

The French pilots' community appears diverse in its technical mastery levels and importance given to the different dimensions of the PWT. With a better understanding of the various profiles, we now wish to assess how this network allows us to overcome technical complexity, leading to both individual and collective appropriation. To this end, we will focus on three different moments of the lifecycle of the PWT: its development, its manufacture during self-building sessions and its use and maintenance.

As we mentioned earlier, PWT were initially solely designed by Hugh Piggott, who then publicly released his recipe without patenting its method. The consequences were a widespread empowering movement and the emergence of a community-based development. Effectively, pilots, supporters and dedicated scientists and researchers organized themselves through online forums and recurring congresses. There is now a continuous process of continuous improvement impacting three aspects. The first one is adapting the design to fit the local needs. An example is the pilots in Madagascar struggling to find appropriate wood for the blades and experimenting with other materials. The second aspect is the improvement of the PWT performance regarding robustness or energy production. As an example, after conducting studies, Kostas Latoufis and his team suggested resizing cables to reduce losses during electricity transportation. The third aspect is the qualitative upgradation of the manufacture manual. In France, Tripalium has a team of volunteers redacting and keeping up to date their own technical handbook, adding more details, explanations, testing and implementing new building processes, which are meant to be easier or less expensive. Through all of these examples, the network appears to have an important role in the development of the PWT, showing transparency over their processes (through forums and congresses), maximizing their efforts towards accessibility (through documentation, reducing costs as much as possible, making open source sizing calculators, etc.) and replicability ("forgiving design" in which mistakes will not drastically affect the quality of the final product (only hand tools required)), while allowing its commercial usability (anyone can sell the product or parts of it or organize self-building sessions). All of these aspects are what constitutes the openness assessment criteria (Bonvoisin et al. 2017), allowing us to consider the PWT as an open-source hardware product.

Pilots' network also has a key role during the self-building sessions, the most common way of building a PWT. In France, these sessions usually last five days, and gather a dozen of trainees to build between one and three wind turbine heads, supported by two or three trainers. The organization places the pilot's freedom at its core and allows everyone to roam and experiment freely with wood working, metal working and electricity installation. In this space, not only do the pilots develop new technical skills, but they also mingle with other pilots sharing some of their beliefs or questionings, while politicizing the relationship to technique. The new encounters are described by the pilots as the main contribution of said sessions, followed by the joyful apprehension of their ability to self-empower through formations and experimentation. Through these sessions, the pilots develop a new perspective on the technique seen as an accessible playground, source of learning and political choices. As described by Aurélien Béranger, this last point is a key experience of the pilots, revealing that any piece of equipment implies a political choice. Throughout the week, the pilots learn to identify and deal with persistent tensions between efficiency, costs, environmental impact, method of production and dependencies (Béranger 2024). A recurring debate is the choice of ferrite or neodymium magnet. The second ones allow for a better energy production but implies financing the high polluting extraction and of rare-earth elements. The collective dimension of the sessions contributes to creating a space with a common culture, dedicated to addressing such tensions and debating them in order to render them meaningful. During these five days, the pilots develop a technical understanding of their PWT allowing its manufacture, use and maintenance as well as its adaptation to their needs, all while sharpening their awareness to underlying political choices and dealing with them.

At last, the pilots' network plays a significant role during the maintenance and use. On a technical aspect, as mentioned in the characterization of the users' network, some pilots decide to delegate parts of, or the whole, maintenance process. In such cases, they rely on mutual aid to troubleshoot and realize the various maintenance operations. It is a current issue for Tripalium to maximize the autonomy of the pilots for this continuous care, and the association created a lot of resources to push forward with the technical empowering. On a more political aspect, the pilots embody shared alternatives way of life. Since the pilots all go through the same phase of self-building, where they face technical and political issues, they share a common experience and sensibility to technical and energy production issues. Moreover, due to the organization of the self-building sessions, most of the PWT are a collective realization where different pilots have manufactured every piece. Therefore, the daily use of the PWT summons a common imaginary ensued by a minimal set of values. Among others, the pilots feel part of a network advocating change through chosen interdependencies with the other members of the community or the local competences that they solicited for the installation. This feeling of community is accentuated by the fact that pilots share a public identity as "small wind turbine ambassadors", whether they claim it or are simply put into it by their neighborhood. The pilots bring new thoughts to their friends, families and local community and become flag bearers of a desirable relative or total autonomy from fossil fuels and will to live sufficient lifestyles. In other words, through the whole collective experience of the PWT, pilots get to experiment, create imaginaries, embodying new future-oriented practices and norms, prefigurate their utopic future (Yates 2015).

These three moments of the PWT lifecycle bring to light the core role of users' network at every step and at various levels. The most evident one is the individual technical appropriation, that the collective approach enables through self-building sessions and community-based development. The idea here is not to say that appropriation would not be possible alone, nor that anybody can easily be able to set up a PWT, but that the users' network acts as a facilitating element lifting barriers in terms of financial, material and cultural resources. On another level, the PWT appears as a collective experience putting into actions beliefs and representations of a future deemed sustainable by the pilots.

8.4. Conclusion

The PWT as a continuous process contributes to defining a common set of minimal values, which can be broken down in at least two dimensions. First, the pilots care for the open-source aspect of their project, allowing them to minimize access costs while maximizing their technical understanding, and allowing them to modify the technology to fit their needs optimally. Secondly, pilots are all looking to empower themselves to choose how their consumed electricity is produced, gaining a relative or total independence from the national grid. As thought by its creator and embodied through several organizations including the French pilots' community, PWT are meant to enable a sufficient lifestyle opposing the macro technical system (MTS) and the idea of a thermo-industrialized society (Gras 1997, 2003). Even though this aspect is not covered in this chapter, it appeared during the interviews that the pilots all shared concerns about the social and ecological implications of energy production. They were especially expressive on their dissatisfaction with the French centralized nuclear production, and the industrial approaches on renewable energy through giant wind turbines. They described these energy production modes as money-driven, not really taking ecological issues into consideration, and neglecting citizen political freedom to choose how and where their energy is produced.

The existence of this common culture shared by all of the pilots contributes to give sense to the PWT use as a way for the pilots to take agency over their energy production and dependencies. The PWT encapsulates a minima alternative to the MTS and creates a narrative for a low-carbon society. This minimal set of values helps us understand who the pilots are, providing insights complementary to the proposed typology.

Beyond the production of said narrative, we also want to draw attention to the new practices emerging around the PWT, considered as a continuous process. Elaborating from the idea that "technologies, infrastructures and practices are interwoven" (Shove 2018), we will first draw the portrait of the PWT socio-technical system through its users, technical dimension, process and infrastructure. The pilots appear to be men with technical background, sharing the open-source and low-carbon future ideals while rejecting the MTS. As a determined technical object, the PWT is unique through its "visibilization" of energy production through its movement, overproduction and low buy rates, time-consuming maintenance, parts made of recycled materials. The building process is also peculiar, relying on self-building and a forgiving design meant to facilitate its creation. In terms of infrastructure, in France, the PWT is nested in Tripalium's and the users' network actions and narrative as well as their horizontal and participative organizations. All of these components are the cogs of the PWT socio-technical system, shaped by and shaping new practices.

These new practices can be described in two dimensions: sufficient and convivial attitudes. The PWT can effectively enable more sufficient lifestyles first through the awareness it brings to energy consumption. Pilots can note energy production in real time and try to maximize its use, introducing load-shifting to their habits. This heightened sensitivity to energy use is also directly connected to a radical shift of paradigm, from efficiency to sufficiency. Before even setting up the PWT, pilots go through a thorough sizing phase. During this stage, they get to radically address what their needs are, and which energy source is the most efficient to answer it, leading to a decrease of electrical devices and global electricity use. The other radical shift is moving away from the MST to embody conviviality. For the pilots, choosing to use a PWT is choosing to empower themselves through self-building, rejecting industrial modes of production and shift from a consumer attitude. Being a pilot implies a long-term commitment to the wind turbine with an extreme importance of care, a maintenance to ensure longevity and efficiency.

Lastly, the choice of the PWT is for the pilots, an affirmative act to achieve autonomy through chosen interdependencies. Insulating from the national grid is only substituting this dependance with one to the wind, their own ability to face lack of or overproduction, mutual aid to maintain their wind turbine, local solidarities in case of unexpected events. Therefore, the PWT can be seen as a convivial tool (Illich 1973; Vetter 2018) from its design to its use as it contributes building communities of pilots, is thought to be as easily manufactured as possible, can easily be adapted to fit local needs and allows for more sufficient lifestyles.

Piggott wind turbine pilots show acute thoughts and advanced practices regarding sufficiency and political dimension of technical choices. They seek autonomy through the conscious choice of their dependencies, highlighted by the PWT as a process. They firmly reject isolationism and seek to regain agency over their daily life, embodying in the present a future built on solidarity and necessary interdependencies to achieve sufficiency.

8.5. Acknowledgments

I would like to express my deepest gratitude to Adrien Prévost for introducing me to the incredible community of Piggott wind turbine enthusiast, his dedication to cross-disciplinary research and for his valuable insights on the Pilots community. I also extend my appreciation to Laure Dobigny and Jean-François Trégouët, who played a core role to help me shape and sharpen my thoughts and ideas. Their curiosity and patience were precious to navigate within academic research. My gratitude also goes towards Benoit Durillon for his precise and guiding proofreading.

Finally, I would like to address a special thanks to Jay Hudnall who allowed me to connect with the French pilots, taught me how to build a PWT and spent long hours answering my numerous questions with good humor.

8.6. References

Béranger, A. (2024). La politisation de la moindre pièce : tensions entre low-tech et high-tech dans les communautés du petit éolien auto-construit. *Réseaux*, 244(2), 223–256.

Bonvoisin, J., Mies, R., Boujut, J.-F., Stark, R. (2017). What is the "Source" of open source hardware? *Journal of Open Hardware*, 1(1), 5.

Gras, A. (1997). *Les Macro-systèmes techniques*. Presses Universitaires de France, Paris.

Gras, A. (2003). *Fragilité de la puissance : se libérer de l'emprise technologique*. Fayard, Paris.

Illich, I. (1973). *Tools for Conviviality*. Harper & Row, New York.

Latoufis, K.C., Pazios, T.V., Hatziargyriou, N.D. (2015). Locally manufactured small wind turbines: Empowering communities for sustainable rural electrification. *IEEE Electrification Magazine*, 3(1), 68–78.

Latzko-Toth, G. and Proulx, S. (2017). Appropriation des technologies. In *Sciences, technologies et sociétés de A à Z*, Bouchard, F., Doray, P., Prud'homme, J. (eds). Presses de l'Université de Montréal, Montreal. [Online]. Available at: http://books.openedition.org/pum/4256 [Accessed 27 July 2023].

Piggott, H. (1988). *A Wind Turbine Recipe Book*. Scoraig Wind Electric, Dundonnell.

Réseau Tripalium (n.d.) L'Éolienne Piggott. *Réseau Tripalium* [Online]. Available at: https://www.tripalium.org/eolienne-piggott [Accessed 1 August 2023].

Semal, L. (2012). Militer à l'ombre des catastrophes : contribution à une théorie politique environnementale au prisme des mobilisations de la décroissance et de la transition. PhD Thesis, Université du Droit et de la Santé – Lille II, Lille [Online]. Available at: https://theses.hal.science/tel-01659916 [Accessed 19 October 2023].

Shove, E. (2018). What is wrong with energy efficiency? *Building Research & Information*, 46(7), 779–789.

Sumanik-Leary, J. (2013). Small wind turbines for decentralised rural electrification: Case studies in Peru, Nicaragua and Scotland. PhD Thesis, University of Sheffield, Sheffield [Online]. Available at: https://etheses.whiterose.ac.uk/6282/ [Accessed 27 March 2023].

Vetter, A. (2018). The matrix of convivial technology – Assessing technologies for degrowth. *Journal of Cleaner Production*, 197(Part 2), 1778–1786.

Yates, L. (2015). Rethinking prefiguration: Alternatives, micropolitics and goals in social movements. *Social Movement Studies*, 14(1), 1–21.

9

Scripting Flexibility in Collective Self-Consumption: Local Limits, Users and Governance

9.1. Collective self-consumption and the challenge of flexibility

Shifting the programming of household appliances from off-peak nighttime hours to the sunniest hours of the day has become one of the key challenges associated with the development of self-consumption, first on the individual scale and now in its collective form, which has been expanding across Europe in recent years. In France, for example, regulatory changes introduced in 2017 have allowed local producers to share renewable electricity with selected consumers via the low-voltage grid, either freely or commercially, under specific conditions (e.g. geographic proximity, smart metering, etc.). These operations have grown rapidly, from 41 in 2020 to nearly 700 in early 2025, and primarily rely on flow-based energy sources such as solar power (Enedis n.d.). Once producers and consumers are gathered within a common legal entity, local electricity is allocated to consumers in real time by public grid operators. Any surplus of local production is injected into the grid and sold, often at a financial loss, while additional energy needs of consumers are covered by conventional electricity suppliers. In entering such initiatives, consumers can benefit from preferential tariffs or find stability by extracting part of their consumption from the international related electricity market. For others, it is also a way to make a commitment to ecological values or local development objectives.

This collective self-consumption framework (CSC) introduces a radically new constraint of simultaneity in the development and integration of renewable energy

---

Chapter written by Jessica ZAPHIROPOULO.

For a color version of all figures in this chapter, see www.iste.co.uk/dobigny/rethinking.zip.

into the electric grid: electricity must be consumed as it is produced. Although still marginal in the overall energy mix, CSC initiatives reflect a broader transformation in the governance of energy systems. Unlike fossil-fueled models, where energy is continuously available and abundant thanks to large technical systems (Gras 1993), CSC requires users to engage more directly with the temporal rhythms and limits of decentralized renewable production, particularly intermittent sources such as solar and wind. In this sense, CSC can be understood as a form of "bounded infrastructure" (Shove et al. 2005, p.9), in which the supply–demand equilibrium is managed differently from large centralized networks due to specific spatial, technical and temporal limits.

One of the main challenges in such systems lies in the limited ability to adjust either production or demand in real time. In both policy and academic debates, this is increasingly framed through the concept of flexibility, presented as a key condition for integrating renewable sources and maintaining grid stability. In this context, CSC has been promoted as a means to enhance this flexibility, in line with European directives encouraging consumer participation (Directive 2019/944). This framing of flexibility centered on users as adaptable agent is mainly rooted in technical and economic approaches, in which flexibility is either a resource to be optimized through market signals or as a set of user behaviors to be adjusted though demand-side management (Coutard and Shove 2019; Blue et al. 2020). More recent scholarship, however, has emphasized the sociotechnical dimension of flexibility, showing how it is and embedded in and co-produced by infrastructures, routines and institutional contexts through a set of practice and material arrangements (Abram and Silvast 2020; Blue et al. 2020). These approaches shift the focus from individual responsiveness to the collective capacity embedded in social rhythms and infrastructure configuration. Building on this sociotechnical perspective, and on Akrich's (1989) work on script which will be detailed later in the chapter, I explore how flexibility is scripted and transformed within CSC projects, and through which logics of role distribution, appropriation of limits and collective negotiation emerge.

Drawing on a comparative analysis of two empirical cases, I pursue a twofold objective: first, how flexibility is materially and institutionally configured through the design and evolution of CSC projects; and second, to develop a more relational and political understanding of flexibility, grounded in local dynamics of coordination and self-limitation. The next section presents the case studies and the methodology of survey. Section 9.3 develops the theoretical ground for the analysis and helps to reframe the challenges around flexibility in CSC. Section 9.4 presents the results and analyzes how assumptions about flexibility are embedded in project design and how they evolve over time, particularly in relation to user engagement. Finally, I use this analysis to develop a more situated and relational understanding of flexibility in light of the existing literature, before concluding with their theoretical and political implications.

9.2. From typology to fieldwork: introducing the case studies

In France, two main types of CSC configurations are currently distinguished by energy sector actors[1]. The first, known as patrimonial operations, is the most widespread. It generally involves public or semi-public actors, such as local authorities or social housing organizations, who group together multiple meters located within the same or adjacent buildings under a single legal entity. In this model, producers and consumers are the same institutional actor, effectively managing energy flows within their own property portfolio in order to gain certain form of autonomy (Poupeau et al. 2022; Debizet 2023).

The second model, which is referred to as open operations, is characterized by more heterogeneous and context-specific arrangements. These initiatives may emerge from preexisting local collectives or be established specifically for energy-related purposes. In both cases, they bring together actors not traditionally involved in the energy sector, such as local businesses, municipalities or residents, who must develop new forms of coordination around energy and resource management. The two case studies presented in this chapter both fall within this category. By focusing on open configurations, where production and consumption are institutionally and spatially separated, it becomes possible to explore how responsibilities for balancing supply and demand are allocated, materialized and transformed over time.

The first project, referred to here as *project A*, began in the summer of 2020. It is led by a social housing organization and the local electricity distribution and supply company in a metropolitan area of approximately 450,000 inhabitants. The CSC project is part of a broader experiment on building energy autonomy. The local energy supply company produces electricity on a building roof with an installed capacity of 209 kWp. A portion of this capacity (44 kWp) is specifically allocated to the self-consumption project. The electricity is sold by the producer to the social housing organization for the common areas, as well as to residents of the building who have opted into the project. The organizing legal entity overseeing the initiative is integrated into the producer's structure (the local energy distribution and supply company). Depending on rental turnover, the project involves between 50 and 80 members.

Project B was launched in the summer of 2021. It is located in a small rural town of approximately 600 inhabitants. The project is part of a broader environmental and social public policy about sufficiency and autonomy. Electricity is produced by a

1 Some actors further distinguish within patrimonial CSC operations between initiatives led by local authorities and social initiatives led by public housing providers, effectively creating a threefold typology. However, as this distinction is not directly relevant to the objectives of this chapter, it has not been adopted here.

local semi-public supply company[2] and the municipality. A portion of this electricity is used directly by the town hall and the school through an individual self-consumption scheme, while the remainder is sold to nine village residents and a local farmer who joined the project. The legal entity managing the CSC is a nonprofit association, primarily governed by consumers.

To explore these two case studies, I adopted a qualitative approach combining multiple sources and methods. Most of the data mobilized in this analysis were collected as part of the ongoing doctoral research conducted between 2020 and 2025. The empirical material includes 35 semi-structured interviews with a range of actors involved in CSC initiatives among which technical intermediaries, energy consultants, project developers, local elected officials, producers and consumers are directly involved in CSC projects. These interviews focused mainly on the perception and narrative of the projects' design, events and evolution, as well as on domestic energy use with the exploration of the notion of limits. These were completed by several weeks of observation in situ in the two CSC projects enabling third parties not directly involved in the projects, and collecting observational data on interactions within meetings, or in public and domestic spaces. Data collected were analyzed inductively using coding method in a computer-assisted qualitative data analysis software (CAQDAS) (Lejeune 2019). The interviews were conducted over a period of several years, sometimes involving the same persons, which helped me isolate specific situations of negotiation around flexibility within the project's life from initial design to later stages of expansion.

The chapter also draws on materials from two collective research projects supported by the French Agency for Ecological Transition (ADEME) and the Urban Planning, Construction and Architecture Plan (PUCA). The first, conducted between 2021 and 2022, offers a broad institutional and regulatory analysis of the emergence of CSC in France based on interviews and institutional document analysis (Poupeau et al. 2022; Poupeau and Lormeteau 2024)[3]. While I was not part of this team, their work provides a useful backdrop for situating the regulatory and historical context in which CSC projects have developed and allows me to give an account of the institutional vision of flexibility inscribed in the CSC framework. In addition, this project offered an in-depth monographic analysis of one of the same cases studies

2 The local semi-public supply company is related to a departmental energy union. In France, most of local authorities delegate the management of public distribution networks, the maintenance of lighting and the development of local energy policies to collective departmental structures (mixed union) that bring together several local authorities in the same area.

3 Emergence et développement de l'autoconsommation collective en France : modèles et perspectives, lead by François-Mathieu Poupeau, LATTS and Blanche Lormeteau, IODE, 2021–2022.

presented here, providing secondary insights in terms of both investigation and analysis. The second project, launched in 2023, focuses on the potential trajectories and reconfigurations of CSC development in the years to come[4]. As a member of this second research team, I actively contributed to empirical and analytical components, which complement the doctoral work while maintaining a distinct focus and critical lens over the studied CSC initiatives.

9.3. Theoretical ground: multi-level scripted flexibility

The analytical approach adopted in this chapter is rooted in Science and Technology Studies (STS), and more specifically in the concept of script developed by Madeleine Akrich. She argues that technical object can be described "as a scenario, a script, defining a space, roles, and rules of interaction between the different actors (human and nonhuman) who will embody these roles", thus sharing skills and attributions within collectives (Akrich 1989, 1991). I propose to apply this concept to CSC initiatives to explore how sociotechnical systems materially and institutionally inscribed a certain vision of and expectations on flexibility in project design and governance. To put it another way, governance, economic models and technical choices help us to understand how responsibilities and roles about uses, behaviors for the supply–demand balance are distributed across the collectives. Following Akrich's work, I want to pay particular attention to how these are subsequently enacted, adapted or resisted by local actors, and especially by consumers.

However, scripts cannot be understood only at the local level. They are embedded in broader institutional architecture. In the case of CSC, many of the expectations around flexibility are defined at the national and even European levels. Introduced in the late 2010s, CSC emerged as part of a broader shift away from feed-in tariffs towards incentives for local self-consumption and from centralized subsidy mechanisms to decentralized energy coordination (Aubert 2020). As presented in the outset, CSC has been framed as a way to activate the flexibility potential of consumers in compensating to a certain extent for the rigidity of the existing centralized system. This model departs significantly from earlier subsidy-based logics and shifts financial risks and operational responsibility from the state onto users and local organizers (Fonteneau 2021), involving them as adaptable agents, capable of aligning their consumption with local renewable production.

Technically, this is mainly embodied in the simultaneity principle, defined in the CSC framework. Electricity sharing is managed in real time, based on 30-minute

4 Le Déploiement de l'autoconsommation collective d'électricité en France (D-ACCeF), lead by Gilles Debizet, Pacte, 2023–onward.

allocation intervals and predefined distribution keys. In CSC projects based on solar photovoltaics, this implies synchronization of consumers with variable solar cycles, shaped by spatial parameters (panel orientation, size, location, etc.), weather conditions and contextual constraints, such as maintenance operations or potential outages. CSC introduces a new material and temporal grounding to energy supply, embeds a strong prescriptive script that expects spatial and temporal coordination, without always enabling it. This creates a structural paradox: while CSC is promoted as a decentralized and participatory model, it relies entirely on the national centralized grid infrastructures. The continuity of fossil-based infrastructures and standardized usage routines tends to decontextualize energy flow and obscure local production users are expected to respond to. For instance, in solar-based CSC systems, optimal consumption timing conflicts with long-standing French load-shifting policies that incentivize nighttime energy use through special tariffs and automated household appliances (e.g. hot water tanks, dishwasher or electric vehicle charging). In this context, flexibility becomes not just a technical function, but a question of visibility, engagement and appropriation of the local production and its limits.

Making the limits of the system actionable, whether they are spatial, temporal or quantitative, requires more than technical devices or price signals. It involves reconfiguring the national sociotechnical scripts through which energy systems are organized and interpreted. In the following section, I explore how local actors have interpreted and adapted these national scripts, sometimes reinforcing them, sometimes subverting them to construct their own ways of making flexibility work.

9.4. From regulation to practice: scripting flexibility in two CSC projects

Although both projects operate under the same regulatory framework and share overarching objectives related to energy sufficiency and local autonomy, they differ significantly in how they translate, adapt and reinterpret the flexibility script. To unpack these dynamics, I analyze each project in turn focusing on three interrelated dimensions: (1) the supply side, where the design of production units shapes the initial framing of flexibility; (2) the user engagement, where expectations are placed on consumers; and (3) the governance structures, which determine how scripts are co-constructed, negotiated or stabilized across actors and time.

9.4.1. *Supply-side flexibility: technology and production design*

The choice of production technology plays a crucial role in shaping how flexibility is initially scripted within CSC projects. Unlike other subsidy-based models of renewable development that prioritize maximizing output for economic rentability and grid integration, these projects in order to reach economic viability aim to align production with local consumption patterns, thereby embedding certain temporal and normative assumptions into the system design.

Until 2021, the absence of national public support mechanisms and access to feed-in tariffs for production included in the CSC's framework pushed project developers to maximize the self-consumption rate (i.e. the share of energy consumed locally), while minimizing surplus energy sold back to the grid at often less favorable rates (Fonteneau 2021). This economic constraint led them to closely align the size of installations with the needs of consumers based on actual or projected consumption data. This logic is particularly evident among social housing providers, such as in *project A*, where CSC operations are often sized on projected consumption data either calculated with the number of residents per household, or with the size of the housing units (Poupeau et al. 2022).

Linking production units with the residential consumers' needs, or at least targeted needs, gives a first hint about the influence demand plays on defining the scope of operations and shaping the script that governs the roles of producers and consumers. However, adapting production to demand is not achieved solely through the sizing of installations. It also involves selecting specific production technologies, location and orientation. In the two case studies presented here, project developers opted for "innovative" technologies, an adjective generally shared by project developers and consumers alike.

In *project A*, solar panels were installed at a tilted angle in order to smooth out the production curve throughout the day (see Figure 9.1). This configuration departs from standard practices in solar architecture, which typically prioritize energy yield and peak output. This choice reflects a concern for avoiding overproduction and matching with tenants' load consumption at the building scale, both in quantity and rhythm, even if it comes at the expense of the global production potential. The building was designed to consume very little energy, thanks to advanced insulation techniques (use of cork, triple glass) expected to outperform the 2012 thermal standards in force during its construction.

The constraint here is not technical per se, but structural: producing more only makes sense if it supports higher self-consumption rates. Otherwise, it reintroduces dependence on external markets and reduces the economic viability of the project. Panel orientation and form were also influenced by the project's ambition for broader building autonomy and including a rainwater recovery system. The setup was intended not only for functionality but also to visibly express energy self-sufficiency (Fonteneau 2020).

Figure 9.1. *Photovoltaic panels on the roof of project A's building (source: author's photograph, September 2024)*

Project B relies on a solar tracker equipped with bifacial panels mounted on a mast, which rotates to follow the sun's trajectory throughout the day (see Figure 9.2). This configuration, like in *project A*, aims to smooth out production over the course of the day. The tracker was installed at the center of the village by the municipality and the producing company. These choices serve a dual purpose: enhancing energy efficiency while also materializing local production in a visible and symbolic and strategic way. By reflecting the sun's position across the day and the seasons, and by flattening in case of strong wind, the tracker offers a public display to raise awareness among consumers of both the rhythms and vulnerabilities of solar production and its constraints:

> So I know that when it's not producing, it's because it's flat and there's too much wind, or there's a technical problem (consumer of the CSC *project B* living near the tracker).

These installations were chosen not only for their functional benefits but also for their symbolic value. They serve as visible "totems" for stakeholders, reinforcing the connection between local production and consumption (inhabitant reporting *project B* leader's words, interview 2022) (Fonteneau 2020).

These supply-side choices illustrate how flexibility is materially scripted from the outset of CSC initiatives, not only through device performance, but through implicit expectations about what counts as adaptation, what timeframes matter and how production should also relate and adapt to residential consumption. This script becomes the baseline upon which other forms of coordination, at the demand and governance level, will later build.

Figure 9.2. *Photovoltaic panels in project B (source: author's photograph, May 2022)*

9.4.2. *Demand-side flexibility: community composition and engagement*

While CSC frameworks are often based on the assumption that users will adapt their consumption patterns to align with production, the two projects studied show that this demand-side flexibility is far from being spontaneous. In both cases, flexibility is not simply a behavioral variable, but a collectively managed and negotiated process that evolves in response to organizational choices such as the selection and management of the consumer community.

In *project A*, after a phase of promoting the housing project and opening it up to applications, the building's residents were chosen among beneficiaries of social housing based on their sensitivity to ecological issues. This required the social housing organization to obtain a special exemption system for the dwelling allocation in agreement with the metropolis and the state departments in order to align with the project's autonomy and sufficiency demonstration goals.

However, this is not representative of other CSC projects managed by social housing providers, as they typically do not select residents based on ecology or energy-related criteria or their wish to join in the CSC project. In such cases, the responsibility for adapting and shifting demand falls more heavily on the organization of the community itself, particularly through the rules governing electricity sharing. As a result, social housing stakeholders often opt against equal sharing in favor of a default system provided by the distribution network operator, which promotes daytime electricity use and maximizes local consumption.

In *project B*, to better align with solar production cycle, the community was intentionally designed to include not only residential consumers, but also consumers with significant daytime energy needs, such as a farmer. In this way, the supply–demand balance is based on an upstream adjustment of consumer profiles rather than expected changes in behavior. As for the electricity sharing rules, they were decided collectively by all consumers. They adopted a two-round allocation system: it begins with equal sharing and is followed by rules prioritizing local economic activities registered at the meter's address. This choice reflects a balance between the objectives of social justice and the need to maintain the balance of the operation.

In both cases, responsibility for demand-side flexibility does not rely solely on load-shifting incentives embedded in the sharing rules, as we may expect. According to several project leaders interviewed, such incentives often prove ineffective, except for a few highly committed users with a strong interest in technology and energy monitoring, matching the "technophile" socio-energy profile described by Stéphane Labranche (2021). Instead, it also lies in the strategic selection of an appropriate consumer profile during the community's initial setup phase.

9.4.3. *Script evolution: the role of demand-side in project development*

Although these sociotechnical scripts about flexibility are primarily designed during the initial phases of CSC projects, they are not fixed. Over time, the roles and responsibilities for balancing supply and demand evolve in response to operational challenges, users' dynamics, and institutional or economic constraints, particularly when projects scale up or reduce their scope.

Despite its initial goal of maximizing self-consumption, *project A* in the social housing building encountered significant barriers. The usual high tenant turnover of social housing complicated efforts to maintain engagement and alignment of consumers with project objectives. Departing members of the building and so of the CSC project reduced the CSC rate, with no guarantee that newcomers would participate. As a result, the initial occupant selection criteria on ecological values were gradually abandoned in favor of a more traditional selection process. Additionally, billing complexities and delays with distribution network operators and national electricity suppliers further obscured the visibility of CSC's benefits for consumers, particularly regarding load shifting. The expectation put on the demand side to match production did not materialize as expected, ultimately destabilizing the project economic viability.

In response, responsibility for balancing supply and demand became increasingly technical, shifting from direct consumer engagement to automated solutions, thus reducing visibility and materiality of local supply limits. Battery storage and automated heating controls were introduced within the building, reducing reliance on consumer behavior and placing flexibility management primarily on technological devices. This situation is reminiscent of the dynamics observed in ecological housing experiments, such as those documented in low-consumption labelled buildings where residents often struggled to appropriate the environmental objectives of their dwelling (Renauld 2013). In both cases, a gap emerges between the project's intended participatory ambition and the actual engagement of occupants.

This shift in responsibility, flexible in *project A*, reflects broader challenges faced by social housing organizations involved in CSC. Some sought alternative strategies, such as expanding the consumer base to include common areas of nearby buildings or local businesses, to enhance self-consumption rate and mitigate individual engagement barriers (manager of CSC projects in a social housing structure, interview 2022).

In contrast, *project B* in the rural area maintained its commitment to sufficiency and autonomy through the active participation of members in negotiating the operational limits on both the supply and demand sides. Encouraged by the project's initial success, members sought to include new consumers. While the original plan was to expand production capacity, legal and financial constraints related to the involvement of a public entity in the project prevented this option. Instead, members opted collectively to reduce their individual consumption to accommodate new participants while adhering to the collectively established electricity distribution balance (Poupeau et al. 2022).

In this context, one of the project members reported that, as of 2022, nine households were participating in the operation and that efforts were underway to reduce individual consumption from the tracker production, both through behavioral changes and by installing individual photovoltaic panels. According to him, these dynamics have contributed to bringing the consumption of locally energy self-generated rate of the collective to between 15 and 20%, with approximately 8–9% of annual surplus remaining unused. On this basis, the possibility of adding two to three new members was discussed within the group, given the margin that still exists before the balance point is reached (consumer of *project B*, interview 2022).

This evolution was driven by two main arguments expressed by members: first, to make the project accessible to a larger number of consumers, given its public status that can be regarded as a public service mission; and second, to maximize the self-consumption rate in order to maintain economic viability. While the producer has not expressed concerns over financial aspects, surplus electricity injected to the grid was perceived by consumers as a "lost" opportunity and inefficiency, both financially and politically, as this locally generated energy was seen as something that should serve the benefit local citizens rather than external suppliers (consumer of *project B*, interview 2022).

In this context, the quantitative limits on energy production, acknowledged and actively taken into account by members upon, were not experienced as a burden but rather as a guiding principle for the social and political management of the operation. It served as a reference for balancing supply and demand and collides with the principle of fostering sufficiency, originally inscribed in the association's statutes. In this sense, the project can be likened to what Giorgos Kallis conceptualizes as a "*collective self-limitation*" initiative: a voluntary, community-based decision to operate within the constraints of the local energy system in order to ensure a fair distribution (Kallis 2021). This was put into practice through regular discussions during general assemblies, the organization of a workshop on how the solar tracker functions and more broadly on energy-saving practices. Over successive annual reviews, members observed a decrease in electricity use from the tracker and expressed their readiness to welcome additional participants, demonstrating a dynamic interpretation of limits not as restrictions to be overcome, but as principles to be shared and collectively negotiated whether on both supply and demand sides.

As CSC projects evolve, the capacity of communities to engage with, renegotiate and adjust limits on both the supply and demand sides remains a critical factor. In cases such as *project B*, this adaptability has allowed sufficiency to become a foundational principle for managing local energy systems. In contrast, in *project A*, the technical and economic constraints of flexibility diluted the initial sufficiency objectives, highlighting the difficulty of aligning flexibility with broader sustainability goals.

9.5. Flexibility beyond individual adaptation

The comparative analysis of *projects A* and *B* shows that flexibility, far from being a uniform or static requirement, is constructed through situated sociotechnical arrangements that evolve over time. While both projects operate under the same legal and infrastructural frameworks, they diverge in how they assign, distribute and renegotiate responsibility for balancing supply and demand. This divergence reflects contrasting interpretations of what flexibility entails, for whom and under what conditions.

In *project A*, flexibility was initially scripted as a shared responsibility between tenants and technology, but it progressively shifted towards technology automation. The challenges of maintaining user engagement, due to housing turnover and weak visibility of energy flows, led to a reconfiguration of the project's internal logic. This trajectory exemplifies a model of delegated flexibility, where behavioral adaptation is minimized and the system is stabilized through infrastructure.

In contrast, *project B* illustrates a model of negotiated flexibility, rooted in collective deliberation and sufficiency-oriented values, understood here not simply as a reduction of energy use, but as a normative and political orientation toward living well within limits (Kallis 2021). From this perspective, sufficiency is less about individual changes than about reconfiguring collective expectations, infrastructures, and social norms. Confronted with institutional constraints that limited production expansion, participants chose to redistribute existing capacity by voluntarily reducing their own consumption. This move, far from being framed as a sacrifice, was perceived as an ethical and political choice grounded in the principle of fair access to locally generated renewable energy. Here, the management of supply and demand relationship is not a side effect of optimization, but a central object of collective governance, through the discussion of self-consumption rates, sharing practice, discussion about community expansion.

These findings suggest that flexibility cannot be fully understood through a lens focused solely on load shifting or individual behavioral change. As Akrich (1991) emphasized, scripts are not merely written in advance. They are reinterpreted, challenged and transformed as technologies, or in that case sociotechnical configurations, become embedded in collective practices. In the case of CSC, flexibility emerges less as a quantifiable capacity than as a relational process involving infrastructures, norms and negotiations. This perspective echoes the work of Blue et al. (2020), who argue that flexibility should be conceived not as an attribute of consumers or devices, but as a multi-temporal, socially entangled phenomenon. While their work highlights the diversity of "flexibilities" in time and practice, the present analysis adds a further layer by emphasizing the political and infrastructural collective nature of these arrangements. In this perspective, flexibility is not solely defined by institutional timeframes, such as the 30-minute intervals used by smart meters for daily production management but is understood over a longer time scale. It is assessed by actors within the broader framework of annual energy and financial balances and reports, positioning flexibility as a long-term process of infrastructure management and evolution.

Finally, this perspective invites us to reexamine the role of limits in energy systems, not as mere constraints to be optimized away, but as generative devices that shape both the organization and meaning of flexibility. In both cases, limits, whether technical, spatial, economic or normative, played a central role in structuring how flexibility was understood and coordinated. In *project A*, limits prompted a shift towards infrastructural delegation, marginalizing users in the process. In *project B*, they served as points of deliberation and collective redefinition, anchoring practices of sufficiency and shared responsibility.

9.6. Conclusion

This chapter has shown that flexibility, far from being a stable or uniform requirement, is situated and negotiated through sociotechnical configurations that evolve over time. Through the comparison of two CSC projects, it became clear that the allocation of responsibilities for balancing supply and demand is not determined solely by regulatory frameworks or technical infrastructures. Rather, the study of the two project trajectories has shown that it is the result of local reinterpretations of national scripts, shaped by material conditions, governance choices and collective norms.

Project A illustrates a trajectory where flexibility becomes increasingly delegated to technical infrastructure, reflecting the difficulties of sustaining user engagement in social housing contexts. Whereas, *project B* exemplifies a form of negotiated flexibility, where energy limits are not simply endured but embraced as a shared principle for collective coordination and negotiation.

In line with scholarship that promotes a sociotechnical approach to flexibility, these findings call for a shift in how flexibility is conceptualized in energy transitions: not as an individual behavioral response or a technical adjustment, but as a relational and political process, embedded in practices, infrastructures and institutions. Flexibility, in this light, is enacted not only through devices and price signals but also through shared meanings, evolving roles and the capacity to make limits visible and actionable within CSC projects. While project leaders may often appear to be the primary actors formulating and directing hypotheses about flexibility, as well as defining and implementing operational limits, our analysis reveals that this responsibility is, in fact, distributed across multiple levels, and that users have some capacity to negotiate, reinterpret and reshape the arrangement over time.

Ultimately, looking at flexibility through the lens of CSC projects offers a valuable entry point for rethinking energy transitions. It draws attention to the relationship between energy sources, production arrangements and everyday consumption practices, and supports a reorientation of energy governance, not around control and optimization, but around coordination, responsibility and collectively negotiated constraints. This is especially critical in decentralized systems, where technical and social scripts must be continuously co-constructed and revised.

9.7. References

Abram, S. and Silvast, A. (2020). Flexibility of real-time energy distribution: The changing practices of energy control rooms. *Revue d'histoire de l'énergie*, 5(2), 1g–17g.

Akrich, M. (1989). La construction d'un système socio-technique. Esquisse pour une anthropologie des techniques. *Anthropologie et sociétés*, 13(2), 31–54.

Akrich, M. (1991). De la sociologie des techniques à une sociologie des usages : l'impossible intégration du magnétoscope dans les réseaux câblés de première génération. *Techniques & culture*, 16. doi: 10.4000/tc.725.

Aubert, F. (2020). "Communautés énergétiques" et fabrique urbaine ordinaire : analyses croisées Allemagne, France, Royaume-Uni. PhD Thesis, Paris-East Créteil University (UPEC), Paris. [Online]. Available at: http://www.theses.fr/2020PESC2016 [Accessed 24 March 2021].

Blue, S., Forman, P., Shove, E. (2020a). Flexibilities in energy supply and demand: Legacies and lessons from the past. *Revue d'histoire de l'énergie*, 5(2), 1–8.

Blue, S., Shove, E., Forman, P. (2020b). Conceptualising flexibility: Challenging representations of time and society in the energy sector. *Time & Society*, 29(4), 923–944.

Coutard, O. and Shove, E. (2019). Infrastructures, practices and the dynamics of demand. In *Infrastructures in Practice: The Dynamics of Demand in Networked Societies*, Shove, E. and Trentmann, F. (eds). Routledge, London.

Debizet, G. (2023). Les communautés énergétiques par-delà le marché unique. *L'Économie politique*, 97(1), 80–91.

Enedis Open Services & Open Data (n.d.). Autoconsommation collective d'électricité en France [Online]. Available at: https://data.enedis.fr/pages/autoconsommation-collective/ [Accessed 27 February 2025].

Fonteneau, T. (2020). Carnets de terrain : le projet de bâtiment autonome ABC comme lieu d'expérimentation de l'autoconsommation collective. Report, Université Grenoble-Alpes, Grenoble.

Fonteneau, T. (2021). Autoconsommation collective ou solidarité nationale? L'adaptation controversée de la tarification du réseau d'électricité pour les autoconsommateurs. *Flux*, 126(4), 52–70.

Gras, A. (1993). *Grandeur et dépendance. Sociologie des macro-systèmes techniques*. Presses Universitaires de France, Paris.

Kallis, G. (2021). Limits, ecomodernism and degrowth. *Political Geography*, 87, 102367.

Labranche, S. (2021). *Énergie et écologie : les sept profils socio–énergétiques*. PUG, Grenoble. [Online]. Available at: https://www.librairieleneuf.fr/ebook/9782706151095/energie-et-ecologie-les-sept-profils-socio-energetiques-stephane-labranche-pug [Accessed 11 June 2025].

Lejeune, C. (2019). *Manuel d'analyse qualitative : analyser sans compter ni classer*, 2nd edition. De Boeck, Louvain-la-Neuve.

Poupeau, F.-M. and Lormeteau, B. (2024). *L'autoconsommation collective d'électricité en France: émergence d'une innovation contrariée*. Mines Paris – PSL, Paris.

Poupeau, F.-M., Lormeteau, B., Marchand, M., Pistoni, R., Dornier, L., Hernigou, N., Barel, S., Brisepierre, G. (2022). Émergence de l'autoconsommation collective d'électricité en France : modèles et perspectives. Report, PUCA, Paris.

Renauld, V. (2013). L'habitat écologique à l'épreuve de son public. *Espace temps* [Online]. Available at: https://www.espacestemps.net/articles/lhabitat-ecologique-a-lepreuve-de-son-public/ [Accessed 10 June 2025].

Shove, E., Chappells, H., Vliet, B.V. (2005). *Infrastructures of Consumption: Environmental Innovation in the Utility Industries*. Routledge, London.

10

Collaborative Energy Management: A Novel Methodology to Reduce Both Energy Price and Environmental Impact of an Energy Community

10.1. Introduction

Integration of renewable energy into the energy mix is an identified strategy for mitigating climate change, following sobriety and efficiency. However, their intermittent nature necessitates the development of new approaches for electricity use and distribution. It is therefore necessary to imagine new organizations and economic models. With an engineering approach, demand-side management becomes crucial to address the constraints on the distribution grid. Optimal energy behaviors are, in this context, often regulated by the price. Moreover, new economic models based on market logic are directed by the deregulation of the energy markets. A more recent legal way for local actors to exchange renewable energy is through energy communities, often combined with collective self-consumption. They open a new approach to energy management based on resource pooling and cooperation.

This chapter explores a new methodology of demand-side management based on a cooperative game. It will be proved that a collective objective based on prioritizing local energy production in community consumption can lead to higher individual costs and environmental impact reduction than individual behavior. The work integrates several motivations of stakeholders in their participation in a collective self-consumption operation. A new methodology based on the life-cycle assessment

---

Chapter written by Adrien BOSSU, Benoit DURILLON and Christophe SAUDEMONT.

For a color version of all figures in this chapter, see www.iste.co.uk/dobigny/rethinking.zip.

(LCA) method is deployed to assess environmental impacts. Ethical distributions of gains based on game theory will also be discussed. Several game-theoretic concepts are explored, opening up philosophical questions such as: should we reward those who participate the most, but who are also often the wealthiest? Should we encourage the most vulnerable? From a technical point of view, other questions are discussed, such as: which behaviors should be rewarded in the context of energy management? Should energy-saving behavior be rewarded? Should flexibility be rewarded?

Readers interested in a more in-depth exploration of the approach, including additional references and technical details, are encouraged to consult the underlying doctoral research, which has been published in the interim (Bossu 2024).

10.2. Context

10.2.1. *Evolution of the electrical grid*

The transition towards decentralized and sustainable energy systems is driven by technological, legal, economic and social innovations. Advances in smart grids, energy storage and demand-side management enable greater flexibility and resilience in electricity distribution (Lampropoulos et al. 2013; Meyabadi and Deihimi 2017). These technological developments support new economic models based on market logic, where self-consumption – both individual and collective – becomes a key strategy for optimizing energy costs and reducing dependence on external suppliers. Legal innovations, including energy communities (Cui et al. 2020; Tushar et al. 2020) and regulatory frameworks for peer-to-peer energy exchange, empower local actors to engage in cooperative management strategies that go beyond traditional market-driven approaches (Moret and Pinson 2019; Reis et al. 2022).

From an economic perspective, collective self-consumption presents an opportunity to lower global energy costs compared to the sum of individual bills. However, this raises fundamental questions about the fair distribution of economic gains among participants. Should benefits be allocated based on energy contribution, financial investment or broader social equity considerations? The role of market mechanisms in structuring energy exchanges – whether through peer-to-peer transactions or community-based markets – adds complexity to these distributional concerns.

In parallel, the environmental implications of energy management strategies must be critically assessed. While the literature predominantly focuses on the impact of direct emissions and climate change mitigation, broader environmental effects, such as resource depletion and land-use impacts, are often overlooked. This gap underscores the need for a comprehensive environmental impact assessment methodology. By integrating life cycle assessment (LCA) into energy community frameworks, a more holistic evaluation of sustainability can be achieved, ensuring that economic and environmental objectives are aligned.

10.2.2. *Objective*

This chapter explores a novel demand-side management methodology based on cooperative game theory, analyzing both economic efficiency and environmental sustainability within energy communities. It raises essential philosophical and technical questions regarding the incentivization of energy-saving behaviors, flexibility and equitable participation. By bridging engineering, economics and environmental science, this study aims to propose a balanced and ethical approach to collective energy management. The objective of this research is threefold:

1) Proposing a collaborative energy management based on pooling resources and flexibility coordination.

2) Proposing a methodology for assessing environmental impacts.

3) Proposing individual gain distribution of a collective self-consumption operation based on collaborative game theory.

10.3. Methodology

10.3.1. *Study case*

Since real-world data was unavailable for this research, a virtual energy community (residential consumers and prosumers with a PV production) was created using simulated consumption data. To ensure realistic data, three methods were considered: a bottom-up approach, which models individual appliances and aggregates their consumption; a top-down approach, which disaggregates total load curves into finer household data; and a hybrid method combining both. The CREST model, a bottom-up approach, was selected due to its accuracy in prior studies (McKenna and Thomson 2020). It was reparametrized to reflect the French energy context, including assumptions about hot water systems and flexibility through electric vehicles. Statistical data from INSEE[1] was used to calibrate household characteristics, ensuring representativeness as explained in a previous research (Durillon et al. 2021).

1 French National Institute of Statistics and Economic Studies.

In parallel, production data were retrieved from the open data website ECO2mix of the French transmission system operator, and local PV production was simulated using the Pvlib library available in Python (Bossu et al. 2024).

10.3.2. *Approach*

This overall optimization process is summarized in Figure 10.1.

The selected optimization process is based on a centralized day-ahead (J-1) supervisor, and is divided into three subcategories:

– a fictitious individual optimization that serves as a baseline for comparison in the subsequent steps;

– a collective optimization that establishes the reference for the optimal strategy to be pursued;

– a gain redistribution mechanism based on game theory.

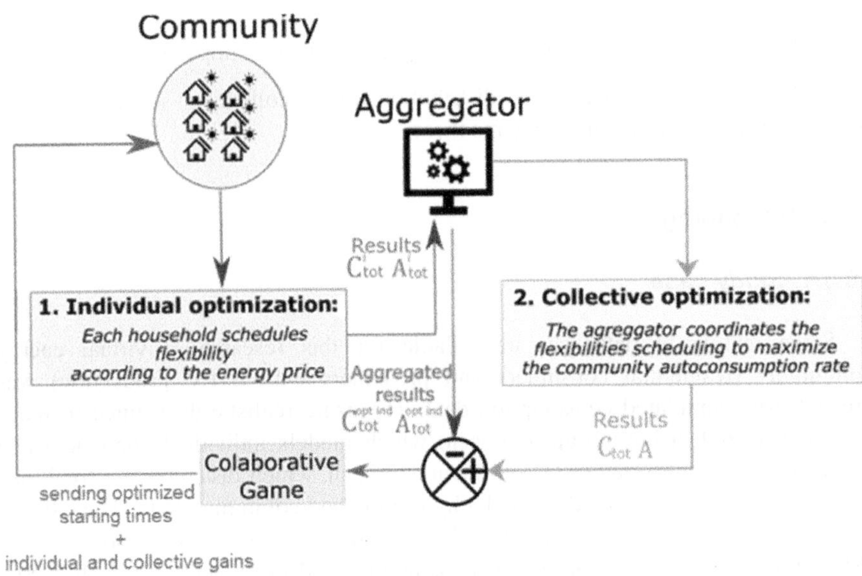

Figure 10.1. *Overall methodology*

10.3.3. *First step: individual optimization*

The first step in the energy management process involves performing an individual optimization for each household within the community. Each household is allocated a share of the photovoltaic production and optimizes its consumption using its available flexibility resources. Each household i adopts a flexibility planning strategy based on their sensitivity preferences towards financial and/or environmental cost $\boldsymbol{\rho}^i = \left(\boldsymbol{\rho}^i_{price}, \boldsymbol{\rho}^i_{env}\right)$.

In this study, two individual sensitivity profiles have been considered: price-sensitive households (optimizing based on the variable $\boldsymbol{C}_{tot}$: total cost of consumed energy of household i), environmentally sensitive households (optimizing based on the variable $\boldsymbol{\tau}_{\text{env}}$: environmental score of energy consumption by household i).

The objective is to determine the maximum potential gains a household could achieve if it were to optimize its consumption independently, based on its preferences (weights $\boldsymbol{\rho}^i$) and its allocated solar energy. The household is thus placed within the framework of a collective self-consumption scheme, as legally defined, with a fixed key distribution for renewable energy allocation. Each household then optimizes its consumption by following the objective function:

$$\boldsymbol{min}\, f^i(\boldsymbol{X}) = \sum \boldsymbol{C}^i{}_{tot}(\boldsymbol{t}) \times \boldsymbol{\rho}^i_{price} + \sum \boldsymbol{\tau}^i_{\text{env}}(\boldsymbol{t}) \times \boldsymbol{\rho}^i_{penv}$$

10.3.4. *Second step: collective optimization*

In parallel with the individual optimizations, the aggregator performs a collective optimization. The aggregator has access to the community's estimated consumption for the following day. Community members declare the appliances they are willing to make available for flexibility optimization. The entire photovoltaic production is considered in this process. The objective of this optimization is to maximize the self-consumption rate A, following the equation:

$$\boldsymbol{min}\, f(\boldsymbol{X}) = -\boldsymbol{A}$$

Computing financial costs, environmental impacts and self-consumption rate, the results of the collective optimization can be determined. By comparing these values to those obtained from individual optimizations, it becomes possible to assess the benefits of the developed energy management system relative to the best outcomes that households could have achieved independently, based on their sensitivity profiles.

The remainder of this chapter will illustrate the advantages of the collective strategy, highlighting the added value of cooperation in energy management. A detailed analysis of these benefits will be provided, demonstrating how the collective approach leads to improved performance in financial, environmental and self-consumption terms compared to purely individual strategies.

10.3.5. *Assessing environmental impacts*

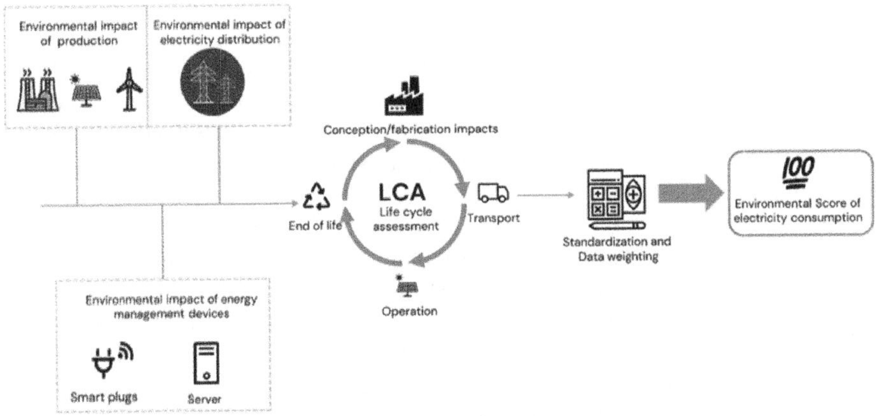

Figure 10.2. *Computation of environmental impacts*

The assessment of the environmental impact of community is required to discuss benefits of self-consuming the local production. Figure 10.2 shows the integration of LCA in evaluating the environmental impact of electricity consumption within the community. LCA is a comprehensive methodology that assesses environmental impacts across the entire lifecycle of a product or system, from processing and manufacturing to end-of-life management. In the context of electricity consumption, LCA considers multiple factors, including the environmental impact of energy production (e.g. emissions from power plants, land use for renewable installations) and electricity distribution (e.g. transmission losses and infrastructure footprint). Additionally, energy management devices, such as smart plugs and servers, introduce their own environmental burdens, which must be accounted for in the broader sustainability assessment.

The LCA process follows a systematic approach, encompassing transportation, operation and standardization of data for accurate impact weighting. By quantifying emissions, resource use and other ecological effects at each stage, LCA enables a standardized environmental score for electricity consumption. This score provides a holistic view of the sustainability performance of an energy community, allowing stakeholders to make informed decisions that align economic incentives with environmental responsibility. The integration of LCA into energy management strategies provides a means, though imperfect and limited, to account for potential unintended environmental consequences of demand-side optimization, aiming for a more balanced and sustainable framework for collective self-consumption.

In this research, environmental impacts are computed into a single score: the issue of a single environmental indicator is complex, and the simplest proposed method is often to assign equal weight to all impacts. Any weighting inherently involves value choices that depend on policies, cultural preferences and other value systems, making consensus on this issue difficult. The approach adopted here is that of the Joint Research Center (JRC), which relies on weighting impacts based on two distinct groups: 2,400 individuals from civil society across six different European countries and 518 experts in LCA (Sala et al. 2018). A modification was then made by merging the last two impacts: the depletion of mineral and metal resources with the depletion of fossil resources. Using this approach, the final environmental scores for each energy source considered are presented in Table 10.1.

Source	**Local PV**	**Grid PV**	**Nuclear Power**	**Hydraulic**	**Gas**	**Fuel**	**Wind power**	**Coal**	**Bioenergy**
Score	1.87	2.69	2.16	1.93	2.23	20.42	1.25	11.59	26.56

Table 10.1. *Environmental scores of each energy source*

10.3.6. *Gain distribution*

To distribute the gain between community members, two key concepts from cooperative game theory are used in this research: the Shapley Value and the Nucleolus. These concepts offer different approaches to distributing gains among participants in a cooperative setting, each rooted in distinct philosophical principles.

The *Shapley Value* is a solution concept that aims to distribute gains based on the average marginal contribution of each participant. In the context of an energy community, this means that each stakeholder's share of the collective benefits is determined by how much they contribute to the overall success of the community. This approach is often seen as meritocratic, as it rewards participants based on their individual efforts and impacts. However, it may not account for the varying capacities and vulnerabilities of different stakeholders, potentially favoring those who are already well resourced. On the other hand, the *Nucleolus* focuses on maximizing the welfare of the least advantaged participants. It achieves this through a recursive process that ensures the worst-treated coalitions receive a fairer share of the gains. The Nucleolus embodies a more collective and egalitarian philosophy, aiming to reduce disparities and promote overall fairness. While it may not reward individual contributions as explicitly as the Shapley Value, it fosters a sense of solidarity and ensures that the community's gains are shared in a way that supports all members, particularly those in need.

These two concepts offer complementary perspectives on gain distribution, allowing for a nuanced approach to energy management that can balance individual merit with collective welfare. For the interested reader, an introduction to these mathematical concepts is provided in the following article: Bossu et al. (2024).

10.4. Simulation and results

10.4.1. *Scenarios*

This study evaluates two optimization scenarios for households within an energy community, each emphasizing different criteria. In Scenario 1, individual households prioritize cost reduction by optimizing their energy usage ($\boldsymbol{\rho\_i}$ = (1,0)), while the community collectively seeks to maximize the autoconsumption rate. In Scenario 2, the focus shifts towards environmental sustainability, where each household optimizes its energy usage to minimize environmental impacts ($\boldsymbol{\rho_i}$ = (0,1)), while the community maintains the same collective objective of maximizing autoconsumption. These scenarios allow for an analysis of the trade-offs between economic and environmental priorities in community energy management. It is important to acknowledge the sensitivity of the environmental results to parameters of the initial LCA. The primary focus should therefore be on the robustness of the methodological approach.

10.4.2. *Energy consumption results*

10.4.2.1. *Scenario 1 results (price-driven optimization)*

In Scenario 1, where households individually optimize price reduction by scheduling flexible appliances during low-cost nighttime grid periods (see Figure 10.3(a)), collective optimization significantly improves outcomes (see Figure 10.3(c)). The self-consumption rate increases by *+46%* when the community coordinates energy use to prioritize local renewable energy (e.g. solar PV). This coordination reduces reliance on grid imports, minimizing financial losses from exported energy. While individual price-focused behavior achieves a *-5% reduction in energy costs*, collective action also lowers environmental impacts by *-12%*, likely due to reduced dependence on high-emission grid sources such as coal and gas. However, individual scheduling during sunny periods without coordination leads to suboptimal self-consumption, as households prioritize nighttime grid energy over local PV use.

10.4.2.2. *Scenario 2 results (environmental impact-driven optimization)*

In Scenario 2, households individually optimize to reduce environmental impacts by scheduling appliances when cleaner energy sources (e.g. wind, solar during the day) dominate the grid mix (see Figure 10.3(b)). Collective optimization further enhances outcomes (see Figure 10.3(c)), boosting the autoconsumption rate by *+10.20%* and achieving a larger price reduction (*-14.20%*) compared to Scenario 1. Despite the focus on environmental goals, collective coordination reduces environmental impacts by *-6.40%*, though this is smaller than Scenario 1's reduction. This paradox arises because individual actions in Scenario 2 already avoid high-impact sources such as fuel (LCA score: 20.42) and coal (LCA score: 11.59), leaving less room for collective improvement. The community's coordination ensures that surplus local PV is prioritized over marginally cleaner grid, balancing cost and sustainability.

10.4.2.3. *Comparative insights and collective benefits*

Both scenarios demonstrate that collective optimization outperforms individual strategies, regardless of initial objectives. Scenario 1's larger self-consumption gains (*+46%* from Figure 10.3(a) to (c)) highlight the inefficiency of uncoordinated price-driven behavior, which prioritizes cheap grid energy over local renewables. In contrast, Scenario 2's smaller autoconsumption boost (*+10.20%* from Figure 10.3(b) to (c))) reflects households already aligning with cleaner energy sources individually. Notably, collective action in Scenario 2 achieves the highest cost reduction (*-14.20%*), as coordination minimizes reliance on expensive, high-impact grid sources. The results underscore that collective strategies harmonize individual goals with community-wide benefits, thereby enhancing both economic and environmental outcomes beyond those attainable through isolated efforts.

10.4.3. *Gain distribution results: focus on Scenario 1*

Scenario 1 illustrates the optimization and the two game theory concepts, with a focus on the financial cost reduction. Table 10.2 (*Individual optimization*) shows the impact of individual optimization on household energy costs, comparing the initial baseline expenses to those after households independently adjust their consumption patterns. In comparison, the average cost for the household in this case is *1.8€*. This demonstrates the advantages of decentralized decision-making, where consumers benefit from lower bills through informed energy scheduling. However, despite these individual savings, the lack of coordination results in inefficiencies at the community level. Without a collective strategy, households may inadvertently increase reliance on the grid at peak times or fail to maximize the utilization of locally produced renewable energy, leading to suboptimal overall outcomes.

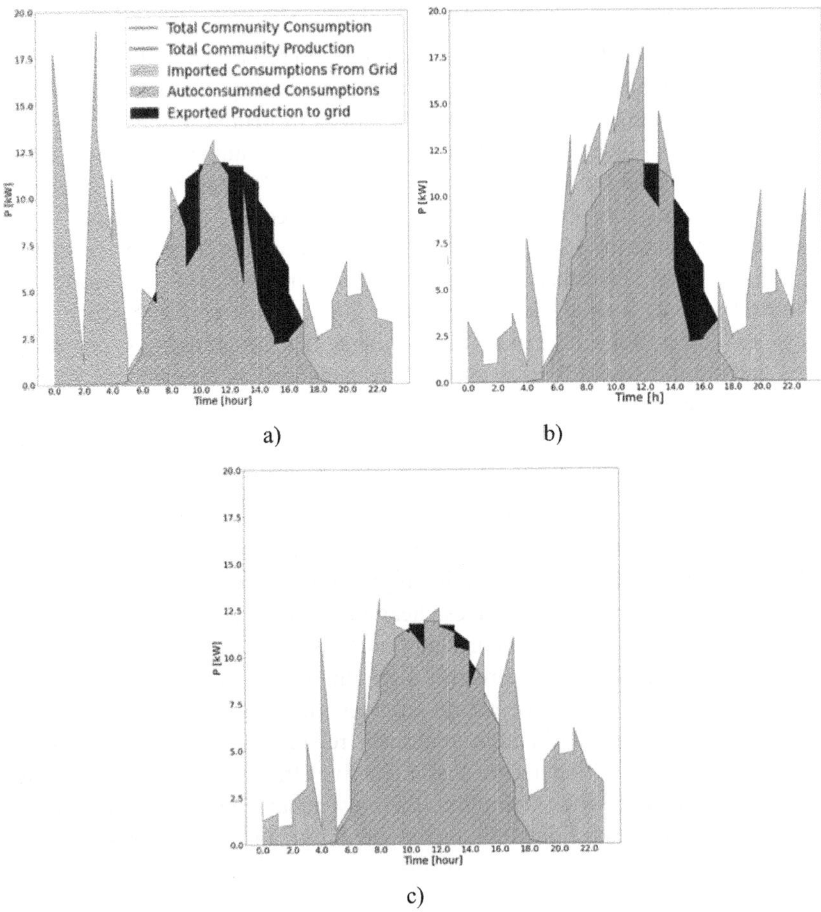

Figure 10.3. *Origin of consumption after individual optimization scenario 1 (a), individual optimization scenario 2 (b) and collective optimization (c)*

Table 10.2 introduces also a cooperative approach (*collective gain distribution*), where collective optimization strategies further enhance cost reductions and ensure a fair distribution of savings. To compare, the average cost for the household in this case is *1.3€*. By coordinating energy use at the community level, total energy costs decrease significantly, maximizing self-consumption and reducing dependence on external sources. The Shapley Value allocates savings based on each household's marginal contribution to the community's cost reduction, preserving variations in costs based on participation levels. Households that contribute more receive greater benefits, reinforcing a merit-based incentive structure. In contrast, the Nucleolus prioritizes fairness by redistributing gains to minimize disparities among participants. This ensures that extreme costs move closer to the community average, effectively fostering equity, particularly for vulnerable households with fewer energy assets.

	Individual optimization		**Collective gain distribution**	
House	Before	After	Shapley	Nucleolus
0	2.40	1.91	1.83	1.64
1	2.72	1.97	1.87	1.69
2	1.48	1.21	0.90	0.94
3	1.54	1.35	1.10	1.07
4	1.78	1.48	1.14	1.21
5	1.51	1.33	1.03	1.05
6	1.67	1.50	1.31	1.23
7	1.58	1.40	1.12	1.14
8	1.76	1.56	1.32	1.29

Table 10.2. *Individual cost (in €) with and without daily individual and collective optimization*

The comparison between the Shapley Value and Nucleolus highlights the trade-off between efficiency and social equity. The Shapley Value aligns with a market-driven philosophy, rewarding active participants and maintaining cost dispersion that reflects individual contributions. This method incentivizes engagement from prosumers and households with flexible loads. Conversely, the Nucleolus follows an egalitarian principle, aiming to reduce disparities by redistributing gains more equitably, thereby protecting vulnerable consumers who may otherwise be disadvantaged. While both approaches outperform individual optimization by enhancing fairness and cost efficiency, the choice between them ultimately depends on the community's priorities; whether economic efficiency or social cohesion.

The results in Table 10.2 underscore that collective action consistently leads to superior outcomes compared to isolated decision-making, demonstrating the potential of cooperative strategies in community-based energy management.

10.5. Conclusion

While the proposed methodology offers a comprehensive approach to energy management in energy communities by integrating cooperative game theory and LCA, several limitations must be acknowledged. One primary limitation lies in the LCA methodology itself, which requires further refinement through more detailed and granular data on various energy components. The environmental impact assessment currently relies on available datasets that may not fully capture the specificities of local energy generation, grid infrastructure and flexibility equipment such as storage systems and smart control devices. Without high-resolution data, the accuracy of environmental impact quantification remains constrained, potentially leading to oversimplifications or misrepresentations of the actual sustainability benefits of different energy management strategies.

Additionally, the use of game theory in optimizing demand-side management introduces computational challenges, particularly in scaling up to larger energy communities. Cooperative game-theoretic models, while effective in ensuring fair and strategic distribution of benefits among participants, require significant computational power as the number of players increases. The complexity of solving these optimization problems grows exponentially with the size of the community, imposing practical limitations on real-time applications. Future work should explore algorithmic improvements or approximation methods to enhance computational efficiency while maintaining the fairness and robustness of the proposed approach.

10.6. References

Bossu, A. (2024). SOREL : supervision optimisée d'un réseau local de distribution au sein d'une communauté énergétique. PhD Thesis, Université de Lille, Lille.

Bossu, A., Durillon, B., Davigny, A., Barry, H., Robyns, B., Belaïd, F., Saudemont, C. (2024). Coalitional game-based gain generation and distribution for collective self-consumption in an energy community. *Mathematics and Computers in Simulation*, 225, 129–147.

Cui, S., Wang, Y.-W., Shi, Y., Xiao, J.-W. (2020). A new and fair peer-to-peer energy sharing framework for energy buildings. *IEEE Transactions on Smart Grid*, 11(5), 3817–3826.

Durillon, B., Salomez, F., Davigny, A., Kazmierczak, S., Barry, H., Saudemont, C., Robyns, B. (2021). Consumers' sensitivities and preferences modelling and integration in a decentralised two levels energy supervisor. *Mathematics and Computers in Simulation,* 183, 142–157.

Lampropoulos, I., Kling, W.L., van den Bosch, P.P.J., Ribeiro, P.F., van den Berg, J. (2013). Criteria for demand response systems. In *2013 IEEE Power & Energy Society General Meeting*, Vancouver.

McKenna, E. and Thomson, M. (2020). CREST demand model. Dataset, Loughborough University, Loughborough [Online]. Available at: https://figshare.com/articles/CREST_Demand_Model_v2_0/2001129 [Accessed 1 January 2023].

Meyabadi, A.F. and Deihimi, M.H. (2017). A review of demand-side management: Reconsidering theoretical framework. *Renewable and Sustainable Energy Reviews*, 80, 367–379.

Moret, F. and Pinson, P. (2019). Energy collectives: A community and fairness based approach to future electricity markets. *IEEE Transactions on Power Systems*, 34(5), 3994–4004.

Reis, I.F.G., Gonçalves, I., Lopes, M.A.R., Antunes, C.H. (2022). Towards inclusive community-based energy markets: A multiagent framework. *Applied Energy*, 307, 118115.

Sala, S., Cerutti, A.K., Pant, R. (2018). *Development of a Weighting Approach for the Environmental Footprint*. Publications Office of the European Union, Luxembourg.

Tushar, W., Saha, T.K., Yuen, C., Smith, D., Poor, H.V. (2020). Peer-to-peer trading in electricity networks: An overview. *IEEE Transactions on Smart Grid*, 11(4), 3185–3200.

11

Dynamic Energy Allocation in a Local Community Integrating User Preferences

11.1. Introduction

The ongoing transformation of energy systems is primarily driven by the objective of decarbonizing society, with the reduction of greenhouse gas (GHG) emissions standing as a central goal. This transition is driven by the electrification of energy end-uses and the increased deployment of renewable energy sources, particularly solar and wind. However, these developments also introduce new challenges that are not only environmental in nature, but also deeply energetic and systemic. While renewables offer low operational emissions, their integration requires addressing variability, grid stability and the coordinated management of decentralized production and consumption. Beyond GHG emissions, the energy transition raises critical concerns about the availability and sustainability of key resources, notably the critical raw materials used in energy storage systems, power electronics and renewable technologies. These considerations highlight that the environmental impact of energy systems must be assessed over their entire life cycle, and that gains in operational efficiency do not automatically guarantee global environmental benefits.

From an energy systems perspective, one of the major challenges lies in designing electrical networks that are intelligent, flexible and inclusive. These networks must be capable of dynamically managing fluctuating renewable production, aligning with real-time consumption patterns, and adapting to the heterogeneous preferences and constraints of diverse actors – including households, businesses and prosumers. Such preferences may reflect varying priorities, such as

Chapter written by Amira DHORBANI, Dhaker ABBES, Kahina HASSAM and Benoît ROBYNS.
For a color version of all figures in this chapter, see www.iste.co.uk/dobigny/rethinking.zip.

cost savings, comfort, autonomy or ecological values, which do not always converge. This work positions itself within this energetic and socio-technical framework. Rather than attempting to directly solve environmental problems, the proposed approach addresses the coordination and optimization challenges inherent to energy management systems.

Durillon (2019) and Durillon et al. (2021) tackled this challenge by developing a methodology to incorporate actor preferences into energy management systems. In the first two works, their approach began with profiling the different stakeholders, considering their energy usage patterns, constraints and priorities. This led to the formulation of a global utility function that captured these preferences using weighting coefficients to represent the relative importance of factors such as cost, comfort and environmental impact. This framework provided a structured way to account for diverse objectives within an energy supervisory system.

Based on this work, other articles such as in Durillon and Bossu (2024) extended this approach by introducing a detailed analysis of environmental impacts through life cycle assessment, addressing concerns such as the environmental costs of communication technologies and the potential for increased energy use. Their research shifted the focus from purely financial optimization to a systemic and multi-criteria approach that integrates environmental impacts into energy management, highlighting the need for tools capable of addressing the complex trade-offs inherent to smart grid operation.

A noteworthy reference on this topic is in Robyns et al. (2024) where authors explored the integration of smart grids and intelligent buildings, emphasizing their roles in promoting sustainable energy practices and facilitating societal shifts toward more efficient energy consumption such as the implementation of renewable energy communities that uses blockchain technology to manage self-consumption and enhance grid flexibility. The study explores how advanced technologies and collaborative strategies can enhance energy management, reduce environmental impact and support the broader goals of the energy transition.

The emergence of peer-to-peer (P2P) energy exchange represents a significant step forward in addressing the diverse priorities of energy stakeholders. By facilitating direct energy transactions between participants – including consumers, producers and prosumers – P2P systems promote greater inclusivity and flexibility.

Inclusivity refers to the ability of all actors, regardless of size or technical capacity, to actively participate in the energy exchange process, ensuring equitable access to local energy resources. Flexibility, on the other hand, relates to the system's capacity to adapt in real time to fluctuations in supply and demand, as well as to the heterogeneous preferences and temporal constraints of the participants. While the concept of P2P energy exchange is compelling, its success hinges on user acceptance, which presents several challenges. User acceptance in this context involves trust in the system, willingness to adopt new technologies and sustained engagement over time. Addressing these factors is critical to ensuring widespread adoption and long-term viability of P2P energy markets.

In their comprehensive review (Sousa et al. 2019), the authors provide a comprehensive analysis of peer-to-peer energy markets, highlighting the significant role of user acceptance in their success. They emphasize that while these markets empower participants through transparency and flexibility, fostering trust and engagement is essential for widespread adoption. One key challenge is bridging the attitude–behavior gap, where individuals' positive attitudes toward these systems do not always translate into active participation. In Ableitner et al. (2020), a real-world P2P energy market involving 37 households and two commercial entities demonstrated a high level of user acceptance and engagement. Participants interacted regularly with the platform, with 79% actively using the web application throughout the 4.5-month study. Also, in Hackbarth and Lobbe (2020), they examined German households' interest in peer-to-peer (P2P) electricity trading, driven by environmental concerns, technical curiosity and a desire for energy independence. It identifies "planners" – future adopters of microgeneration technologies – as key participants, rather than current prosumers. Key motivators include community sharing and innovative pricing, while complexity and unfamiliarity pose barriers. The study suggests targeting planners with tailored and transparent strategies to enhance participation and support energy transition.

Many other studies have also used P2P energy trading to explore various aspects of decentralized energy management, including economic efficiency, grid stability and environmental benefits. For example, in Kong et al. (2022), the authors provided a comprehensive framework for P2P energy trading that aligns with global efforts to promote renewable energy and reduce greenhouse gas emissions. In Suthar and Pindoriya (2024), the article proposes a co-optimization model for P2P energy trading that incorporates uncertainties in renewable generation and enforces distribution network constraints.

A methodology is presented in this chapter for dynamic energy allocation, which incorporates user preferences into energy management. The system architecture is first defined, detailing the interaction between buildings, electric vehicles and battery storage. A utility-based framework is then introduced to quantify user priorities and enable an adaptive energy distribution strategy. To optimize energy flow while accounting for technical constraints and economic factors, a rule-based decision-making approach is developed. The effectiveness of the proposed method is assessed through simulations, demonstrating its capacity to balance multiple and sometimes conflicting objectives.

11.2. System description and problem formulation

To evaluate the proposed approach under realistic conditions, we considered the use case of the grid demonstrator located at Lille Catholic University in France. This example incorporates two rooftop photovoltaic systems with capacities of 185 kWp and 28 kWp, paired with a 203 kWh lithium-ion battery storage unit. The infrastructure also includes multiple electric vehicle charging stations, each providing 22 kW of charging power, alongside four commercial buildings. Additionally, the setup features a connection point to the main distribution grid. This configuration provides a pertinent case study as it effectively meets a functional local energy community with diverse participants, combining generation, storage, consumption and grid interaction. Each of these users has their own preferences which guide the allocation of energy according to their priorities. A visual description of this demonstrator can be seen in Figure 11.1.

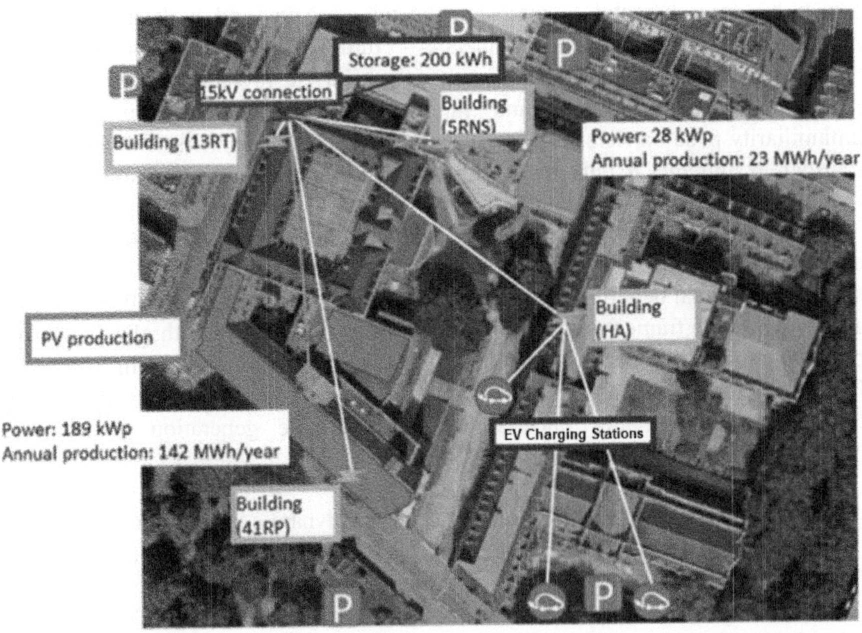

Figure 11.1. *Lille Catholic University demonstrator grid (Stephant 2021a)*

11.2.1. *Utility functions and constraints*

Each user i has a utility function Ui, representing their satisfaction by modeling individual objectives mathematically. These functions incorporate personalized preferences through manually set weighting coefficients (α, β, γ) that are chosen by the users themselves. These coefficients rank the user's priorities, focusing on, respectively, maximizing cost efficiency, consuming local PV production and ensuring comfort. The cost efficiency term is calculated based on the power drawn from the grid, PV-generated power and battery storage, with their respective costs C_{grid}, C_{pv} and C_s. The exact methodology used to determine these coefficients, as well as the detailed formulation of the cost efficiency term and the underlying assumptions in its computation, are thoroughly described in Stephant (2021b) and Robyns et al. (2024). These references provide an in-depth justification for the choice of parameters, the mathematical framework supporting the model and the empirical validation of the proposed approach.

For example, in a building, the utility function balances these priorities. The user can set a higher α_b if minimizing costs is critical, meaning the system will prioritize drawing cheaper energy sources. A higher β_b would push the system to use as much PV-generated energy as possible. Finally, a high γ_b would ensure that the building's energy needs are fully met, prioritizing user comfort over cost efficiency.

In the case of EVs, the utility function would prioritize ensuring the battery reaches a desired state of charge (SOC) by the end of the charging session. A high α_{EV} could reduce charging costs by optimizing when the vehicle charges, while β_{EV} would ensure the EV uses renewable PV energy when available. A high γ_{EV} would focus on fully charging the EV within the available time, ensuring it meets the user's transportation needs.

By manually setting these coefficients, each user has the flexibility to tailor their energy usage according to their own priorities, allowing for a personalized and optimized balance between cost, renewable energy usage and comfort in the energy management system:

i) Buildings: the system includes four buildings, each with varying energy demands Pb that can be met through grid power, PV-generated power or battery storage. The total demand is defined as P_{totalb}. The following utility function is proposed for buildings:

$$U_b = \alpha_b \cdot \mathrm{Cost}_b + \beta_b \cdot \mathrm{UsePV}_b + \gamma_b \cdot \mathrm{Comfort}_b \qquad [11.1]$$

where:

$$\mathrm{Cost}_b = C_{grid} \cdot P_{\mathrm{grid}b} \cdot \Delta t + C_{\mathrm{pv}} \cdot P_{\mathrm{pv}b} \cdot \Delta t + C_s \cdot P_{sb} \cdot \Delta t$$

where C_{grid}, C_{pv} and C_s represent the cost per unit energy from the grid, photovoltaic system and storage system, respectively.

– $UsePV_b = \frac{P_{pvb}}{P_{totalb}}$, which represents the ratio of PV-supplied power to the building's total power consumption.

– $Comfort_b = \frac{P_b}{P_{bmax}}$, where P_{bmax} represents the maximum power demand of the building. This metric reflects the extent to which the energy needs of the building are satisfied during each time step. A comfort value equal to 1 means that the building is receiving all of the power it requires to operate normally. A value below 1 indicates a shortfall in energy supply, suggesting that certain energy-dependent services may be reduced, delayed or unmet, potentially affecting user satisfaction.

– $\alpha_b,\beta_b,\gamma_b$: building preferences.

– P_{gridb},P_{pvb},P_{sb}: power drawn by the building from the grid, power supplied to the building from the PV system, and power provided to the building from the energy storage system, respectively, in each time step Δt.

ii) EVs: for an EV user connected to the charging station, the following function is proposed:

$$U_{EV} = \alpha_{EV} \cdot Cost_{EV} + \beta_{EV} \cdot UsePV_{EV} + \gamma_{EV} \cdot Comfort_{EV} \quad [11.2]$$

where:

$$– Cost_{EV} = C_{grid} \cdot P_{gridEV} \cdot \Delta t + C_{pv} \cdot P_{pvEV} \cdot \Delta t + C_s \cdot P_{sEV} \cdot \Delta t$$

where C_{grid},C_{pv} and C_s represent the cost per unit of energy from the grid, photovoltaic system and storage system, respectively. Δt is the time step used in the simulation.

– $UsePV_{EV} = \frac{P_{pvEV}}{P_{totalEV}}$, where $P_{totalEV}$ is the total power consumed by the electric vehicle from all sources.

– $ComfortEV = \frac{SOC_{EV}}{SOC_{EVmax}}$, where SOC_{EV} is the state of charge of the electric vehicle at arrival time, and SOC_{EVmax} is the maximum state of charge for the EV battery.

– P_{gridEV}, P_{pvEV}, P_{sEV}: power drawn by the EV from the grid, power supplied to the EV from the PV system and power provided to the EV from the energy storage system, respectively.

For each EV, certain constraints must be considered. According to the current case study, discharging EV to meet other demands is not allowed, and the charging power is limited by the station's maximum capacity but only applies if the EV is connected to the charging station. This is represented by the Boolean variable $\delta_{EVi}(t)$, which indicates whether the EV is connected at time t (1 if connected, 0 otherwise):

$$0 \leq P_{EVi}(t) \leq P_{EVmax} \cdot \delta_{EVi}(t) \quad \forall i \qquad [11.3]$$

In addition, the state of charge (SOC) has upper and lower limits, which are characteristics of the vehicle's battery:

$$SOC_{EVi,min} \leq SOC_{EVi} \leq SOC_{EVi,max} \quad \forall i \qquad [11.4]$$

These constraints ensure that the battery operates within safe and functional boundaries. The $SOC_{EVi,min}$ prevents deep discharges that could damage the battery or reduce its lifespan, while the $SOC_{EVi,max}$ avoids overcharging.

For each EV connected, we have the state of charge defined as follows:

$$SOC_{EVi}(t+1) = SOC_{EVi}(t) + \eta_{VE} \cdot \frac{P_{EVi}(t) \cdot \Delta t}{C_{EVi}} \cdot \delta_{EVi}(t) \qquad [11.5]$$

where η_{VE} is the efficiency of the electric vehicle under load and C_{EVi} is the capacity of the vehicle's battery.

iii) Battery storage: we consider the storage system as a full actor within the energy community, meaning it plays an active role in energy exchanges and decision-making. The battery storage unit charges from the grid or PV and discharges to meet the energy needs of buildings and EVs. To optimize its operations, we define a specific utility function for this user, similar to how human users optimize for comfort. In this case, the battery's utility function represents its operational preferences, such as maintaining an optimal SOC, which ensures longevity and efficiency. While the concept of "comfort" traditionally applies to human satisfaction, for the battery, it reflects its need to remain within its preferred operational range to perform optimally:

$$U_s = \alpha_s \cdot Cost_s + \beta_s \cdot UsePV_s + \gamma_s \cdot Comfort_s \qquad [11.6]$$

where:

– $Cost_s = C_{grid} \cdot P_{gridCharge} + C_{pv} \cdot P_{pvCharge} + C_s \cdot P_{sdischarge}$

C_{grid}, C_{pv} and C_s represent the cost per unit energy from the grid, photovoltaic system and storage system, respectively.

$$-\text{UsePV} = \frac{P_{pvCharge}}{P_{totalCharge}}$$

where $P_{totalCharge}$ is the total power used to charge the battery from all sources

$$-\text{Comfort}_s = \frac{SOC_s}{SOC_{smax}}$$

where SOC_s is the state of charge of the storage system at a given time, and SOC_{smax} is the maximum state of charge for the storage system.

- $P_{gridCharge}$, $P_{pvCharge}$, $P_{sdischarge}$: power used to charge the battery from the grid, power used to charge the battery from the PV system and power discharged from the battery to meet energy needs, respectively.

- $P_{pvCharge}$, $P_{totalCharge}$: power drawn from the PV system to charge the battery, and total power used for charging from all sources, respectively.

It is important to note that when the battery discharges, it releases energy to meet demand and gain income. This process is represented by a negative value of $P_{sdischarge}$. Moreover, similarly to the EV, the SOC has boundaries:

$$SOC_{smin} \leq SOC_s \leq SOC_{smax} \quad [11.7]$$

The power of the battery storage P_s also has operational boundaries, ensuring that it operates within safe and efficient limits:

$$P_{min} \leq P_s \leq P_{max} \quad [11.8]$$

Similarly to electric vehicles, the battery's state of charge is defined by:

$$SOC_s(t+1) = SOC_s(t) + \eta_s \cdot \frac{P_s(t) \cdot \Delta t}{C_b} \quad [11.9]$$

where η_s is the efficiency of the storage system; during charging, η_s is equal to 1, and during discharging, it is equal to 0.98.

We also have C_b that represents its capacity.

11.2.2. *Global constraints*

In addition to all of these constraints, the integration of multiple energy sources and demands necessitates a careful balance of power within the energy community. The total power generated from the PV panels P_{pv}, the power drawn from the grid P_{grid} and the power from the battery storage unit P_s must collectively meet the energy demands of all users in the system, including the power consumption of the buildings P_{b_j} and the electric vehicles P_{EV_i}. This requirement is represented by the following power balance equation at each time step t:

$$P_{pv}(t) + P_{grid}(t) + P_s(t) = \sum_{i=1}^{6} P_{EVi}(t) + \sum_{j=1}^{3} P_{bj}(t) \qquad [11.10]$$

To further ensure the stability and reliability of the energy system, it is also crucial to limit the power drawn from the grid to within subscribed limits in order to ensure that the power demand from the grid P_{grid} does not exceed the maximum subscribed power $P_{subscribed\ grid}$.

$$P_{grid}(t) \leq P_{subscribed\ grid} \qquad [11.11]$$

11.3. Methodology

In the context of our energy management system, rules are used to ensure that energy demands are met while optimizing the use of renewable energy and maintaining system constraints. The process of making decisions involves a number of logical checks or conditions which control how power allocations happen within the system. Energy can come from PV panels, the grid or the discharge of the battery, depending on these conditions. A key innovation in our work is the incorporation of preference coefficients ($\boldsymbol{\alpha}$, $\boldsymbol{\beta}$, $\boldsymbol{\gamma}$) into the decision-making process. These coefficients allow each user – whether a building, electric vehicle or storage system – to express their priorities across cost minimization, renewable energy usage and comfort.

This formulation provides a flexible and transparent mechanism to tailor energy allocations according to individual needs. However, this modeling choice also introduces certain limitations. It presumes that users are both willing and able to understand the implications of their preferences and accurately translate them into quantitative weights. In practice, the effectiveness of this approach depends on user awareness, comprehension of the system's functioning and the stability of their declared priorities over time. While the current method ensures that user inputs are respected in the allocation process, further research is needed to evaluate whether these preferences align with broader community goals or optimal system behavior.

11.3.1. *Rule-based rule description*

As shown in Figure 11.2, our energy community consists of multiple types of users. These users will provide their preference coefficients, which will be used in the dynamic allocation algorithm.

The energy dispatch process dynamically allocates available power among different users based on their predefined preference coefficients ($\boldsymbol{\alpha}$, $\boldsymbol{\beta}$, $\boldsymbol{\gamma}$), which represent cost minimization, renewable energy utilization and comfort priorities, respectively. At each step, the system first assesses the total available energy from photovoltaic (PV) production, battery storage and the grid. If PV production exceeds the total demand, it is allocated proportionally to users based on their normalized $\boldsymbol{\beta}$ coefficients, ensuring that those who prioritize renewable energy receive a higher share. Normalization ensures a fair distribution by first summing all $\boldsymbol{\beta}$ coefficients across users to establish a reference, then dividing each user's $\boldsymbol{\beta}$ coefficient by this sum to calculate their proportional weight. This normalized weight is then multiplied by the total available PV energy to determine each user's allocated share. Next, the system transitions to mode 1, where it checks for any surplus PV energy. If surplus energy is available, it is allocated to charge electric vehicles (EVs) based on their normalized $\boldsymbol{\beta}$ coefficients. If no EVs are connected, the remaining PV energy is stored in the battery. If no surplus PV energy remains for EVs, the system switches to mode 2.

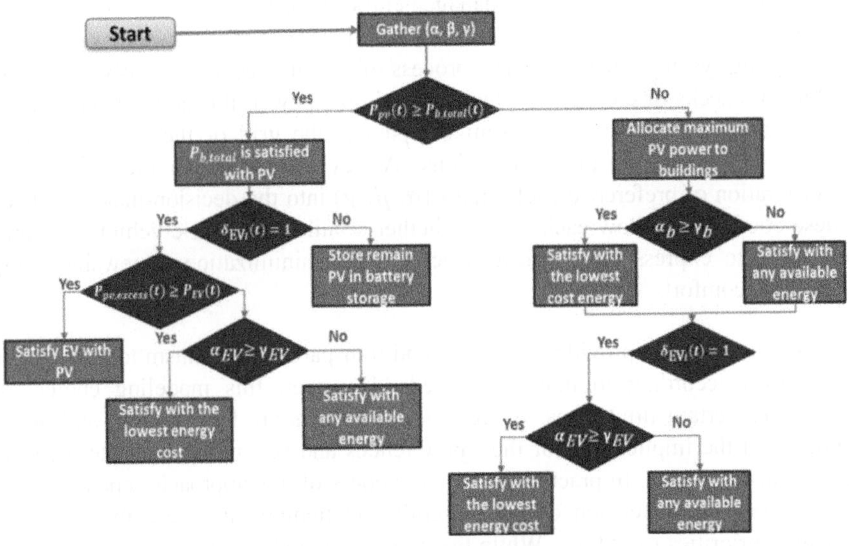

Figure 11.2. *Global system description*

If PV production is insufficient to cover all of the buildings' demand, the system switches to mode 2, where the remaining energy requirements are met by either discharging the battery or drawing power from the grid, prioritizing the lowest-cost option based on real-time grid and storage prices. The allocation of grid and storage power follows a similar proportional distribution, using users' $\boldsymbol{\alpha}$ and $\boldsymbol{\gamma}$ coefficients to prioritize those seeking cost minimization while maintaining system constraints such as maximum subscribed grid power and state-of-charge limits. This preference-based allocation ensures that energy distribution is both fair and efficient, aligning with individual user priorities while optimizing overall system performance.

11.4. Simulations and results

The optimization simulation is conducted over a full day, specifically on June 10, 2019, starting at midnight (00:00) and ending 24 hours later (24:00). Table 11.1 summarizes the key economic parameters used for energy sources and grid interaction in the simulation.

Parameter	Value
Grid energy cost C_g	0.05–0.15 €/kWh (time-varying)
PV energy cost C_{PV}	0.10 €/kWh (fixed)
Storage energy cost C_S	0.35–0.40 €/kWh (time-varying)
Subscribed grid power limit $P_{subscribed}$	580 kW

Table 11.1. *Economic parameters used in the simulation (June 10, 2019)*

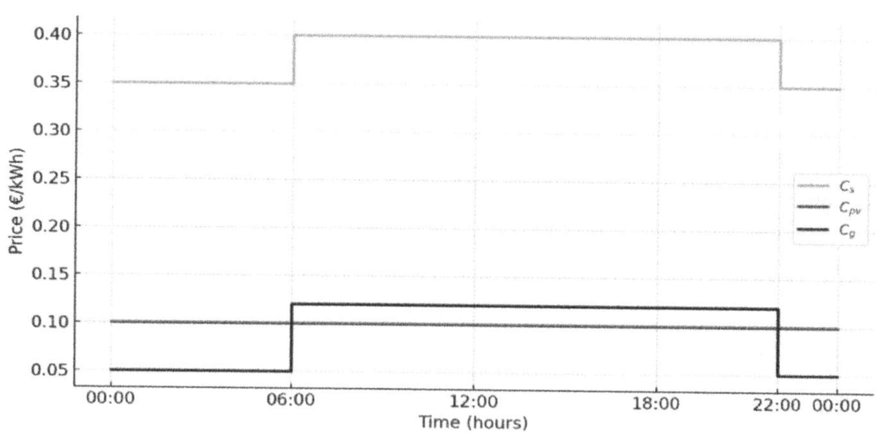

Figure 11.3. *Prices over 24 hours*

The variation of these prices over time is visualized in Figure 11.3, while Figure 11.4 shows the photovoltaic production curve over the same period.

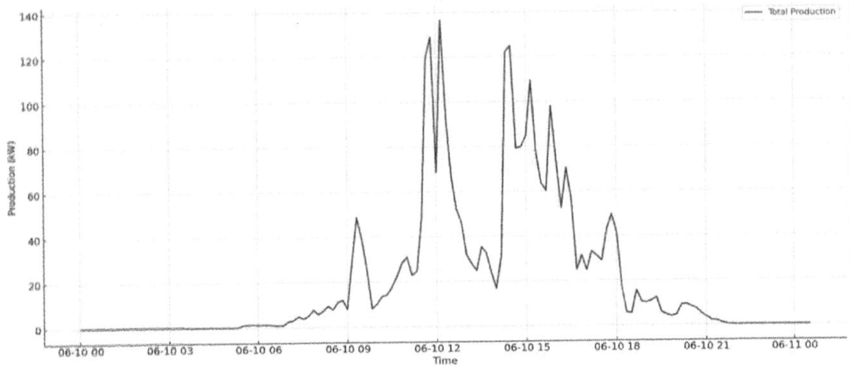

Figure 11.4. *PV production*

11.4.1. *Physical parameters of different users*

This section presents the physical parameters used to model the different actors. In total, there are four electric vehicles, one battery, and four buildings. The characteristics of the EVs and the battery are summarized in Tables 11.2 and 11.3. The data for these simulations were generated using Python, allowing for the testing and validation of the proposed energy management system under various scenarios.

The underlying scripts may be made available upon request for research purposes.

These scenarios were tested to evaluate the robustness of our algorithm. We distinguish four specific cases:

– an economic scenario where actors aim to reduce their costs or maximize their profits (denoted as I);

– a scenario where users prioritize their comfort (denoted as II);

– a scenario in which actors primarily focus on exchanging energy within the community (denoted as III);

– a scenario that incorporates different objectives (denoted as IV).

Characteristics	EV1	EV2	EV3	EV4
t_{arrival}	9:15	10:30	11:00	10:00
$t_{departure}$	12:30	19:00	16:30	18:00
$SOC_{EVi,\mathrm{min}}$	20%	45%	40%	25%
$SOC_{EVi,\mathrm{max}}$	100%	100%	100%	100%
$SOC_{EVi,required}$	60%	80%	80%	85%
$P_{EV\mathrm{max}}$ (kWh)	22	22	7.2	22
η_{VE}	80%	80%	80%	80%

Table 11.2. *Physical parameters of EVs*

Characteristics	Battery
SOC_{smin}	10%
SOC_{smax}	100%
SOC_s	35%
P_{s_charge} (kW)	40
$P_{s\_discharge}$	-80
C_S (kWh)	203

Table 11.3. *Physical parameters of battery storage*

User	Scenario I	Scenario II	Scenario III	Scenario IV
Building 1	0.8, 0.1, 0.1	0.1, 0.1, 0.8	0.1, 0.8, 0.1	0.1, 0.8, 0.1
Building 2	0.8, 0.1, 0.1	0.1, 0.1, 0.8	0.1, 0.8, 0.1	0.6, 0.2, 0.2
Building 3	0.8, 0.1, 0.1	0.1, 0.1, 0.8	0.1, 0.8, 0.1	0.3, 0.4, 0.3
Building 4	0.8, 0.1, 0.1	0.1, 0.1, 0.8	0.1, 0.8, 0.1	0.2, 0.7, 0.1
EV1	0.8, 0.1, 0.1	0.1, 0.1, 0.8	0.1, 0.8, 0.1	1/3, 1/3, 1/3
EV2	0.8, 0.1, 0.1	0.1, 0.1, 0.8	0.1, 0.8, 0.1	1/3, 1/3, 1/3
EV3	0.8, 0.1, 0.1	0.1, 0.1, 0.8	0.1, 0.8, 0.1	1/3, 1/3, 1/3
EV4	0.8, 0.1, 0.1	0.1, 0.1, 0.8	0.1, 0.8, 0.1	1/3, 1/3, 1/3
PV	0.8, 0.1, 0.1	0.1, 0.1, 0.8	0.1, 0.8, 0.1	1/3, 1/3, 1/3
Battery	0.8, 0.2, -	0.5, 0.5, -	0.2, 0.8, -	0.5, 0.5, -

Table 11.4. *User scenarios* (α, β, γ)

11.4.2. *Results*

11.4.2.1. *Impact of user preferences on energy allocation*

In this study, a Python-based model was developed to simulate and optimize the energy management of a system based on logic rules. The model was initialized with the parameters defined earlier. To analyze the impact of user preferences on energy allocation, we focus specifically on scenario IV, as it represents a combination of different objectives for buildings. This scenario allows us to study how the trade-offs between cost minimization, renewable energy utilization and comfort preferences influence energy distribution decisions.

As shown in Figure 11.5, power profiles for buildings over a 24-hour period highlight energy consumption patterns categorized into PV power used, grid power used and storage power used.

– *Load*: represents the total power demand of each building at each time step, generated arbitrarily.

– *PV power used*: indicates the amount of power supplied by the photovoltaic (PV) system.

– *Grid power used*: shows the amount of power drawn from the grid.

– *Storage power used*: represents the power supplied from the storage system.

The load profiles of the four buildings vary considerably throughout the day, reflecting their typical usage patterns with distinct peak times and overall demand levels. PV usage follows a consistent daily pattern, starting in the morning, peaking around midday and decreasing toward the evening, effectively meeting the buildings' energy needs during sunlight hours.

The electricity grid is mainly used when photovoltaic power is insufficient, for example, early in the morning, late in the evening or during cloudy periods, with notable peaks during periods of high demand to guarantee a stable supply. The red lines highlight the role of the storage system in balancing supply and demand, intervening during peaks in demand or when there is insufficient photovoltaic power and energy from the grid.

Efficient charging and discharging of the storage system reduces dependence on the grid and makes the most of the available photovoltaic energy.

To ensure a reliable energy supply, it is essential to examine how demand is met at each time step. This makes it necessary to analyze the power balance, which provides a detailed view of how energy sources interact to satisfy consumption dynamically.

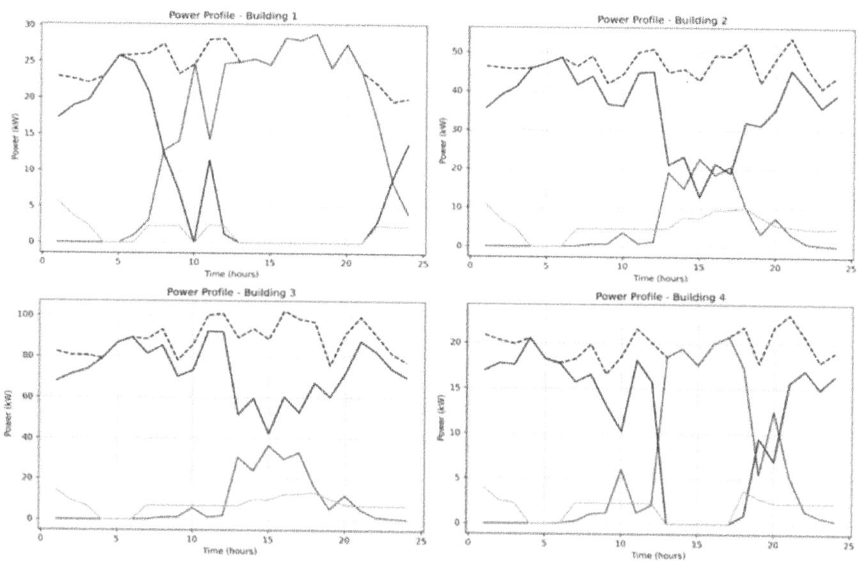

Figure 11.5. *Power profile of buildings*

The power balance for each building, shown in Figure 11.6, provides a comprehensive view of how energy demand is satisfied throughout the day due to different preference coefficients (α, β, γ).

Building 1 ($\alpha = 0.1$, $\beta = 0.8$, $\gamma = 0.1$) relies primarily on PV energy, significantly reducing its grid dependency, while storage usage remains minimal due to its low γ value. In contrast, building 2 ($\alpha = 0.6$, $\beta = 0.2$, $\gamma = 0.2$) exhibits strong grid dependency, using little PV. Storage use is moderate, indicating some buffering but an overall priority for grid consumption.

Building 3 ($\alpha = 0.3$, $\beta = 0.4$, $\gamma = 0.3$) shows a more balanced energy distribution, with PV contributing significantly and storage being actively used to reduce grid dependence, particularly in the evening, demonstrating the effect of a higher γ value in shifting energy supply.

Lastly, building 4 ($\alpha = 0.2$, $\beta = 0.7$, $\gamma = 0.1$) follows a pattern similar to building 1 but has slightly greater grid reliance due to its lower β value.

Across all buildings, higher β values promote PV self-consumption, higher γ values facilitate better storage utilization and dominant α values lead to a cost-optimized energy mix, where the selection between grid, PV and storage varies dynamically based on availability and real-time cost considerations.

This analysis also highlights how even subtle differences in coefficients lead to measurable changes in source prioritization, illustrating the method's sensitivity to user preferences.

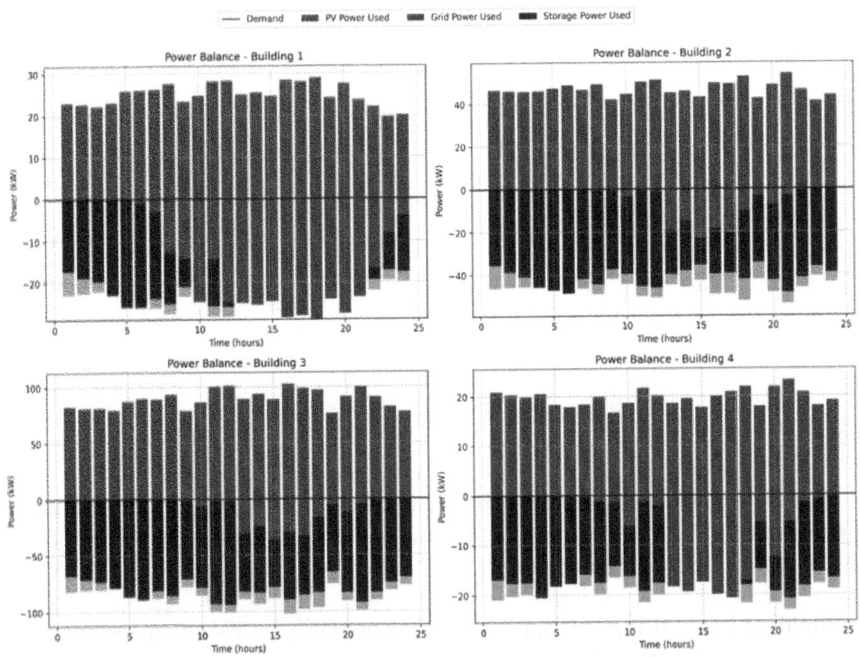

Figure 11.6. *Power balance for buildings*

To better understand user preferences, we created a pie chart illustrating the percentage of energy used by different sources (see Figure 11.7).

Building 1, which has a higher beta coefficient so a higher preference for using more photovoltaic (PV) energy, has the highest PV energy usage among the buildings at 60.1%.

Figure 11.8 shows the SOC evolution for the EVs over a 24-hour period. Each EV starts with a different initial SOC, reflecting their respective minimum SOC constraints, and only begins charging when connected to the charging station within their predefined arrival and departure times.

The SOC profile for battery storage in Figure 11.9 shows how the battery charges and discharges to balance the energy needs of the buildings throughout the day. Early in the morning, the SOC is low, reflecting minimal PV generation. As the PV generation rises, the SOC increases, particularly from 6 am to 6 pm, indicating that excess solar energy is being stored.

This matches the buildings' power profiles (see Figure 1.5) where solar power is heavily used. The dips in SOC, especially in the morning and late afternoon, occur when the battery discharges supply power, filling in the gaps when solar power drops or demand increases.

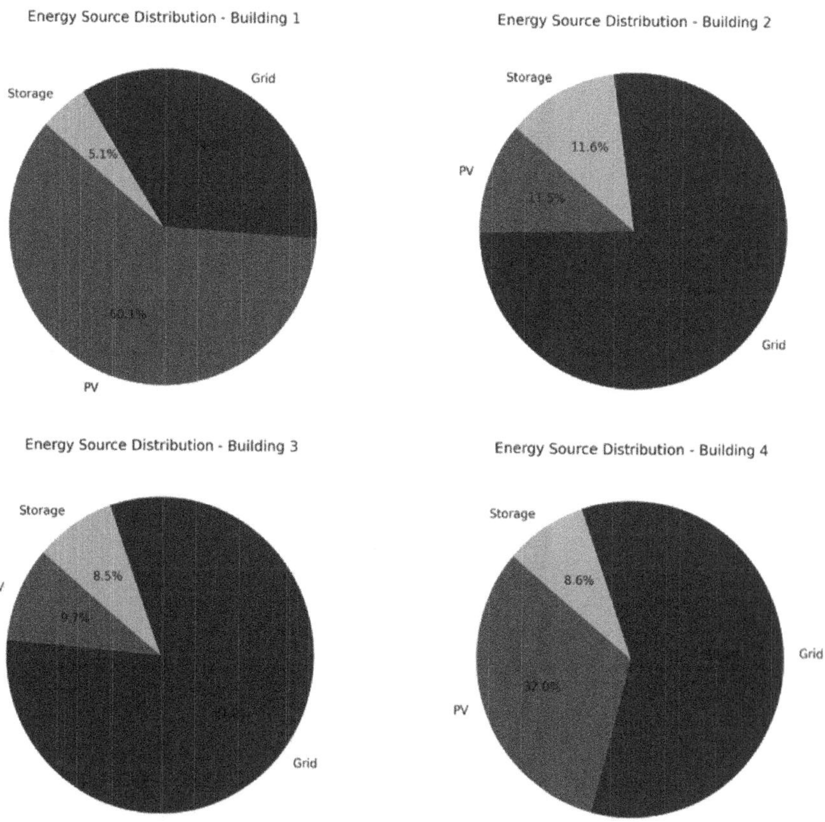

Figure 11.7. *Energy source distribution for buildings*

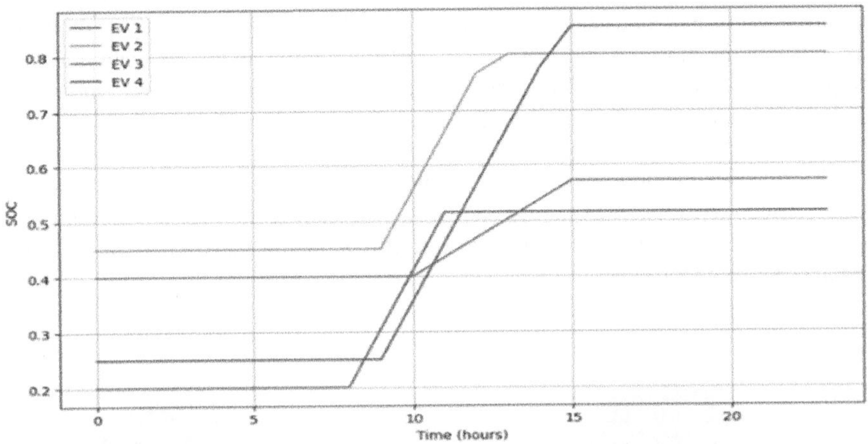

Figure 11.8. *EV SOC*

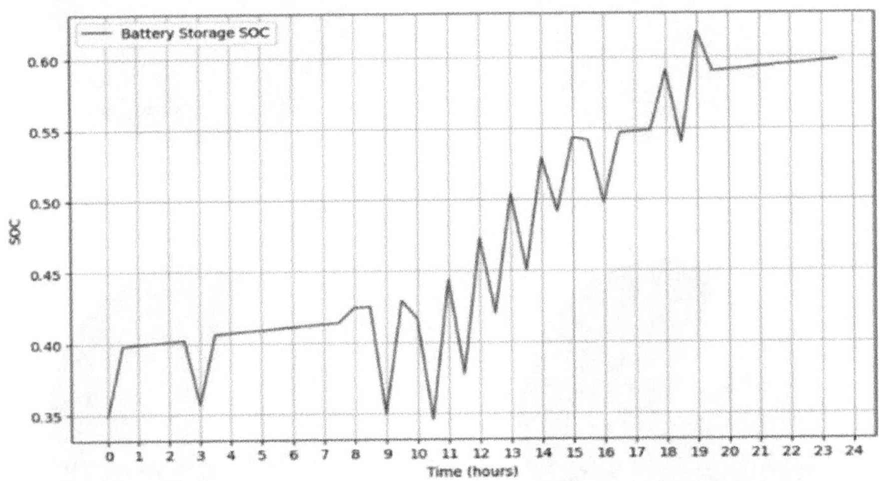

Figure 11.9. *Battery SOC*

11.4.2.2. *Cost paid indicator*

Table 11.5 gives the details of costs paid by all of the users, grouped by type (electric vehicles, buildings, photovoltaic generators, battery). A positive value indicates income, while a negative value indicates expenses. The four scenarios reflect different energy management priorities, directly influencing cost distribution among users.

Scenario I is a cost-minimization strategy, where users prioritize reducing expenses by optimizing energy allocation from the most economical sources. This results in the lowest total system cost (-€396.74), as energy consumption is carefully managed to minimize overall cost.

In comparison, scenario II (comfort maximization) increases costs to -€430.88, ensuring that all user demand is fully satisfied, regardless of economic efficiency. This increase in cost suggests a shift toward using more grid energy or less restrictive energy allocation, prioritizing user satisfaction over savings.

Scenario III, which maximizes renewable energy usage, further raises costs to -€458.63, the highest among all scenarios. This indicates that while PV use is maximized, it does not necessarily translate to lower costs for users. The increased expenses suggest that while PV energy is prioritized, it may not always be available at the right time, leading to a reliance on more expensive alternatives such as battery discharges.

Scenario IV adopts a multi-objective strategy; this scenario reflects a realistic trade-off where different users follow separate priorities – some minimizing costs, others ensuring demand satisfaction and others maximizing PV energy use.

Scenario	Scenario I	Scenario II	Scenario III	Scenario IV
Building 1	-85.64	-84.47	-103.14	-66.05
Building 2	-139.48	-144.17	-164.47	-133.34
Building 3	-244.33	-259.69	-284.32	-253.07
Building 4	-61.04	-62.20	-72.45	-53.42
EV1	-3.59	-3.93	-6.97	-1.21
EV2	-4.70	-4.67	-7.14	-2.34
EV3	-4.21	-4.98	-7.27	-2.03
EV4	-4.74	-5.36	-7.59	-3.54
PV	174.36	162.38	222.02	83.26
Battery	-23.38	-23.79	-27.29	-13.18
Sum	-396.74	-430.88	-458.63	-444.92

Table 11.5. *Cost results*

11.4.2.3. *Constraints*

The power balance is visually represented through bar plots (see Figure 11.10) where total demand is shown as positive bars and total supply as negative bars. As we can see, the power balance is respected for each time step. All of the constraints such as SOC for EVs and storage are also respected.

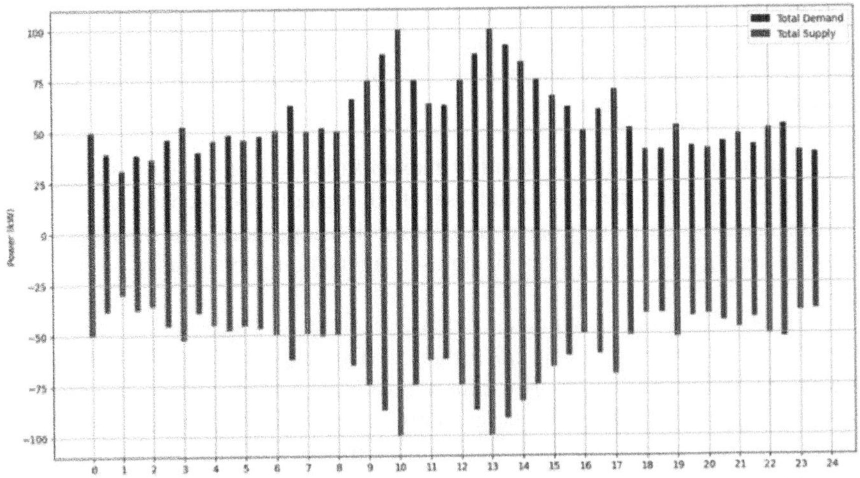

Figure 11.10. *Power profile of buildings*

In addition to maintaining power balance, it is also crucial to assess the system's robustness by ensuring that grid consumption remains within predefined limits under various scenarios. This prevents potential grid overloads and ensures operational reliability.

As shown in Figure 11.11, the total grid consumption across all scenarios consistently stays below the imposed grid limit of 580 kW on a summer day. Similarly, even during a winter day, the grid consumption remains within this limit, as shown in the figure.

These results demonstrate that the system operates efficiently across different seasons.

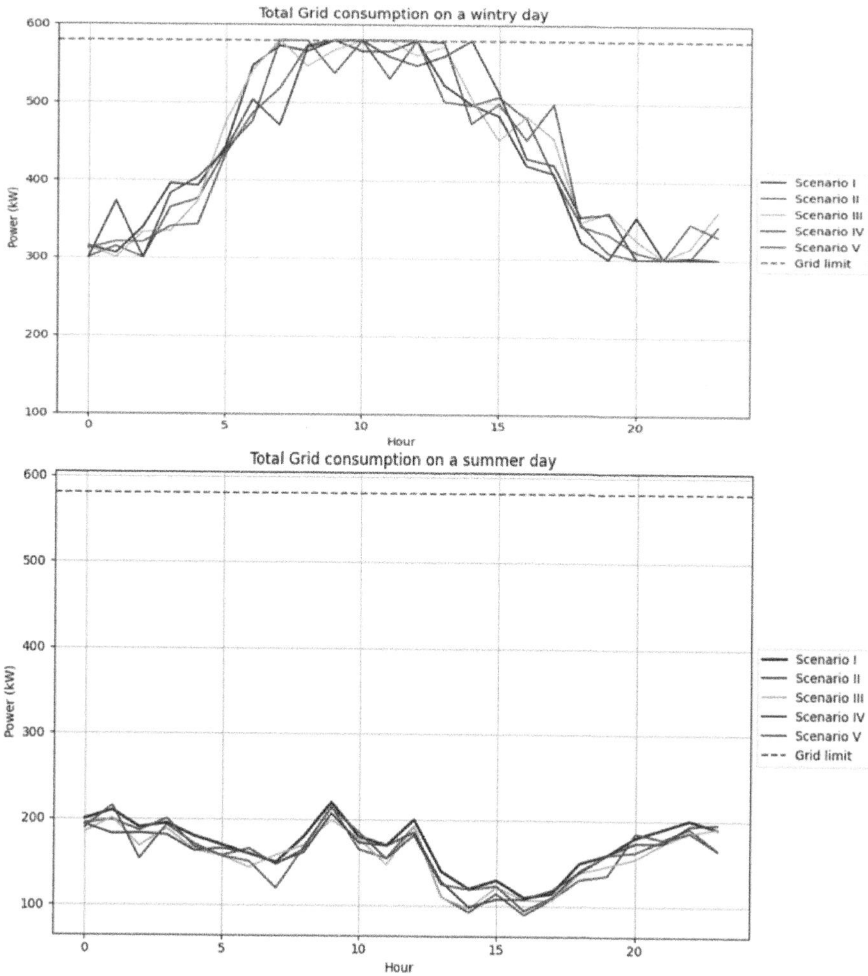

Figure 11.11. *Grid limit respect*

11.5. Conclusion

This chapter proposed a dynamic rule-based framework for energy allocation in a local community, integrating user preferences through utility functions that balance cost, renewable energy use and comfort. The results demonstrated that varying the preference coefficients (α, β, γ) significantly influences energy distribution. The system respected all operational constraints and adapted well under different user scenarios and seasonal conditions. However, limitations include static user preferences, lack of uncertainty modeling and absence of learning capabilities, which restrict the model's adaptability and scalability. Future work should focus on incorporating demand and production forecasting, integrating adaptive or learning-based mechanisms, and extending the utility framework to include equity or carbon minimization goals. Deploying the method via smart contracts could also support real-time automation and transparency. Overall, the approach offers a robust, interpretable foundation for managing decentralized energy systems while aligning allocations with stakeholder values:

– The proposed framework ensures fair energy distribution by dynamically adapting to individual user preferences.

– Simulations confirm that high β values favor PV self-consumption.

– Scenario-based analysis reveals clear trade-offs between cost savings, comfort and sustainability goals.

– The system always respects critical constraints such as grid power limits and SOC ranges.

11.6. Acknowledgments

This work has been achieved within the framework of the EE4.0 (Energie Electrique 4.0) project. EE4.0 is co-financed by the European Union with the financial support of the European Regional Development Fund (ERDF), French State and the French Region of Hauts-de-France.

This work is financed by the French ANR through the AI_Engineering_PhD@Lille project.

11.7. References

Ableitner, L., Tiefenbeck, V., Meeuw, A., Wörner, A., Fleisch, E., Wortmann, F. (2020). User behavior in a real-world peer-to-peer electricity market. *Applied Energy*, 270, 115061.

Durillon, R. (2019). Modélisation de l'acceptation et de l'implication de producteurs et consommateurs d'énergie électrique en vue d'une stratégie de supervision énergétique multi-acteurs. PhD Thesis, Université de Lille, Lille.

Durillon, B. and Bossu, A. (2024). A review of environmental considerations in energy management systems for smart grids. *Renewable and Sustainable Energy Reviews*, 203, 114739.

Durillon, B., Salomez, F., Davigny, A., Kazmierczak, S., Barry, H., Saudemont, C., Robyns, B. (2021). Consumers' sensitivities and preferences modelling and integration in a decentralised two levels energy supervisor. *Mathematics and Computers in Simulation*, 183, 142–157.

Hackbarth, A. and Lobbe, S. (2020). Attitudes, preferences, and intentions of German households concerning participation in peer-to-peer electricity trading. *Energy Policy*, 138, 111238.

Kong, K.G.H., Lim, J.Y., Leong, W.D., Ng, W.P.Q., Teng, S.Y., Sunarso, J., How, B.S. (2022). Fuzzy optimization for peer-to-peer (P2P) multi-period renewable energy trading planning. *Journal of Cleaner Production*, 368, 133122.

Robyns, B., Dobigny, L., Abbes, D., Durillon, B., Barry, H., Saudemont, C. (2024). *Smart Grids and Buildings for Energy and Societal Transition*. ISTE Ltd, London, and John Wiley & Sons, New York.

Sousa, T., Soares, T., Pinson, P., Moret, F., Baroche, T., Sorin, E. (2019). Peer-to-peer and community-based markets: A comprehensive review. *Renewable and Sustainable Energy Reviews*, 104, 367–378.

Stephant, M. (2021a). Distributed optimization of energy profiles to improve photovoltaic self-consumption on a local energy community. *Simulation Modelling Practice and Theory*, 108, 102242.

Stephant, M. (2021b). Increasing photovoltaic self-consumption with game theory and blockchain. *EAI Endorsed Transactions on Energy Web*, 8(34), e1.

Suthar, S. and Pindoriya, N.M. (2024). Chance-constrained co-optimization of peer-to-peer energy trading and distribution network operations. *Sustainable Energy, Grids and Networks*, 38, 101344.

12

Awareness and Perception of Collective Self-Consumption in the University Community

12.1. Introduction

Collective self-consumption (CSC) is a model of energy production and consumption in which multiple consumers (e.g. households, businesses or communities) share locally produced renewable energy, typically generated from sources such as solar power, wind turbines or other decentralized renewable systems. The key idea is that the energy produced by the community is consumed locally, allowing for more sustainable energy practices and promoting energy independence. Thus, sustainable development and collective self-consumption are deeply interconnected, especially within the context of energy systems.

As the world increasingly focuses on addressing climate change, promoting renewable energy and achieving energy independence, CSC presents a significant opportunity for sustainable development. However, for collective self-consumption to be successful, the behavior of individuals and communities plays a key role.

Collective self-consumption offers an innovative and sustainable model for local energy production, but its success depends heavily on individuals' behaviors and attitudes. Research indicates that a combination of economic, social, psychological and regulatory factors influences the decision to engage in collective energy projects. By understanding these drivers and barriers, policymakers and project

---

Chapter written by Hanene BETTAIEB, Adrien BOSSU, Benoit DURILLON, Arnaud DAVIGNY, Christophe SAUDEMONT and Hervé BARRY.

organizers can design more effective and inclusive energy-sharing schemes that appeal to a broader range of participants, ultimately contributing to a more sustainable and resilient energy system.

Therefore, CSC stands at the crossroads of three fundamental concepts:

– The concept of energy communities, based on the collective collaboration of local actors (citizens, businesses, communities) in the production, management and consumption of energy. The objective is to promote the energy transition by enabling local actors to play an active role in energy management, encouraging self-production and pooling of energy resources. This may include renewable energy production (solar, wind, etc.), demand management, energy storage and distribution via local networks or "smart grids".

– Smart energy networks (smart grids): essential in this context, as they allow for the optimized management of energy through real-time communication and control technologies. These networks not only include physical infrastructures (solar panels, batteries, wind turbines), but also digital systems that coordinate decision-making among the different actors in the energy community.

– Human behavior: a crucial indicator in the efficiency and sustainability of these energy communities. The integration of human behavior in energy management is concerned with how social choices, preferences and interactions influence collective decisions. For example, how community members respond to economic incentives or energy-sharing rules, or how social dynamics influence participation in collective initiatives.

12.2. Literature review

The integration of human behavior in local energy network management is a major challenge both scientifically and socio-economically. Energy communities offer a unique opportunity to experiment with new models of governance, social inclusion and energy management using smart grid technologies and decentralized infrastructure. However, the main disadvantage of these solutions is that the way savings are shared among community members can influence their decision to participate.

CSC is a model that simultaneously involves technical (management of production and consumption), social (sharing energy) and economic (distribution of benefits) aspects. To better understand these different aspects, we propose a literature review that explores the key themes and results of research on collective self-consumption. There are different motivations for participating in CSC, such as economic, environmental and social. One of the main drivers of participation is the possibility to reduce electricity costs. Studies indicate that participants can reduce their electricity bills by consuming locally produced energy, often at a lower price than traditional suppliers (Sorrell 2015). In this economic context, the prospect of a long-term return on investment through renewable energy production is also a motivating factor. However, barriers such as the initial cost of installation or management fees can discourage some individuals, especially if public incentives are not available.

Many participants are motivated by the environmental impact of their actions: collective self-consumption reduces dependence on fossil fuels and promotes cleaner energy production (Dall-Orsoletta et al. 2022). Research shows that individuals are more likely to participate in collective self-consumption projects when they perceive their participation as a means of supporting a sustainable energy model. From a social perspective, CSC is seen as an opportunity to strengthen social ties and community cohesion. Several studies point out that the community aspect and mutual support among members play a major role in the decision to participate (McCauley and Heffron 2018). The ability to collectively manage energy production and consumption can enhance individuals' sense of autonomy and control over their energy choices, allowing them to move away from dependence on large energy companies (Wolsink 2012).

Despite these motivations, different barriers hinder the development of this energy policy. A lack of understanding of technologies related to energy production and management (e.g. solar panels, energy storage, smart grids) can deter individuals from joining collective self-consumption projects. In addition, some participants may perceive collective energy management as risky due to concerns about the fairness of the distribution of energy produced or the stability of the system (Abada et al. 2025). Another major obstacle is the initial investment required for infrastructure, such as solar panels, storage systems or energy management devices. Costs can be reduced by government support, but this largely depends on local policies. In some regions, laws and regulations regarding collective self-consumption are unclear, which can create a sense of uncertainty among potential consumers (Inês et al. 2020). The lack of clear standards or policies on issues such as surplus management or redistribution of savings can slow adoption.

To overcome these barriers, awareness and communication are essential. Highlighting the economic, environmental and social benefits of collective self-consumption can encourage participation. Effective information campaigns can help reduce uncertainty and foster project buy-in (Bourne 2016). In this context, we aim to study the behavior of individuals in relation to this energy-sharing model. While CSC remains a promising model for the energy transition, its adoption depends on several factors related to the behavior of individuals alongside economic, environmental and social motivations, all of which play a crucial role.

12.3. Methodology: field data collection method

Among the new and emerging practices of energy transition, individual and CSC stand out as promising approaches. For individuals participating in such initiatives, awareness of these policies is essential. This awareness can be fostered through multiple actions, including education and training, on the one hand, and the implementation of public information measures – targeting children, young people and adults – on the other.

However, the key question remains: are energy users and stakeholders aware of the various actions related to energy management and environmental protection? Therefore, this project aims to study the level of awareness of individuals about collective self-consumption policies. Based on a new energy-sharing method using collaborative play (Bossu 2024; Bossu et al. 2024), this study aims to explore the acceptability of this system among stakeholders through data analysis.

Through this survey, we propose a synthesis of the answers of different topics treated throughout the questionnaire such as basic information, awareness and acceptability of the proposed energy-sharing method. Therefore, we closely identify the different profiles of prosumers to better inform energy control strategies considering consumer behavior.

Our study targets students and staff in the university community during 2024 and aims to analyze how to integrate sustainable development, especially climate change action, and the influence on their daily behavior by studying their awareness.

This survey was conducted among students, researchers, teachers and staff of the Junia (Graduate Scholl of Engineering) and UVSQ (Université de Versailles Saint-Quentin-en-Yvelines) universities, France, to assess their perception of renewable energy and their attitude toward individual and collective self-consumption. So, this study targeted 85 respondents spread over two different cities: Lille (Hauts de France) for Junia and Guyancourt (Ile de France) for UVSQ. The sample included a diverse range of participants, including engineering and social sciences students, researchers, teachers and administrators. Participants answered a questionnaire that combined closed and open questions, allowing for both quantitative and qualitative data to be collected (see Figure 12.1).

This is a survey that was conducted by a face-to-face questionnaire as this method ensures more reliable responses. To increase the number of respondents, a portion of the questionnaires were also distributed via email.

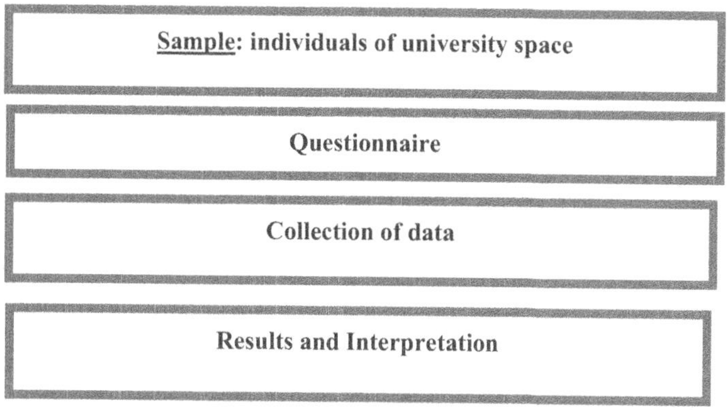

Figure 12.1. *Steps of survey*

Questionnaire design and methodology

In designing the questionnaire, several key considerations were addressed:

– *Comprehensibility*: would the respondent be able to understand and answer the questions?

– *Willingness*: would the respondent be motivated to participate and provide accurate responses?

– *Sufficiency*: would a single question yield the necessary insights, or would further elaboration be required?

– *Relevance*: was each question essential to the study's objectives?

To ensure clarity and accessibility, the questionnaire was structured with concise, straightforward and comprehensive questions. It comprises two main sections:

1) *General and demographic information*: this section gathers data on respondents' age, income, name, field of study, socio-professional category (student or teacher), housing status (tenant or owner) and geographical location.

2) *Energy consumption behaviors and policy awareness*: this section investigates household energy habits and knowledge of individual and collective self-consumption (ISC and CSC) policies.

Additionally, a portion of the questionnaire is dedicated to assessing individuals' satisfaction with CSC, particularly following their participation in a CSC initiative, as well as their acceptance of new energy-sharing systems.

Once the data is collected, recorded and coded, we proceed with a systematic analysis of the results (section 12.3). This involves descriptive statistical interpretation and discussion of key findings. Based on these insights, we formulate recommendations and strategies to enhance policy implementation and improve stakeholder engagement in CSC initiatives (section 12.4).

12.4. Results and interpretations

The following section presents the results of the study, which examines individuals' behaviors and knowledge of collective self-consumption (ACC), as well as their willingness to adopt the proposed energy-sharing system.

12.4.1. *Descriptive statistics*

Gender	Man	58%
	Woman	42%
Institution	UVSQ	60%
	JUNIA	40%

Table 12.1. *Gender and institution of individuals*

According to Table 12.1, 58% of respondents are men. A gender parity was achieved by chance in the sample. Between the two universities, we have more respondents coming from UVSQ (60%), with a social and economic profile, than those from Junia (40%), with a technical profile.

12.4.1.1. *Knowledge ISC, CSC and energy community*

Survey questions focused on the knowledge of individual self-consumption, collective self-consumption and energy community among respondents. The results are shown in Table 12.2.

Knowledge ISC	Yes	24%
	No	76%
Knowledge CSC	Yes	20%
	No	80%
Knowledge energy community	Yes	16%
	No	84%

Table 12.2. *Knowledge ISC, CSC and energy community (%)*

The results show that about 80% of respondents do not know either ISC or CSC or energy community, indicating a lack of information or visibility on these terms and policies for energy transition. The minority (20%) who know these terms are technical profiles (coming from Junia).

Installation of photovoltaic panels	Yes	2%
	No	98%

Table 12.3. *Installation of photovoltaic panels*

Out of 85 individuals surveyed, we found only two people who installed photovoltaic panels at home (Table 12.3). This result intrigues us regarding the knowledge and awareness of energy transition techniques in the university space.

12.4.1.2. *Acceptability and new methods of energy sharing*

The following question sought to study the acceptability of individuals for the energy-sharing system proposed by a research work cited above:

> The difficulty of producing and consuming local photovoltaic energy is that this production is intermittent and depends on sunlight. This means that solar panels do not produce consistently, and the amount of energy produced in a day varies considerably. The challenge is to try and adapt energy consumption to production by shifting appliances during the day such as washing machines or Electric Water Heater.

If you are offered the opportunity to join a program aimed at providing solar energy to your neighborhood, you or your neighbors would produce energy made available to everyone, with the sole intention of amortizing the investment in the installation over its lifetime. A coordinator would suggest preferred times to turn on appliances whose use you believe can be postponed during the day (no effort is required to reduce energy consumption). The objective is to coordinate all of the devices that your neighborhood would consider able to shift (foyer by foyer) to consume the entire energy produced by photovoltaic installations. You will be rewarded for your participation with a reduction on your bill at the end of the month.

Would you agree to join such a system?

☐ Yes, I agree

☐ Rather agree

☐ Neither agree nor disagree

☐ Rather not in agreement

☐ Not all agreed

The results are presented in Table 12.4.

Acceptability of new method of energy sharing	Yes, I agree	24.706%
	Rather agree	32.941%
	Neither agree nor disagree	24.706%
	Rather not in agreement	3.529%
	Not all agreed	8.235%
	No answer	5.882%

Table 12.4. *Acceptability of new collaborative game energy-sharing system (Bossu et al. 2024)*

Around 58% of respondents are between very much and rather agree with this energy-sharing system, with about 25% expressing strong agreement. This suggests that, although the agreement is present, there is also a significant proportion of people who remain reserved (around 42%), of whom about 25% are indifferent due to lack of information on the energy transition. The results showed that respondents are not familiar with renewable energy-sharing technologies such as ISC, CSC and energy communities. Many UVSQ students consider these concepts to be perceived as complex, technical or too specialized for the average person to understand. Without clear and simple explanations, people may feel disconnected or intimidated by the subject.

Among the responses of individuals are those who think that the lack of CSC operations and energy-sharing systems implemented in their daily lives leaves them not realizing the potential benefits.

12.5. Conclusion

The results of this study revealed that respondents have limited knowledge of renewable energy-sharing technologies, specifically ISC, CSC and energy communities. It is clear that there is a significant need for better education and awareness regarding the long-term benefits of collective self-consumption. Another finding is that individuals in the economics academic space are not familiar with these technologies and are not interested because of the complexity of terms for them. This requires encouraging communication between different employees by adopting the interdisciplinary in a more effective way.

The study also emphasizes the importance of moving from the simple proposal of actions to the active participation of individuals in the practice of these actions. This requires individuals to take responsibility for their participation. In addition, promoting public commitment to the energy transition requires social and territorial engineering to ensure inclusiveness and engagement. This study is awareness, information and communication implemented through the questionnaire which remains a tool to discover and address concepts related to ACC, especially for individuals with an economic profile. Ultimately, the study underscores the importance of raising awareness not only to encourage individual action but also to foster collective participation in energy transition initiatives. A more coordinated approach, involving education, communication and active engagement, is crucial for the successful implementation of collective self-consumption.

To better understand individuals' behaviors toward the system-sharing technology and broaden the analysis, offering more precise recommendations, this study could be complemented by the addition of a qualitative econometric model that would quantify the key factors influencing the adoption of collective self-consumption and provide predictions on potential participation. The goal would be to model the probability of an individual adopting the energy-sharing system proposed in this study, based on various explanatory variables (demographic characteristics, environmental attitudes, economic factors, etc.).

12.6. Acknowledgments

This work has been achieved within the framework of the EE4.0 (Energie Electrique 4.0) project. EE4.0 is co-financed by the European Union with the financial support of the European Regional Development Fund (ERDF), French State and the French Region of Hauts-de-France.

12.7. References

Abada, I., Ehrenmann, A., Lambin, X. (2025). Risk-sharing in energy communities. *European Journal of Operational Research*, 322(3), 870–888.

Bossu, A. (2024). Sorel : supervision optimisée d'un réseau local de distribution au sein d'une communauté énergétique : prise en compte des critères électriques, environnementaux, économiques et sociétaux. PhD Thesis, Université de Lille, Lille.

Bossu, A., Durillon, B., Davigny, A., Barry, H., Robyns, B., Belaïd, F., Saudemont, C. (2024). Coalitional game-based gain generation and distribution for collective self-consumption in an energy community. *Mathematics and Computers Simulation*, 225, 129–147.

Bourne, L. (2016). Targeted communication: The key to effective stakeholder engagement. *Procedia – Social and Behavioral Sciences*, 226, 431–438.

Dall-Orsoletta, A., Cunha, J., Araújo, M., Ferreira, P. (2022). A systematic review of social innovation and community energy transitions. *Energy Research & Social Science*, 88, 102625.

Inês, C., Pontes Luz, G., Marín-González, E., Gährs, S., Hall, S., Holstenkamp, L. (2020). Regulatory challenges and opportunities for collective renewable energy prosumers in the EU. *Energy Policy*, 138, 111212.

McCauley, D. and Heffron, R. (2018). Just transition: Integrating climate, energy and environmental justice. *Energy Policy*, 119, 1–7.

Sorrell, S. (2015). Reducing energy demand: A review of issues challenges and approaches. *Renewable and Sustainable Energy Reviews*, 47, 74–82.

Wolsink, M. (2012). Undesired reinforcement of harmful 'self-evident truths' concerning the implementation of wind power. *Energy Policy*, 48, 83–87.

PART 3

Social and Controversy Networks

13

Fracking in the UK: Dependent Stakeholders and Relational Legitimacy

13.1. The social license to operate of fracking in the UK

The relationship between industry and local community stakeholders in sites of contestation and flux, such as the oil and gas extractive industries, is well documented. However, several scholars have called for research concerning how such dependent stakeholders, notably the local community stakeholders, are defined, formed and comprised. This chapter is based on research focused on the case of fracking in the United Kingdom. Drawing on ethnographic data, which included 12 semi-structured interviews with participants from two cohorts of stakeholders, namely the industry and the community, stakeholder observations were thematically analyzed. Triangulation was achieved with academic survey data, relevant government documents, and local and national media reportage.

This research investigates the concept of the social license to operate (SLO) within the context of dynamic and dependent stakeholder relationships in contested environments, focusing on the fracking industry in the UK. The SLO refers to the acceptance or approval of a company's activities by local communities and stakeholders, particularly in extractive industries where public contention often arises (Boutilier and Thomson 2011; Demuijnck and Fasterling 2016). While the SLO has been explored as a strategic tool for legitimacy, its relevance extends to any business activity engaging community stakeholders (Morrison 2014).

The findings show that the neglect of relational legitimacy in the SLO of fracking in the UK leads to the formation, definition and self-identification of local community stakeholders into much more powerful and salient actors than is often

---

Chapter written by Matthew O'Meara WALLIS.

assumed in the literature, and indeed by the stakeholders themselves. The study highlights the problem of focusing on institutional legitimacy to the detriment of relational legitimacy, emphasizing that "dependent" stakeholders engage in value judgments critical for businesses' social license to operate.

Local community stakeholders are traditionally seen as dependent, holding urgent and legitimate claims but lacking power (Van Buren 2010). Wagner et al. (2012) describe such stakeholders as over reliant on organizations to meet their needs. However, this dependency does not preclude agency, as communities can challenge this power dynamic, especially when their perceived well-being is at stake. The extractive industry's tendency to focus on institutional legitimacy often leads to eroded trust and degraded relationships with these stakeholders, highlighting the importance of relational legitimacy (Kougiannou and Wallis 2020) in the SLO.

Despite its relevance, the SLO has received limited scholarly attention beyond its use as a corporate strategy (Holt and McNulty 2008). Relational legitimacy, informed by value judgments and community engagement, is a key dimension of the SLO and cannot be fully controlled by firms. Instead, it requires meaningful dialogue and collaboration with stakeholders (Buhmann 2016; Painter and Deslandes 2017). This study contributes to the literature by examining such relational aspects of the SLO by exploring how local community stakeholders perceive their relationships with extractive firms and how these perceptions influence relational legitimacy. Existing studies on fracking have focused on environmental risks and public health but often overlook the ethical and relational dimensions of stakeholder interactions (Clarke et al. 2015; Cotton 2017). In much of the social geography literature, especially around mining and other extraction activities, such local communities are viewed as altruistic actors in contestation with monolithic corporate interests. Already with such a perception, there is less trust and relational legitimacy embedded (Kougiannou and Wallis 2020). Using a case study of fracking in the East Midlands, the research adopts a qualitative approach to capture the nuanced and evolving nature of these relationships.

13.2. Defining the local community stakeholder

Defining stakeholders is not straightforward (Willis 2012), as the "community" can consist of individuals or organizations who claim to represent an interest, being characterized by either geography, interaction or identity. For the manager of stakeholder relations, context is important, and it is often unclear who has either the formal or informal authority or the resources to engage in specific processes. In the management literature, definitions of the "community" are relatively evenly split between individuals of the general public and the definition of community as groups or organizations (Dunham et al. 2006). In this research, we approach the "community" in terms of stakeholder theory. Instead of community relations being the focus, the context of a single-issue event that includes geography, interaction and identity points to the dependent stakeholder of the "community". This local stakeholder is dependent: lacking in agency, power and salience in the stakeholder relationship with the firm.

This is identified as virtuous – and outside the scope of corruption to its truth by the maximizing of profit values held in contrast by the extractive industry (and firm) in question. In the literature, and from the firm's perspective, the environmental activist or community activist is either a derivative stakeholder, or not one at all (Kaler 2003; Phillips 2003). This means they are low-power or dependent stakeholders, and the extractive industry firm usually considers the local community stakeholder as therefore lacking in salience. Painting the local community stakeholder in these terms is also attractive for the local communities themselves, allowing them to engage in a dialogue as the underdog – the David versus Goliath narrative. The local community stakeholder "others" in ethical encounters the industry while shaping the stakeholder relationship, and this is furthermore a key part of how the stakeholder self-identifies.

13.3. Institutional legitimacy

As the local community stakeholder has little salience or "power, legitimacy, or urgency" (Mitchell et al. 1997), understandably, the firm turns to whom or what may be salient. In the terms of the case study, this is the institutional legitimacy possessed by the government, shareholders, planning officers and technicians involved in the process of fracking.

One of the manifestations and ways to empirically explore the SLO is by examining legitimacy perceptions in stakeholder relationships (Kougiannou and Wallis 2020). The SLO also invokes legitimacy, which dates to the inception of organization theory (Deephouse and Suchman 2008; Gehman et al. 2017). Suchman describes SLO as: "a generalized perception or assumption that the actions of an entity are desirable, proper, or appropriate within some socially constructed system of norms, values, beliefs, and definitions" (Suchman 1995, p. 574).

Fracking is regularly subjected to legitimacy evaluations, with stakeholders themselves also being evaluated as to whether they have legitimate claims based upon "contract, exchange, legal title, legal right, moral right, at-risk status, or moral interest in the harms and benefits generated by company actions" (Agle et al. 1999, p. 508).

13.4. Relational legitimacy

Reviewing the literature, questions that emerge include what the composition of "dependent" stakeholders is and whether stakeholder relationships with power asymmetries are virtuous, vicious, both, or neither. Highlighting the composition of dependent stakeholders and where and what values might reside in networked stakeholder relationships, we emphasize the role of relationality in both virtue ethics and stakeholder theory. Relationality encompasses the interconnectedness and interdependence of individuals, organizations and stakeholders.

It is the acknowledgment that ethical decision-making and leadership effectiveness are inherently relational and, indeed, embodied processes (Painter and Werhane 2023). Ethical decision-making is not solely an individual endeavor, but it is influenced by relationships and interactions. Relational legitimacy emphasizes the importance of ethical dialogue, collective decision-making and shared values among organizational members. Therefore, relationality can be regarded as closely aligned with stakeholder theory. After all, stakeholder theory posits that organizations should consider the interests of all stakeholders. Relationality expands how various organizations, such as healthcare organizations (Werhane 2000), build and maintain positive relationships with various stakeholders.

The importance of recognizing the "other" in ethical encounters while shaping the discourse on relationality in business ethics is a key part of how the stakeholder self-identifies. Trust is central to relationality, as it fosters cooperation and open communication among stakeholders. Building and maintaining trust is crucial for ethical behavior within organizations. The SLO is created and maintained gradually, over time, as an organization builds trust with the community. Furthermore, relationality is a key dimension of ethical leadership, which involves being attentive, responsive and accountable to the diverse and dynamic relationships that constitute organizational life (Painter 2008). Relational legitimacy challenges the dominant models that are based on individualism, rationality and hierarchy, and proposes a more inclusive, embodied and collaborative way of leading. Pérezts et al. (2020) suggest that relationality is grounded in the local realities and cultures of different stakeholders as well as open to learning from other perspectives and traditions.

13.5. Degraded stakeholder relationships and relational legitimacy

This case study shows that the local community did not wish to endure the uncertainties that fracking might entail, such as seismic activity, injustice through natural rights violations, diminished natural resources of some kind or threats to their health. More prosaically, but perhaps more honestly and in plainer language, they were also worried about the industrialization of the countryside and its impact on their countryside lifestyle, and indeed, what might happen to their house prices (Boudet et al. 2014). According to the case study, the fracking firm in question, however:

> As part of its commitment to open and transparent communications [GasInc] has undertaken extensive community engagement alongside this application including setting up a community liaison group which was convened in June 2014.

Unfortunately for GasInc, whatever its strategy was with this group in terms of a local stakeholder relationship, it failed.

> You know with their community engagement and there it's how they try and want to win, win over communities by saying oh look at all the good we're doing in their in their community… Problem is of course they didn't actually liaise at all except for a few meetings. They never got round to delivering owt (sic) that was promised did they? (P08).

Understandably, perhaps considering power, legitimacy and urgency, GasInc neglected this stakeholder to pursue institutional legitimacy in the planning process. Therefore, for many local community stakeholders, it was only when planning permission was granted, with its subsequent media coverage, that they had been truly engaged and understood the risks for the first time. P01 – a journalist research participant – described the febrile feelings in the village:

> Then suddenly, you know, local people were getting very alarmed about what might happen in their village… here was a public meeting and it was quite fractious from… from what I gather and people were locally quite angry. I think the company had probably not expected this sort of response.

This confirmed the sense by research participants that the dependent stakeholder of the local community arguably has little power or influence over the firm's actions. The stakeholders have conflicting values, and there is an asymmetry in power between these stakeholders when power is conceptualized (Lukes 1974).

There is a conflict between the values and practices of these two stakeholders, which impacts the power in the stakeholder relationship. To reiterate Freeman (1984) inversely, this relationship got off on the wrong foot and therefore ruined the chance for mutual trust, understanding and collaboration. Without a positive and transparent connection with its stakeholders from the outset, numerous disadvantages can (and did) ensue.

Ineffective communication and engagement at the beginning of the stakeholder relationship as demonstrated then prevented shared understandings of expectations, values and goals. The interests of the fracking firm in question GasInc were misaligned to its local community stakeholders. This disordering of the relationship hindered collaboration and maximized conflict. So much so, that it prevented the business from navigating an existential challenge effectively.

In other words, it lost relational legitimacy. The relationship between the community stakeholder and GasInc was degraded severely in the perceptions of the local residents. The dependent stakeholders of the local community felt that GasInc and the oil and gas industry's relationship with the government led to a stance on shale gas that undermines the validity of the local stakeholder relational legitimacy in terms of the institutional legitimacy of the planning applications, planning hearings and decisions made by the local county council. As one local resident put it: "It goes to the heart of whether there is any point in having a planning Committee at all in ***ham if in the end, it's going to be overruled." (P09). This is a degradation initially in the relationship between the local community and the firm, but it also degrades the relationship between the local community and their democratic representatives. Institutional legitimacy trumps relational legitimacy, but this has numerous unintended consequences.

The relationship between the case study fracking firm and local communities is initially neglected and, therefore, becomes antagonistic rather than cooperative. In the local community, especially, this view was bolstered by a framing of powerlessness in the planning system, which led to distrust in the firm. While some local people claimed to trust GasInc, many found their ways of engaging with the public both disingenuous and ineffective – especially as part of a stakeholder relationship. The day-to-day interaction between residents and GasInc was one of the most important factors in determining the stakeholder relationship between them. This potential relationship was neglected entirely by GasInc in favor of focusing scarce resources on the planning process and institutional legitimacy. They provided little to the local community stakeholder except for words – and scant words at that – in the eyes of the local community. Any additional worries of the local community, such as the long-term impact of industrialization of the countryside, and indeed the viability and accountability of the fracking business, would go unsatisfactorily unanswered.

This provided an information vacuum that led some research participants who were more informed to begin questioning the effectiveness of regulatory and monetary mechanisms that were tasked with overseeing the industry. Opponents of GasInc and fracking generally had stepped into this breach. Indeed, much of this newer sophisticated information had been given by these opponents of GasInc to the local community, in addition to that self-obtained research online. GasInc was notably absent in mitigating or moderating these anxieties with a robust stakeholder relationship with their local community.

13.6. The SLO in stakeholder relationships

One councilor is on record at the planning meeting as saying:

> The fact is that the Council or the members of the Committee have been given legal advice which has tied the members' hands, I'm afraid. And, basically, it made it a waste of time for any of us **to be** here for the last few days.

However, once the advice that the councilors originally received was put in writing and made public, and alternative legal advice was sought, they did not put the councilors' options in quite as stark terms.

In preparation for the appeal, a statement of common ground was supposed to be prepared in cooperation between GasInc and the ***ham CC. As some councilors recall, the draft statement that was presented to them listed as areas of agreement those aspects of shale gas activities that the councilors could not formally object to because of the limitations of the planning process. This was due to the "exploratory" pipeline being granted permission rather than actual fracking.

Nevertheless, they did believe that those aspects of shale gas development could adversely impact the local community. However, the social impact of risk was discounted as irrelevant. This potential misunderstanding regarding the possible sources of harm is consistent with research that shows how state agencies sometimes distort local stakeholder concerns in official proceedings.

Many residents expressed their disappointment to the researcher in various settings, in that they were permitted to present their point of view to the councilors. However, there was no room for discussion between the councilors and members of the public. The residents followed the rules of the planning process. However, they firmly believed that the scientific risk assessment, as well as technical and planning aspects that were taken into consideration when determining the fracking applications, were insufficient grounds for making good decisions. The stakeholder relationship is immediately seen to be severed or broken here, despite the local information meetings – as GasInc and the local stakeholders never genuinely engaged in dialogue outside formalized settings such as planning hearings. Within those settings, it was not much better as the stakeholder relationship was just as neglected, leading to further perceptions of injustice and degrading or diminishing the importance of the stakeholder relationship.

Scholars from a range of different scientific disciplines have also pointed out that a social understanding of risk should supplement the technical understanding of feasibility. Local stakeholders should take part in the process, which also involves a broader deliberation of possible alternatives. It goes on to state that these dimensions correspond to the three roles of the planning system. However, some residents have pointed out that the planning reports on the basis of which the fracking decisions at the ***ham CC were to be determined were deficient.

The deficiency was that they only addressed the environmental and economic dimensions of fracking, thus failing to take into account risks borne and rights violated by the local community stakeholders.

The NPPF (Department for Communities and Local Government, 2012) also seems to have established a broader set of considerations that should be taken into account in planning. It states that the three core and interconnected dimensions of sustainable development are economic, social and environmental.

As P09 put it during a hearing: "this is the elephant in the room", and added:

> In recent years, planning has tended to exclude rather than include people and communities… [We] should have been given the same consideration within the social element of the NPPF jointly and simultaneously with the economic and environmental considerations, especially within the planning reports.

This research reveals that local residents are concerned about the social role of planning being overlooked, which has resulted in the process becoming merely a vehicle for facilitating shale gas development, regardless of local apprehensions. In such cases, power appears to be concentrated in the hands of the most influential stakeholder, the fracking firms.

This finding validates the local community stakeholders' suspicion of regulatory capture, where business interests seem to align with those of the government. Furthermore, there is a belief that restricting local groups by delineating what is considered a legitimate concern and an appropriate objection to fracking has undermined the strength of their arguments.

Such constraints are perceived as deliberate attempts to minimize the impact of objections on shale gas development projects. Stern (1991) found that, as a response to highly uncertain impacts (such as in the case of fracking), policymakers tended to underestimate the risks and privilege single-point estimates to more accurate and more complex ranges of possibility (this, of course, includes qualitative studies such as this research).

The findings of this study confirm the view that the lack of consideration for the local community in the case of planning permission for fracking can influence the outcomes of planning and political decision-making concerning fracking in the East Midlands and beyond. This disadvantages the local community stakeholders through further disempowerment. This further diminishment of power was felt as a procedural injustice that needed remediation by the vulnerable and dependent stakeholders. The institutional license to operate was becoming increasingly untethered from the social. The SLO is in trouble locally, and GasInc did little to mitigate this through any attempted engagement with the local stakeholders. While GasInc retained institutional legitimacy, it had completely lost its relational legitimacy and, with it, the SLO.

13.7. Conclusions from findings: trust and justice in stakeholder relationships

The multiple ways that power and justice dynamically interact within the local community–fracking firm stakeholder relationship emerge as a key theme in the analysis, especially through the lens of trust/distrust. While power and justice are distinct concepts, their interplay is often observed in practice (Kougiannou and Wallis 2020) as it has in this research. In the realm of stakeholder relationships, the ability to influence the allocation of resources and opportunities typically lies with the most powerful or salient stakeholder – be it the firm or the local stakeholder. Their value judgments are instrumental in either advancing or impeding their objectives. The stakeholder classification framework by Mitchell et al. (1997) identifies three attributes that stakeholders may possess in relation to focal organizations: power, legitimacy and urgency. According to this model, stakeholders with legitimate and urgent claims, yet lacking the power to enforce them, are termed "dependent stakeholders" (Wagner et al. 2012). The emerging awareness among local community stakeholders of their role within this framework, and their subsequent actions to address this positioning, is a key contribution. Namely, the coalition forming and retaking of agency and power building the dependent stakeholder engaged in.

With the nature of its operations and especially the extractive industry's impact on the environment, firms such as GasInc are regularly subjected to legitimate evaluations from multiple stakeholders, including national and local governments. The local community is represented by its government as well as claims to be a stakeholder in its own right beyond government (Mitchell et al. 1997; Bowen et al. 2010). The oil and gas industry is usually represented as the most powerful and salient stakeholder (Mitchell et al. 1997) when it comes to power and its exercise. On the other hand, the typical representation of the local community stakeholder in such extractive industry scenarios characterizes it as dependent (Wagner et al. 2012;

Greenwood and Van Buren 2010). As such, it is subject to its rights being violated and possessing less agency. For example, Kougiannou and Wallis (2020) observed in their coal industry case study that this sector serves as a compelling example of power asymmetry and hypocrisy. These stakeholders, often vulnerable to the effects of corporate hypocrisy, play a crucial role in making legitimacy evaluations (Bowen et al. 2010). In this case study, such evaluations led to resistance and rejection of the dependent role in the stakeholder relationship. This was especially prompted by the degraded nature of relational legitimacy in the first place.

Greenwood and Van Buren (2010, p. 425) highlight the key role organizational trustworthiness could play in the study of organization–stakeholder relations, arguing that "trust is a fundamental aspect of the moral treatment of stakeholders within the organization-stakeholder relationship". They argue that trust is the only option for dependent stakeholders when interacting with an organization. In trust and justice studies, Lewis and Weigert (2012) extrapolate where local stakeholders' distrust in a fracking firm might stem from. There is a profound sense of justice being violated, which encompasses concerns regarding the unfair distribution of risks and benefits (distributive justice), lack of transparent and inclusive decision-making processes (procedural justice), and feelings of disrespect and non-recognition (interactional justice). GasInc was often associated with various risks, such as environmental pollution, health hazards and potential disruptions to the local community's way of life, all of which give rise to a sense of distributive injustices. This imbalance fosters distrust, as stakeholders feel that the firm's operations benefit a few at the expense of many, violating the principle of fairness in the distribution of both risks and rewards. This is a core component of justice, which is the process by which decisions are made.

When local stakeholders are excluded from the decision-making process, or when the process is opaque and not consultative, there is a violation of procedural justice. The firm's failure to engage with the community, and disregard for local input into the decision-making process, exacerbate feelings of injustice.

The perceived absence of procedural fairness erodes trust, as stakeholders may perceive the firm as operating in disregard of democratic principles and the rule of law. Interactional justice relates to how the firm engages with stakeholders during interpersonal interactions. GasInc's neglect and deterioration of its communications have breached principles of interactional justice, which has led to a further erosion of trust.

Thematic analysis serves to reinforce the depiction of the local stakeholders as vulnerable or dependent. Moreover, the analysis discloses that such a characterization does not signify the conclusion of the dynamic interplay. In fact, the local community, to quote one activist "flipped the table on them". Local community stakeholders in alliance with "professional campaigners" using distributed leadership (Uhl-Bien 2006) presented a formidable and ultimately more salient grouping with a strategic and political exercise of power that shaped and shapes the value (MacIntyre 2016) and vice (Shklar 1984) of stakeholder relationships. Namely, its success in shutting down GasIncs operations and a total ban (or moratorium) on fracking operations nationwide.

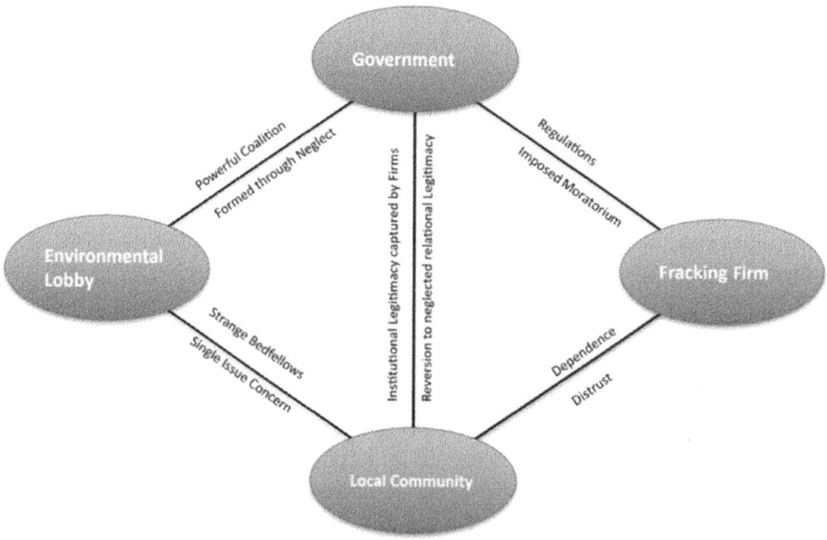

Figure 13.1. *Relational webs: institutional power and legitimacy flows*

13.8. Discussion

This study reveals that the composition and formation of local community stakeholders in the fracking industry are far more dynamic and complex than traditionally portrayed. Local community stakeholders, often dismissed as dependent and low-power entities, demonstrated significant agency by mobilizing collective resources, forming alliances with activists, and leveraging social media to amplify their concerns. These coalitions allowed them to shift from being perceived as passive actors to influential stakeholders capable of challenging corporate practices. GasInc's oversight in recognizing this complexity led to its mismanagement of stakeholder relationships. By focusing on institutional legitimacy rather than understanding the evolving composition of the community, GasInc failed to anticipate the strategic alliances and distributed leadership that empowered local stakeholders. The emergence of these coalitions emphasizes the importance of acknowledging the fluidity of stakeholder roles and the transformative potential of collective action.

GasInc's actions, such as using euphemistic terminology such as "exploratory drilling" to downplay fracking activities, were perceived as insincere and manipulative. These actions eroded trust and reinforced the community's negative value judgments, framing GasInc as a corporate antagonist.

The study highlights the importance of aligning corporate actions with community values to foster mutual respect and trust, which are foundational to effective relational engagement. We find that SLO is multi-dimensional, encompassing trust, justice and relational engagement. It is earned through transparent, equitable and inclusive interactions that prioritize the well-being and values of all stakeholders. Distributive justice requires a fair allocation of risks and benefits; procedural justice demands transparency and meaningful participation in decision-making; and interactional justice calls for respectful and honest communication.

GasInc's approach failed to uphold these principles, leading to the erosion of its SLO. By prioritizing institutional legitimacy and regulatory compliance over relational legitimacy, the company neglected the relational dynamics necessary to build trust. Community perceptions of hypocrisy – where corporate rhetoric on community engagement was not matched by action – further undermined GasInc's credibility. This breakdown in relational engagement fostered active resistance, culminating in the loss of the SLO and the formation of a powerful coalition opposing GasInc's operations. Additionally, this study challenges the conventional characterization of local community stakeholders as merely dependent or low-power entities. Building on Mitchell et al.'s (1997) stakeholder salience framework, it demonstrates the fluidity of stakeholder roles and the potential for these groups to leverage alliances and social capital to exert significant influence. The research extends previous work by Kougiannou and Wallis (2020) on relational legitimacy, showing how local stakeholders use value-based judgments to critique corporate practices and reshape power dynamics.

By framing stakeholder relationships as dynamic and context-dependent, the study highlights the importance of ethical dialogue and shared values in building sustainable and inclusive business practices. This approach moves stakeholder theory beyond transactional models, offering insights into how relational legitimacy can address the ethical and operational challenges faced by extractive industries. The case of GasInc has practical implications as well, highlighting the necessity of prioritizing relational legitimacy in corporate practices. Companies must engage in authentic dialogue with stakeholders, ensuring that their voices are heard and respected. Transparent communication, equitable risk-sharing and inclusive decision-making are essential to building trust and maintaining the SLO. The study also cautions against organizational hypocrisy, where discrepancies between rhetoric and actions degrade trust and stakeholder relationships.

This study highlights the critical importance of relational legitimacy in sustaining the SLO within the fracking industry. Redefining the composition and agency of local community stakeholders challenges the old dyadic "spoke-and-wheel" model of stakeholder management, and which stakeholders are prevalent or not. Such traditional classifications were normatively weak and underscores the need for ethically grounded corporate practices that can be based on relational legitimacy. The findings emphasize the transformative potential of inclusive and transparent engagement strategies, offering valuable lessons for theory and practice in stakeholder management.

13.9. References

Agle, B.R., Mitchell, R.K., Sonnenfeld, J.A. (1999). Who matters to Ceos? An investigation of stakeholder attributes and salience, corpate performance, and Ceo values. *Academy of Management Journal*, 42(5), 507–525.

Boudet, H., Clarke, C., Bugden, D., Maibach, E., Roser-Renouf, C., Leiserowitz, A. (2014). "Fracking" controversy and communication: Using national survey data to understand public perceptions of hydraulic fracturing. *Energy Policy*, 65, 57–67.

Boutilier, R.G. and Thomson, I. (2011). Modelling and measuring the social license to operate: Fruits of a dialogue between theory and practice. *Social Licence*, 1, 1–10.

Bowen, F., Newenham-Kahindi, A., Herremans, I. (2010). When suits meet roots: The antecedents and consequences of community engagement strategy. *Journal of Business Ethics*, 95(2), 297–318.

Buhmann, K. (2016). Public regulators and CSR: The 'Social Licence to Operate' in recent United Nations instruments on business and human rights and the juridification of CSR. *Journal of Business Ethics*, 136(4), 699–714.

Clarke, C.E., Hart, P.S., Schuldt, J.P., Schuldt, J.P., Evensen, D.T.N., Boudet, H.S., Jacquet, J.B., Stedman, R.C. (2015). Public opinion on energy development: The interplay of issue framing, top-of-mind associations, and political ideology. *Energy Policy*, 81, 131–140.

Cotton, M. (2017). Fair fracking? Ethics and environmental justice in United Kingdom shale gas policy and planning. *Local Environment*, 22(2), 185–202.

Deephouse, D.L. and Suchman, M. (2008). Legitimacy in organizational institutionalism. In *The Sage Handbook of Organizational Institutionalism*, Greenwood, G., Oliver, C., Suddaby, R., Sahlin, K. (eds). Sage Publications, London.

Demuijnck, G. and Fasterling, B. (2016). The social license to operate. *Journal of Business Ethics*, 136, 675–685.

Dunham, L., Freeman, R.E., Liedtka, J. (2006). Enhancing stakeholder practice: A particularized exploration of community. *Business Ethics Quarterly*, 16(1), 23–42.

Foucault, M. (1984). *The Use of Pleasure: The History of Sexuality*, Volume 2, Hurley, R. (trans). Pantheon Books, New York.

Freeman, R. (1984). Strategic management: A stakeholder theory. *Journal of Management Studies*, 39(1), 1–21.

Gehman, J., Lefsrud, L.M., Fast, S. (2017). Social license to operate: Legitimacy by another name? *Canadian Public Administration*, 60(2), 293–317.

Greenwood, M. and Van Buren, H. (2010). Trust and stakeholder theory: Trustworthiness in the organisation–stakeholder relationship. *Journal of Business Ethics*, 95(3), 425–438.

Holt, R. and McNulty, T. (2008). Securing the licence to act: A foundational capability. *Journal of Strategy and Management*. 1(1), 72–92.

Kaler, J. (2003). Differentiating stakeholder theories. *Journal of Business Ethics*, 46(1), 71–83.

Kougiannou, N.K. and Wallis, M.O.M. (2020). Chimneys don't belch out carnations! The (in) tolerance of corporate hypocrisy: A case study of trust and community engagement strategies. *Journal of Business Research*, 114, 348–362.

Lewis, J.D. and Weigert, A. (2012). The social dynamics of trust: Theoretical and empirical research, 1985–2012. *Sociological Theory*, 30(4), 364–386.

Lukes, S. (1974). *Power: A Radical View*. Macmillan, London.

MacIntyre, A. (2016). *Ethics in the Conflicts of Modernity: An Essay on Desire, Practical Reasoning, and Narrative*. Cambridge University Press, Cambridge.

Mitchell, R.K., Agle, B.R., Wood, D.J. (1997). Toward a theory of stakeholder identification and salience: Defining the principle of who and what really counts. *Academy of Management Review*, 22(4), 853–886.

Morrison, J. (2014). The social license. In *The Social License: How to Keep your Organization Legitimate*. Palgrave Macmillan, London.

Painter, M. (2008). Systemic leadership and the emergence of ethical responsiveness. *Journal of Business Ethics*, 82, 509–524.

Painter, M. and Deslandes, G. (2017). Reconceptualizing CSR in the media industry as relational accountability. *Journal of Business Ethics*, 143(4), 665–679.

Painter, M. and Werhane, P.H. (2023). *Leadership, Gender, and Organization*. Springer, Berlin.

Pérezts, M., Russon, J.A., Painter, M. (2020). This time from Africa: Developing a relational approach to values-driven leadership. *Journal of Business Ethics*, 161(4), 731–748.

Phillips, R. (2003). *Stakeholder Theory and Organizational Ethics*. Berrett-Koehler Publishers, Oakland, CA.

Shklar, J.N. (1984). *Ordinary Vices*. Belknap Press of Harvard University Press, Cambridge, MA.

Stern, P.C. (1991). Learning through conflict: A realistic strategy for risk communication. *Policy Sciences*, 24(1), 99–119.

Suchman, M.C. (1995). Managing legitimacy: Strategic and institutional approaches. *Academy of Management Review*, 20(3), 571–610.

Uhl-Bien, M. (2006). Relational leadership theory: Exploring the social processes of leadership and organizing. *The Leadership Quarterly*, 17(6), 654–676.

Van Buren, H.J. (2010). Taking (and sharing power): How boards of directors can bring about greater fairness for dependent stakeholders. *Business and Society Review*, 115(2), 205–230.

Wagner, M., Emerson, H., Alves, M.R. (2012). A model for stakeholder classification and stakeholder relationships. *Management Decision*, 50(10), 1861–1879.

Werhane, P.H. (2000). Business ethics, stakeholder theory, and the ethics of healthcare organizations. *Cambridge Quarterly of Healthcare Ethics*, 9(2), 169–181.

Willis, P. (2012). Engaging communities: Ostrom's economic commons, social capital and public relations. *Public Relations Review*, 38(1), 116–122.

14

Articulação: The Ecological Networking of the Brazilian Agroecological Movement

14.1. Introduction

Agroecology refers to forms of agriculture that produce ecologically by relying on the systemic properties of a given agroecosystem (Altieri 1995). But this now very popular term also refers to the dense network of social organizations that promote this alternative agricultural model. Brazil has a well-established national agroecological movement engaged in both promoting ecological agronomic practices and shaping agricultural public policies.

With *The Appropriation of Nature: Essays on Human Ecology and Social Relations*, the British anthropologist Tim Ingold (1986) has been trying to build bridges between these two dimensions. Drawing inspiration from James Gibson's "ecological approach to visual perception" (1979) and from the study of fungi, he "refuses to make a radical separation between humans and nonhumans" and likes to "model one through the other" (Descola and Ingold 2013, p. 19). Thus, building on an idea first proposed by Haudricourt (1962), the way people conceive and practice agriculture should influence their relationships and the types of social relations that characterize a given culture. Taking this idea one step further, we can imagine that the way agroecosystems are conceived and managed daily could similarly influence how people relate to each other and organize at different spatial scales, from the plot and the household to broader territorial levels.

To address this hypothesis, it is essential to model ecological networks using life and complexity sciences while incorporating the sociology of social movements to

---

Chapter written by Sébastien CARCELLE and Mariana BENITEZ.

describe human organizations. We will therefore present these two approaches to Brazilian agroecology, focusing on agroecological coffee production. Beyond a simple analogy, we aim to develop a new understanding of social networks, focusing on the concept of "articulation".

14.2. From ecological networks

14.2.1. *Agroecosystems as complex networks*

Cultivated plants have been historically managed and domesticated within biodiverse systems composed of plants, animals and other living beings. Indeed, even after decades of policies and programs pushing for their transformation into industrialized monocultures (Kloppenburg 2005), there are still dozens of examples of diversified *agroecosystems* being practiced all over the world: rice fish culture in China, Shimbwe Juu Kihamba Agroforests in Tanzania, Ghout Oasis in Algeria, Milpa in Mesoamerica, olive agroforestry in the Mediterranean region and among many others. Even in extensive monocultures, there are organisms besides crops, such as herbivores, pollinators or soil bacteria, that can be of central importance for production.

Agroecosystems thus include cultivated plants, associated uncultivated organisms and the interactions among them, as well as their interactions with people and the environment. These interactions can be understood from an ecological perspective, which helps identify and study relations of mutualism, facilitation, competition, predation and symbiosis, among others. Interestingly, in agroecosystems many of these interactions are often regulated by human labor, through specific practices of agricultural management, which integrate the ecological knowledge, needs and cultural preferences of generations of peasants and farmers. For instance, many of the arrangements in which crops are grown help minimize plant competition for sunlight and space (Vandermeer 2009).

Interactions within agroecosystems are intricate and give rise to properties that cannot be attributed to individual elements of the system. These so-called emergent properties include some of the most important features of agroecosystems for humans, such as overall productivity and the agroecosystem's response to pests and diseases. The emergence of these properties is one of the defining characteristics of complex systems. Another hallmark of complex systems, also exhibited by agroecosystems, is their nonlinear and often unpredictable response to changes (Vandermeer and Perfecto 2017).

14.2.2. *The coffee network*

From an ecological and complex-system perspective, agroecosystems can be conceptualized as networks in which nodes represent species or functional groups (e.g. primary producers, herbivores or predators), and edges stand for the ecological interactions between nodes (Bohan et al. 2013; Benítez et al. 2014). Directed edges indicate the direction of relationships by distinguishing the source from the target. These relationships can be positive if they promote the growth of the target population (e.g. facilitation, symbiosis or mutualism) or negative if they reduce it (e.g. competition, predation or parasitism). Network models enable us to both visualize multiple interactions in a single formal representation and learn about the structure and dynamics of these otherwise hard-to-grasp agroecosystems. For instance, just by analyzing the patterns of edges among nodes, we can infer what kind of perturbations make a given network more vulnerable or, on the contrary, what nodes or links promote the persistence of the network in the face of different types of perturbations (Bohan et al. 2013; Benítez et al. 2014; Allen et al. 2022). Moreover, we can study how the size of the population changes in response to their interaction with other nodes (i.e. analyze the network dynamics in time or in space). Performing dynamic studies under different simulated conditions can help us explore management scenarios. However, beyond predicting precise quantitative outputs, network models allow us to better understand agroecosystems in their complexity and, as Richard Levins (2015) has remarked, help nourish the common sense that underlies agricultural practices.

Coffee is a widely cultivated crop and is usually cultivated in regions of high biological and cultural diversity. Although not native to the Americas, coffee is now a vital source of livelihood for millions of families across Latin America, including in Brazil (Volsi et al. 2019). Like many crops, coffee is grown along a gradient of management intensification, which is associated with a gradient of biodiversity, from highly diverse shaded plantations to large, shadeless monocultures (Moguel and Toledo 1999). The group led by Ivette Perfecto and John Vandermeer has studied ecological interaction around coffee plants in a diversified shaded plantation, home to approximately 90 tree species. After decades of empirical documentation, these interactions have been integrated into ecological networks that have helped better our understanding of this agroecosystem (Perfecto et al. 2014; González et al. 2021).

Figure 14.1 depicts a recently updated version of the ecological network associated with coffee agroecosystem (González et al. 2021). This image allows us to immediately visualize the intricate relationships among insects, fungi, birds and other organisms that interact with coffee plants in various ways. We can also observe that some nodes exhibit greater "centrality", which measures the importance or influence of a node within the network. Another interesting feature of this network is that it includes not only direct interactions but also a type of indirect interactions, known as high-order interactions (HOIs), in which an element of the network modulates the interaction between other nodes. In the network model, pseudo-nodes are introduced to represent the events, processes or actions through which these HOIs happens. These interactions shape community dynamics in ways that cannot be predicted by studying only paired interactions and are now considered crucial for understanding ecosystem stability and responses to environmental change (Kéfi et al. 2012). In the coffee network, second- and third-degree HOIs have been reported as enhancing certain aspects of the agroecosystem's robustness, particularly in response to species removal (González et al. 2021).

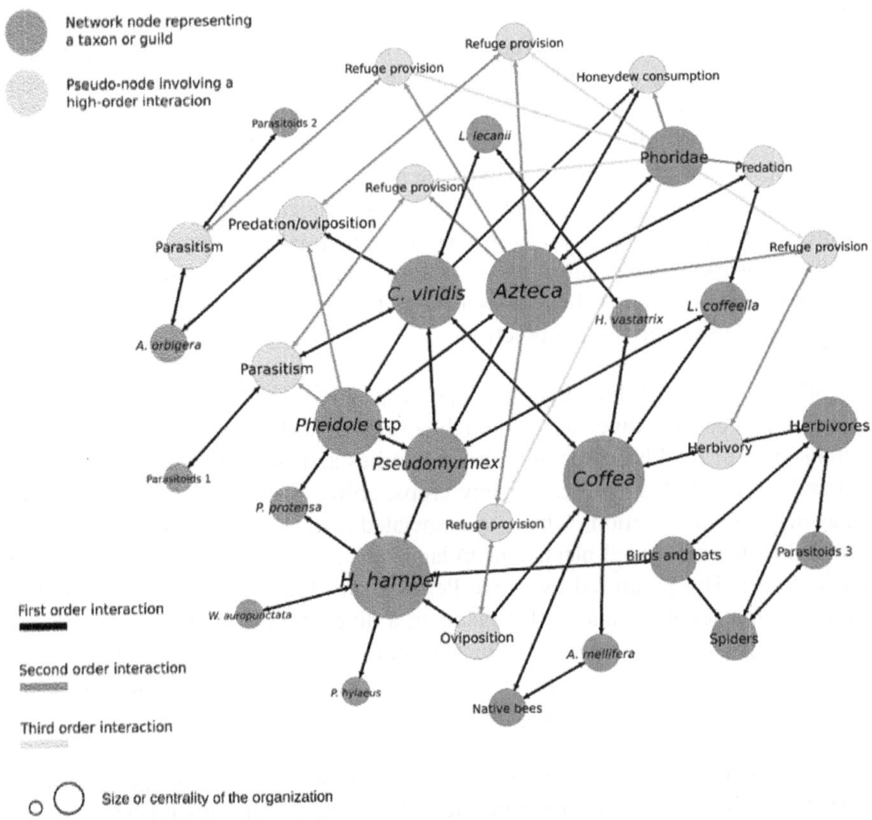

Figure 14.1. *The coffee agroecosystem model. For a color version of this figure, see www.iste.co.uk/dobigny/rethinking.zip*

14.2.3. *A way of thinking about the world*

One key lesson from network modeling in agroecosystems is that their most important features emerge from multiple, often subtle interactions. In particular, the coffee network has been studied in the context of pest control, yielding valuable insights. On the one hand, these studies support the idea that no organism is a pest per se, i.e. organisms such as the coffee rust or the berry borer only become a pest under certain circumstances in which their population is not regulated by other elements of the network. On the other hand, it is almost impossible to point to a specific element of the network that can control a given potential pest. It is the whole network's complex interactions that can explain the coexistence of all of these organisms and the fact that, under certain conditions, none of them become pests (Perfecto et al. 2014; González et al. 2021).

Understanding agroecosystems as complex networks helps us consider unforeseen consequences, nonproportional relationships between causes and effects and distributed causality. It serves as an antidote to the magic bullets that industrialized agriculture often promises: the single coffee variety that will resist a given disease, the new product that will eradicate a specific pest and so forth. In contrast, this way of thinking provides "tools for thought" in Waddington's sense (1974). It prepares us for surprises and supports agricultural managements based on species coexistence and biodiversity. These lessons are of major interest for understanding the social organization of agroecology in Brazil.

14.3. To agroecological articulations

14.3.1. *A movement of movements*

Although in Brazil there is talk of an agroecological social movement (Wezel et al. 2009), this movement has no clear boundaries and unites a diverse range of actors and institutions that collaborate to promote an alternative agricultural model. As a result, this vast network lacks a clear structure or fixed organizational chart, although a form of coordination has gradually emerged as agroecology has become institutionalized in the scientific and political landscape over the past 30 years. Thus, coordination at the federal level is mainly carried out by the *Articulação Nacional de Agroecologia* (ANA) created in 2002, which is divided into articulations in the different states of the country. The term *Articulação* is used to describe this collective organization that brings together numerous entities of different sizes and statuses (unions, cooperatives, technical centers, food or health research institutes, producer associations, NGOs, etc.) at the national, regional and local levels. It presents itself not as an entity but as "a space for articulation and convergence between movements, networks and organizations of Brazilian civil society engaged in concrete experiences of promoting agroecology, supporting family farming and building sustainable alternatives for rural development" (ANA)[1]. Other similar forms of collective organization exist in Brazil, such as the *Articulação do Semárido Brasileiro* (ASA), which was founded in the 1990s and has numerous connections with the agroecology movement. Every four years, the ANA organizes the National Agroecology Meetings (ENA), which are preceded and followed by regional agroecological meetings (ERA). These events bring together family producers, technical organizations and NGOs promoting agroecology, as well as students and academics.

Agroecology in Brazil also has a scientific dimension, and in 2004, the *Associação Brasileira de Agroecologia* (ABA) was founded on the occasion of several concomitant events in Porto Alegre (RS): the II Brazilian Congress of Agroecology (CBA), the V International Seminar on Agroecology and the VI State Seminar on Agroecology. Since then, the ABA has organized the CBA every two years, which is also the opportunity of numerous regional and local preparatory meetings with entities such as the *núcleos de estudos em agroecologia* (NEAs) usually based in higher education institutions. As a scientific association, it also promotes agroecology in universities through the publication of an interdisciplinary journal, the *Revista Brasileira de Agroecologia*, launched in 2006. As the history of the ABA's creation shows, the Brazilian agroecology movement is shaped by these regular events, which can bring together several thousand people, as well as by frequent local meetings with just a few dozen participants. These meetings, circumscribed in time, provide a tangible reality to the movement and act as connectors between actors and institutions, ultimately giving rise to new entities and organizations.

Agroecology serves as a rallying call, bringing together actors from diverse backgrounds. These meetings provide a space where new themes and slogans emerge and spread, occasionally influencing public policy (Sabourin et al. 2018). More than a movement in the strict sense of the word, it is best understood as a network interface. Agroecology can therefore be thought of as a "political operator" (Descola 2019) enabling the connection and disconnection of different entities and rural social movements, such as the *Movimento dos Trabalhadores sem Terra* (MST), *Via Campesina*, or even trade union confederations (CONTAG, CUT, etc.) and organizations defending the rights of indigenous peoples and traditional communities (CONAQ, etc.). All of them are united in their resistance to a dominant model of monoculture – both literal and figurative – that pervades Brazilian public space and territory. They constantly seek to include new allies in a form of convergence of struggles, including urban movements, where agroecology aligns

1 See: https://agroecologia.org.br/o-que-e-a-ana/.

with other causes such as the defense of minorities and the recognition of gender identities.

14.3.2. *Articulation versus network*

Inspired by the modeling of an agroecosystem presented earlier, Figure 14.2 offers a rough graphic representation of the agroecological movement. The dark grey circles represent the entities that make up this movement, varying in size and nature. The light grey circles represent agroecological events, processes and actions which are of varying sizes and just as important as the entities whose interaction they enable or modulate. In addition, the links between the different actors in the movement also vary in nature and can be classified as rank 1, if direct, or rank 2, if indirect. Extending the comparison with agroecosystems, we can consider that the strength and resilience of this movement – which has been able to resist the dismantling of agroecological public policies since 2016 – is certainly due to the density of the indirect or intermittent ties (Granovetter 1973; Gefen and Laugier 2020) that connect these entities. These interconnections provide the movement with great agility and flexibility while making responses to change less predictable.

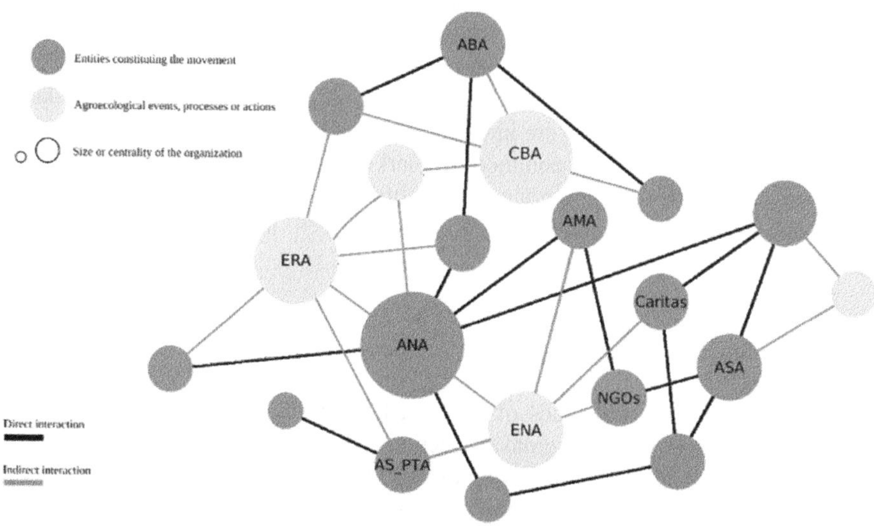

Figure 14.2. *Simplified representation of the Brazilian agroecological movement network*

What stands out in agroecological meetings, as well as in the preparatory and review sessions for these events, is the very inclusive and horizontal nature of the relationships that prevail. For instance, agronomists or scientists never position themselves as experts in front of farmers. Decisions – especially regarding the choice of themes and organizational matters – are always made collectively. Everything is debated and chosen together, giving this vast collective an "acephalous" character, to borrow a classic anthropological term (Evans-Pritchard 1968). Of course, an outside observer who actively participates will eventually identify the influencers and leaders – those whose words carry more weight and who guide collective action – although they never openly claim this role. This fluidity of relationships and style of leadership is reflected in social networks such as WhatsApp, which are widely used in Brazil. These digital spaces facilitate the circulation of information within the vast movement, helping to maintain connections between actors and ensure the homogenization of discourse, despite geographical distances.

Another notable characteristic of this movement is its distinctly feminine character, with women making up a large proportion of the teams. Gender awareness is strong, with frequent references to the patriarchal, macho and racist aspects of Brazilian society (Hillenkamp 2022). In this context, the agroecological social movement exemplifies a different kind of networking, one that is more feminine, inclusive, polycentric and acephalous. This is also reflected in the term "articulation" (*Articulação*), while the notion of "network" (*rede*) would be more masculine, hierarchical and possibly even military. Anthropologists have already applied this concept to describe other rural realities in Latin America, such as Enrique Mayer's work in the Andes: "Hence, the household members are articulated peasants, in the sense that they are interrelated ('articulated') with other households, their communities, and the commodity markets" (Mayer 2002, p. xiii). This definition underscores the importance of entities of different sizes – individuals, communities and even markets – as we have seen in the case of Brazilian agroecology.

14.4. Conclusion: the ecological art of networking

In his work at the crossroads of biology, humanities and social sciences, Tim Ingold has developed an original conception of the environment, aiming "to conceive of it no longer as what surrounds us (what is "out there" and not "in here") but as a zone of interpenetration within which our lives and those of others intertwine into a homogeneous whole" (Descola and Ingold 2014, p. 19). Such an "ecology of life" allows us to consider how each person carries with them wherever they are a certain type of connection to the environment, particularly with the plants, animals and surroundings in which they grew up or are accustomed to living. So, when Brazilian women farmers, living several thousand kilometers apart, meet up at an agroecological gathering, they bring with them, in a more or less visible and conscious way, their entire (agro)ecosystem. It is not on this "cosmopolitical" aspect

(Carcelle 2024), however essential, that we wanted to focus on this work. Rather, our focus is on the very form of the organization of the agroecological movement. Ingold himself is very skeptical of the concept of "network", preferring the notion of "meshing" to describe social organizations as well as individual activities. For our part, we emphasize the notion of "articulation", which is another way of conceptualizing networks, based on a term used by the actors themselves in agroecology.

Without falling into the trap of trying to superimpose models and schematizations derived from biology onto social facts – of which Ingold himself is critical – we have sought to draw inspiration from recent contributions in the sciences of complexity applied to agroecosystems to renew our approach to alternative social movements. Like an agroecological coffee plot with numerous interspecies interactions, the agroecological movement is realized above all in the meeting spaces between the different entities that comprise it. If in an agroecosystem we can observe relationships of predation as well as symbiosis, in the Brazilian agroecological movement, which is itself an interface of networks, these interactions tend to be horizontal and inclusive, revealing a certain art of networking that corresponds to a deep expectation of many of our contemporaries.

14.5. Acknowledgements

The authors acknowledge the editors for their proposal to be part of this book. S.C. was supported by ART-Dev CIRAD. M.B. was supported by UNAM PASPA – DGAPA.

14.6. References

Allen, W.J., Bufford, J.L., Barnes, A.D., Barratt, B.I., Deslippe, J.R., Dickie, I.A., Goldson, S.L., Howlett, B.G., Hulme, P.E., Lavorel, S. et al. (2022). A network perspective for sustainable agroecosystems. *Trends in Plant Science*, 27(8), 769–780.

Altieri, M. (1995). *Agroecology: The Science of Sustainable Agriculture*. Westview Press, Boulder, CO.

Bohan, D.A., Raybould, A., Mulder, C., Woodward, G., Tamaddoni-Nezhad, A., Bluthgen, N., Pocock, M.J., Muggleton, S., Evans, D.M., Astegiano, J. et al. (2013). Networking agroecology: Integrating the diversity of agroecosystem interactions. *Advances in Ecological Research*, 49, 1–67.

Carcelle, S. (2024). *Cosmopolitique de l'agroécologie. Enquête au nord du Minas Gerais (Brésil)*. Mimesis, Milan.

Descola, P. (2019). De la nature universelle aux natures singulières. Quelles leçons pour l'analyse des cultures ? In *Les natures en question*. Odile Jacob, Paris.

Descola, P. and Ingold, T. (2014). *Être au monde. Quelle expérience commune ?* Presses Universitaires de Lyon, Lyon.

Evans-Pritchard E.E. (1968). *Les Nuer. Description des modes de vie et des institutions politiques d'un peuple nilote*. Gallimard, Paris.

Gefen, A. and Laugier, S. (eds) (2020). *Le pouvoir des liens faibles*. CNRS Éditions, Paris.

Gibson, J. (1979). *The Ecological Approach of Visual Perception*. Mifflin and Company, Houghton.

González, C.G., Van Cauwelaert, E.M., Boyer, D., Perfecto, I., Vandermeer, J., Benítez, M. (2021). High-order interactions maintain or enhance structural robustness of a coffee agroecosystem network. *Ecological Complexity*, 47, 100951. doi: 10.1016/j.ecocom.2021.100951.

Granovetter M.S. (1973). The strength of weak ties. *American Journal of Sociology*, 78(6), 1360–1380.

Haudricourt, A.G. (1962). Domestication des animaux, culture des plantes et traitement d'autrui. *L'Homme*, 2(1), 40–50.

Hillenkamp, I. (2022), Dans la forêt, le genre. Agroécologie et féminisme dans un environnement sous tension au Brésil. *Journal des Anthropologues*, 168–169, 79–95.

Ingold, T. (1986). *The Appropriation of Nature: Essays on Human Ecology and Social Relations*. Manchester University Press, Manchester.

Kéfi, S., Berlow, E.L., Wieters, E.A., Navarrete, S.A., Petchey, O.L., Wood, S.A, Boit, A., Joppa, L.N., Lafferty, K.D., Williams, R.J. et al. (2012). More than a meal… integrating non-feeding interactions into food webs. *Ecology Letters*, 15(4), 291–300.

Kloppenburg, J.R. (2005). *First the Seed: The Political Economy of Plant Biotechnology*. University of Wisconsin Press, Madison.

Levins, R. (2015). *Una pierna adentro, una pierna afuera*. CopIt ArXives, Mexico City.

Mayer, E. (2001), *The Articulated Peasant. Household Economies in the Andes*. Westview Press, Boulder, CO.

Moguel, P. and Toledo, V.M. (1999). Biodiversity conservation in traditional coffee systems of Mexico. *Conservation Biology*, 13(1), 11–21.

Perfecto, I., Vandermeer, J., Philpott, S.M. (2014). Complex ecological interactions in the coffee agroecosystem. *Annual Review of Ecology, Evolution, and Systematics*, 45(1), 137–158.

Sabourin E., Le Coq J.-F., Fréguin-Gresh S., Marzin J., Bonin M., Patrouilleau M.M., Vázquez, L., Niederle, P. (2018). Public policies to support agroecology in Latin America and the Caribbean. *Perspective*, 45, 1–4. doi: 10.19182/agritrop/00020.

Vandermeer, J.H. (2009). *The Ecology of Agroecosystems*. Jones & Bartlett Publishers, Burlington.

Vandermeer, J.H. and Perfecto, I. (2017). *Ecological Complexity and Agroecology*. Routledge, New York.

Volsi, B., Telles T.S., Caldarelli C.E., Gabardo da Camara, M.R. (2019). The dynamics of coffee production in Brazil. *PloS One*, 14(7). doi: 10.1371/journal.pone.0219742.

Waddington, C.H. (1974). Tools of thought about complex systems. *Ekistics*, 28, 9–18.

15

Toward an Ecology of Architectural Experimentation Beyond Environmental Regulatory Standards

Science produces knowledge to mobilize the world, technology extends and multiplies action, and art evokes aesthetic emotion. Architecture seems to embody all of these at once. It is knowledge, it is technology, it is the pursuit of beauty[1]

Michel CALLON

15.1. Introduction

In 2021, the European Commission launched the "Fit for 55" package as part of its broader strategy to combat climate change. This legislative initiative sets a legally binding goal for the EU to cut net greenhouse gas emissions by at least 55% by 2030. Achieving these EU-wide targets raises important questions about how local governments can reconcile top-down climate directives with bottom-up, context-specific and experimental approaches.

In France's construction sector, the gradual increase of thermal and environmental regulations (RT 2012 and RE 2020) has led to a proliferation of performance certification protocols in the building sector. These protocols aim to promote an exemplary model of environmental efficiency, integrated into the various stages of project design.

Chapter written by Hector DOCARRAGAL MONTERO.

1 Callon (1996, p. 31).

In response to this increasing dominance of environmental standards, a new local architectural trend is emerging – one that explores innovative, frugal, transgressive and often "out-of-standard" ecological practices. Initiatives such as La Preuve par 7, Bellastock, Le Grand 8, Méandre Etc, Plateau Urbain and Le Hangar Zéro exemplify emerging creative possibilities within experimental frameworks. Drawing on what Michel de Certeau termed "tactics of resistance", these practices challenge the prevailing model of "green" standards in favor of alternative approaches to designing environmentally efficient buildings. However, in the absence of a suitable regulatory framework, these projects often evolve intuitively, operating on "trickery" (Violeau 2004, pp. 147–166). and inventive margins of normative structures in order to accelerate implementation.

The growing interest of the French State in such experimental strategies is reflected in the institutionalization of new protocols that allow exemptions from standard building regulations, thus encouraging innovation. These can already be observed in laws such as Law on the Freedom of Creation, Architecture and Heritage (LCAP)[2] and Law for a State at the Service of a Trusted Society (ESSOC)[3], which aim to "simplify construction rules" to "facilitate project realization and foster innovation[4]". This experimental framework lends legitimacy to the ongoing debate about the need for a legal system that can rethink environmental performance beyond the prescriptive boundaries of the *Code de la construction eut de l'habitation* (building code for housing), the foundation of French building regulation. However, what are the implications when the law itself legitimizes exemption – effectively turning deviation from established norms into a new standard practice?

To address this question, this chapter examines whether recent policy developments genuinely support a non-conventional, "out-of-standard" architectural practice – one more capable of responding to the complex social, technical and environmental challenges of the ecological transition. To this end, this chapter proposes a rigorous methodology for analyzing the multidimensional character of architectural projects. First, it traces the genealogical emergence of public support for architectural innovation, particularly through the validation of experimental frameworks such as the LCAP and ESSOC laws. Second, it introduces an original ethnographic methodology – CAST (Context, Actors, Strategies, Theme) – designed to facilitate a multidimensional reading of architectural practices and to recognize the legitimacy of unconventional approaches that extend beyond the scope of these legislative initiatives. Finally, this chapter demonstrates how these experimental practices seek to redefine the material foundations of an ecology of

2 Law on the Freedom of Creation, Architecture, and Heritage of July 7, 2016.

3 Law on the State in the Service of a Trust-Based Society of August 11, 2018.

4 Ministry in Charge of Housing, DGALN/DHUP/QC1, 2020.

"out-of-standard" experimentation in architecture, ultimately challenging and redefining established environmental paradigms in construction across various European contexts[5].

15.2. Genealogy of the regulations framework for architecture: between overabundance and transgression of environmental standards

In 1987, following the initial formulation of the concept of "sustainable development" in the United Nations' Brundtland Report, *Our Common Future*, which was later reaffirmed in various Earth Summits and agreements (Rio, 1992; Johannesburg, 2002; Paris, 2015), a regulatory framework was established to systematically guide the sustainable growth of nations. In the building sector, this encouraged the enhancement of the energy performance of buildings to reduce the environmental impact of construction.

In France, from the laws and measures introduced under the Grenelle Environmental Forum (2009–2010) to the adoption of the Energy Transition for Green Growth Act (LTECV) followed by the National Low-Carbon Strategy (SNBC) in 2015, the strengthening of standards and regulations in the construction sector, notably with the latest environmental regulation, RE 2020, represents the concrete implementation of this global ecological framework. This regulation critically impacts the practice of architectural projects by enforcing the newly established environmental standards and advocating for environmental efficiency based on measurable and predictable reports.

Today, such regulatory frameworks require architects to predefine the spatiality of their projects based on imposed and standardized technical solutions. In this context, the growing recognition of the ecological, technical and social complexities of environmental transition calls for a critical reassessment of the negative impacts these regulations have on architectural design.

In 2015, the conference series "Hors la loi, pour dépasser la loi", organized at the Maison de l'Architecture Île-de-France during the COP21 summit in

5 In 2024, I successfully defended my doctoral thesis, titled *The Transgression of Standards in Architecture and the Evolution of Environmental Regulations: The Impact of the LCAP and ESSOC Laws,* at Université Paris 1 Panthéon-Sorbonne. My research seeks to develop an ethnographic methodology for analyzing the multidimensional nature of architectural projects, thereby legitimizing "out-of-standard" practices challenging and transcending established environmental frameworks.

Paris[6], questioned the effectiveness of environmental norms, regulations and standards that shape architectural practice, advocating for their reconsideration and transcendence.

This raises a crucial question: are the numerous regulations and legislative initiatives aimed at fostering an ecological transition in the construction sector truly commensurate with the challenges of climate change as identified by the international scientific community (COP: Conference of the Parties and IPCC: Intergovernmental Panel on Climate Change)[7]?

Beyond environmental standards – often derived from international agreements and difficult to adapt to local ecological contexts – it is crucial to recognize the capacity of designers, such as architects and urban planners, as well as users, to propose inventive, experimental and at times complex or problematic responses. These evolving practices – centered on bio-based materials, reuse and the frugality of traditional and vernacular techniques – are already challenging and reshaping the prevailing environmental standards in architecture. Ultimately, these approaches reflect a complex, emerging ecological imagination, offering an alternative to the efficiency-driven model promoted by the dominant environmental engineering paradigm within the architectural profession.

In response to these architectural concerns, the French government is seeking to encourage regulatory "exemptions". The enactment of the LCAP in 2016 and the ESSOC in 2018 are significant examples, demonstrating the extent of these architectural explorations. Notably, under the 2018 ESSOC law, the government introduced a validation protocol for "solutions with equivalent effect" (SEE) to existing regulations. This framework enables architects to validate alternative operational methodologies that deviate from conventional rules, which are often seen as inadequate for addressing contemporary environmental challenges.

Although the French government acknowledges the urgency and complexity of revising regulatory frameworks, current public policy still struggles to accommodate the spontaneous and intricate operational methods of architects and designers working in experimental contexts (see Figure 15.1). Several factors explain this difficulty.

---

6 A series organized by the Institut pour la Conception Éco-Responsable du Bâti (ICEB) and the Collectif Démarche Durable (CO2D), held from September 28, 2015, to December 8, 2015 (Online). Available at: www.asso-iceb.org (Accessed June 2022).

7 The latest IPCC report, released on March 20, 2023, confirms the alarming outlook on climate change. In France, the construction sector alone is responsible for approximately 123 million tons of CO_2 emissions per year, accounting for 44% of the country's total emissions (Online). Available at: www.ecologie.gouv.fr (Accessed 10 February 2025).

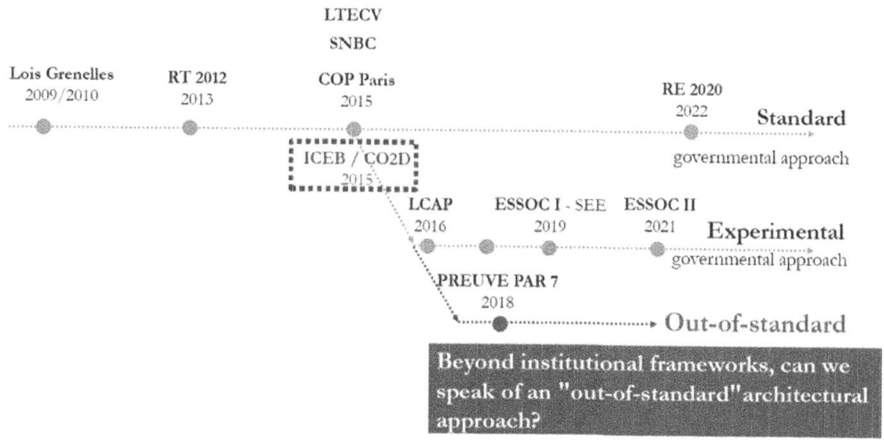

Figure 15.1. *Genealogy of the regulations framework for architecture in France. For a color version of this figure, see www.iste.co.uk/dobigny/rethinking.zip*

First, contemporary architectural design and construction unfold within a highly intricate environment. Projects must navigate not only technical and physical constraints but also social and cultural dimensions. This complexity introduces additional uncertainty into the design process, which increasingly depends on interdependent production chains and ongoing negotiations between public and private actors. Within this evolving landscape, new forms of "out-of-standard" architectural experimentation are emerging. These practices challenge the bureaucratic underpinnings of the LCAP and ESSOC laws, as well as the regulatory frameworks they have helped establish. Their emergence invites critical reflection on the relevance of applying uniform green engineering standards to architecture.

Second, while France's RE 2020 regulation is intended to offer a comprehensive framework for assessing environmental performance – by considering buildings as integrated systems of materials and energy flows – it also places heavy emphasis on quantifiable metrics, particularly carbon assessments aligned with the EU's *Fit for 55* climate targets. However, this focus should be understood within a broader historical trajectory of regulatory development, including earlier thermal regulations (e.g. RT 1974), major legislative reforms (such as Grenelle 2010 and LTECV 2015) and standardization processes rooted in industrialization (e.g. AFNOR 1926; CSTB 1947; HQE 1996). Together, these milestones have shaped the conceptual and regulatory environment in which today's architects operate.

Finally, could these emerging "out-of-standard" practices – grounded in the complex realities of social, technical and environmental conditions – offer a more effective path for ecological transition than the generic, prescriptive model, exemplified by RE 2020 and its historical lineage of rigid regulations?

Rather than attempting to regulate unconventional practices through ineffective exemptions within traditional frameworks, it is crucial to develop a more adaptive approach – one that embraces the fluidity of decision-making, inherent uncertainties and the ongoing integration of regulatory standards into construction objectives. In this context, identifying the "out-of-standard" elements in architectural design is key to crafting a more comprehensive strategy that uncovers the creative nuances of regulatory contexts. To truly understand the unpredictable nature of this reality, we propose formalizing an original, tailored methodology that fosters the rediscovery of the fertility of the contextual, the specific and the local, while aligning with the complexity of today's environmental challenges in construction.

15.3. Multicriteria analysis of experimental practices in architecture using CAST ethnographic method

The analytical study of architecture overlooks the concrete nature of the environment[8]

Christian NORBERG-SCHULZ

This section draws on an ethnographic study of architectural projects that diverge from standardized procedures. Conducted as part of my doctoral research, the study applies a specialized methodology to examine 27 contemporary architectural projects, either recently completed or still underway at the time of analysis. In the context of the ecological transition in France's construction sector, these unconventional practices appear to be deeply rooted in their specific local conditions, rather than strictly adhering to regulatory frameworks such as RE 2020 or experimental provisions such as ESSOC and LCAP.

To grasp the operational logic of these projects, it is helpful to turn to sociologist Daniel Bertaux, who argues that in experimental fields, no single project classification can fully capture the nuances of a specific issue. He proposes that architects' decisions regarding regulatory norms are guided by a principle of "differentiality" (Bertaux 1997, p. 23) – the notion that individuals in the same institutional role may approach their duties in vastly different ways due to distinct personal structures. So, how can we develop an operational methodological framework that accounts for this complex and heterogeneous reality?

8 Norberg-Schulz (1981, p. 5).

To address this challenge, empirical data from the 27 projects were collected and structured using the narrative interview method, or "life story" approach, which draws on the tradition of biographical sociology as developed by Bertaux (1997). This approach was further supported by the development of an original multicriteria classification system, organized around four key analytical axes, designed to assess the diverse contextual factors that influence experimental design processes:

– *C*: *Context* (Experimental Framework).

– *A*: Actors (Stakeholders).

– *S*: Strategy for challenging Norms.

– *T*: Theme (Environmental Theme).

The interdependencies among the various experimental variables demonstrate that a project cannot be reduced to a single analytical criterion. Rather, it must be understood within a structured operational framework capable of systematically organizing and classifying experimental data (see Figure 15.2). Our methodology draws inspiration from studies in the anthropology of Science and Technology Studies (STS), the Actor-Network Theory (ANT)[9] developed in France by Madeleine Akrich, Michel Callon and Bruno Latour in 2006.

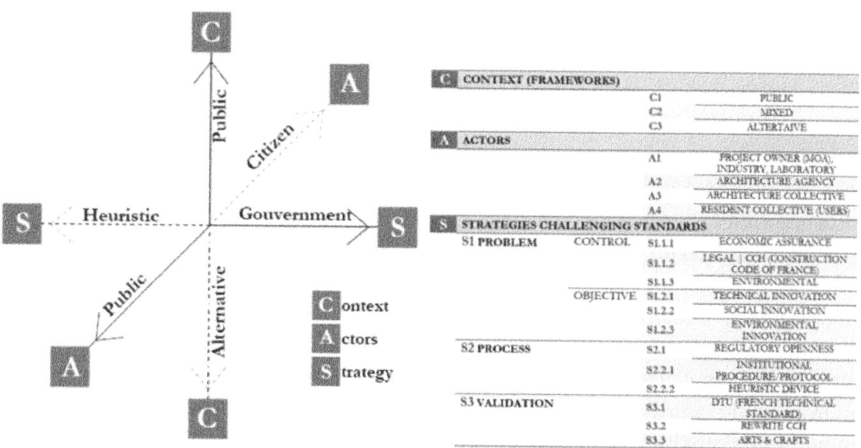

Figure 15.2. *Summary of CAS data: context – actors – strategy. For a color version of this figure, see www.iste.co.uk/dobigny/rethinking.zip*

9 Actor-Network Theory emerges within the broader renewal of the sociology of science and technology. From the 1970s onward, this "new" sociology has challenged the classical approach, which views science as a rational activity that generates paradigms through the systematic testing of working hypotheses.

Applying the ANT method in the CAST creation enables an empirical analysis of decision-making within the working environment, examining how this context influences the design process, its content and its evolution. In our study, we conceptualize this work as a system where an input resource (INPUT – regulatory expertise) produces an output resource (OUTPUT – new experimental performance). The data generated through the CAST multicriteria interact with and influence this linear INPUT/OUTPUT sequence, forming a complex network of interdependencies (see Figure 15.3).

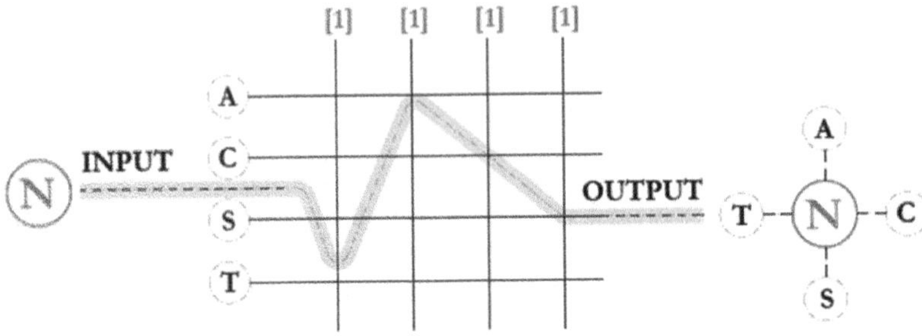

Figure 15.3. *Network CAST diagram (© Hector Docarragal Montero). For a color version of this figure, see www.iste.co.uk/dobigny/rethinking.zip*

The CAST framework introduces a novel epistemological system for analyzing the complexities of architectural design processes in relation to environmental contexts. Moreover, the interdependencies among various experimental coordinates underscore that a project cannot be defined by a single analytical criterion. Instead, it must be understood within an operational framework of systematically organized and classified experimental data. This specific representation of the project process – as a sequential chain of "micro-processes" – aims to develop a methodological tool for dynamically analyzing architectural design processes. These representations have supported the documentation of extensive empirical investigations, including the "life stories" developed for each of the 27 projects. They have also played a vital role in shaping a coherent evaluation framework capable of assessing the environmental impact of experimental approaches while enabling comparative studies – one of which is presented here.

As an illustration, this chapter applies the CAST framework to compare the approaches of two projects. These projects were selected for their unconventional trajectories in responding to standardized regulatory frameworks such as RE 2020, as well as the experimental provisions of the ESSOC and LCAP laws. Accordingly, we take a closer look at the day-to-day practices involved in the "Le Onze" residential project (No. 01) in Chartres and the "La Vallée" school project (No. 02) near Paris. The comparative analysis of these two cases aims to establish an initial, consistent evaluation framework for measuring the environmental impact of experimental uses of recycled aggregates in concrete (OUTPUT) while comparing these practices to the requirements of the NF EN 206/CN standard regarding the composition of recycled aggregates (INPUT). We begin by presenting a summary table of data *[CAST]* from the two multidimensional production contexts.

– *[C] Experimental Framework*: public competition for both projects, with the "Permis d'Innover" (AMI, LCAP law framework) for Project No. 01 and Public competition (MOP)[10] for Project No. 02.

– *[A] Actors*: an ecosystem of stakeholders, including public entities (Municipal governments and constructors), research laboratories (CERIB for Project No. 01), as well as an architectural firm (Samuel Demas for Project No. 02).

– *[S] Strategy for challenging norms*: LCAP law experimental framework for Project No. 01 and a creative interpretation of DTU standards for masonry walls in Project No. 02.

– *[T] Environmental Theme*: use of concrete incorporating recycled aggregates from construction sites for both projects.

Then, through a sequential analysis of the experimental data [C, A, S, T], we demonstrate that each experimental approach follows a distinct trajectory, within which the EN 206/CN standard is reconsidered in context. From this perspective, the experimental approach initiated by CERIB (Project No. 01), although left incomplete within the public exemption framework (LCAP law) *[C1]*, found partial application in the local real estate market of Chartres with the construction of the "Le Onze" residence, delivered in 2020 *[C3]*. The use of recycled aggregate concrete in this project paves the way for new collective planning scenarios and contributes to structuring a local economic dynamic. Similarly, the "La Vallée" school complex project (Project No. 02), led by the municipality of Châtenay-Malabry *[A1]* and Eiffage's deconstruction division within a public framework *[C1]*, under the direction of the Samuel Delmas architecture firm *[A2]*, exemplifies a rethinking of the project's initial parameters, resulting in an unexpected yet more contextually fitting solution. The facade's expressive design, originally intended to be built with rammed earth but ultimately realized in dry-packed concrete incorporating 100% recycled aggregates and earth, strikes a balance between technical innovation and a regionally integrated approach.

These experimental approaches align with the environmental objectives of RE 2020 by promoting reversible construction strategies that support life cycle analysis (LCA) and carbon footprint reduction. However, an analysis of the two

10 Public competitions: Appel à Manifestation d'interet (AMI) and Maitrise d'ouvrage publique (MOP).

diagrams (see Figure 15.4) through a narrative thread reveals that beyond regulatory frameworks, experimental work – by generating a flow of empirical data unique to each exploratory context – can more effectively reduce environmental impact, lower costs and promote a circular economy compared to the standard approach.

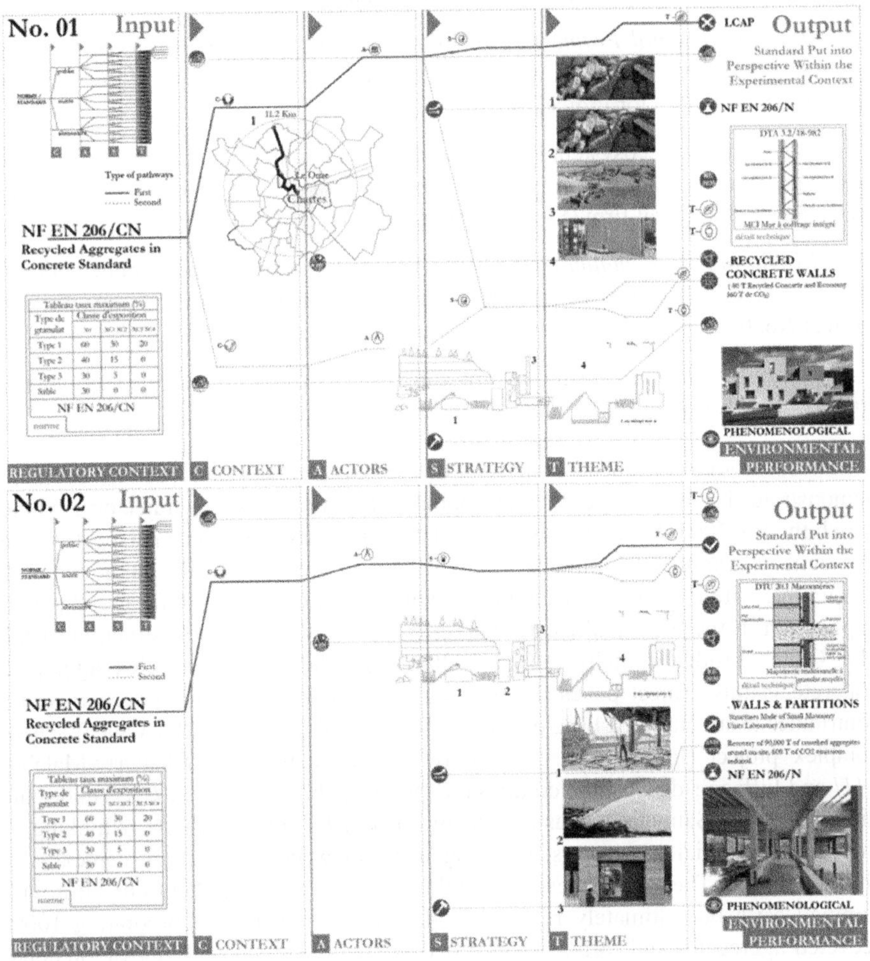

Figure 15.4. *Comparative representations of environmental approaches based on CAST data (Project no. 01 © Arnaud Lombard; Project no. 02 © a+ samueldelmas architects). For a color version of this figure, see www.iste.co.uk/dobigny/rethinking.zip*

Ultimately, unlike bureaucratic exemption mechanisms – such as the "*Solution d'Effet Équivalent*" under the ESSOC law – or the environmental regulatory ambitions of RE 2020, which often fail to accommodate the spontaneous experimental dynamics of innovative projects, the "out-of-standard" projects analyzed using the CAST method embrace the inherent diversity of exploratory contexts, adapting to the specific variables of each project. Integrating project methodologies within a multidimensional analytical framework enables the validation of more complex, context-specific solutions. Thus, this methodology opens new pathways for architectural innovation in response to the urgent challenges of ecological transition.

15.4. From multicriteria analysis to comparative synthesis

To conclude, this study presents a summary of findings based on ethnographic monitoring of "out-of-standard" practices. In addition to the two projects discussed in the previous section – Cerib (Project No. 01) and Samuel Delmas (Project No. 02) – this section presents six selected experimental trajectories, each analyzed using the CAST framework. All six recent architectural initiatives share a unifying characteristic: the innovative use of bio-based, low-carbon ecological materials, implemented within a circular economy paradigm *[T]*. These case studies were carried out across three distinct experimental contexts – public, hybrid (public–private) and alternative frameworks *[C]*. Each project is analyzed through the lens of its principal actors *[A]*, drawing on personal "life story" narratives to reveal how these initiatives developed exemplary experimental practices that critically engage with and challenge conventional environmental standards *[S]*.

In the public context, we identified two notable projects:

– *(Project No. 01)*. The research project on "green concrete" led by the Industrial Technical Center for Concrete Research (CERIB) as part of the 2018 "Permis d'Innover" public competition (AMI).

– *(Project No. 02)*. The "green concrete" project by the architectural firm Samuel Delmas, also developed for the public competition (MOP law) for the "La Vallée" public school in Châtenay-Malabry.

In a mixed partnership context, we selected:

– *(Project No. 03)*. The renovation of the Halle de Colombelles, led by the architectural firm Encore Heureux and the engineering office Albert & Compagnie.

– *(Project No. 04)*. The renovation of the Maison des Canaux carried out by the cooperative architecture firm SCOP Grand 8 in Paris.

Finally, within an alternative context, we chose:

– *(Project No. 05).* The thermal renovation of a building using wood and straw, led by the architectural firms Trait Vivant and LandFabrik, located on Rue de la Convention in Paris.

– *(Project No. 06).* The Chamarel cooperative housing project in Vaulx-en-Velin.

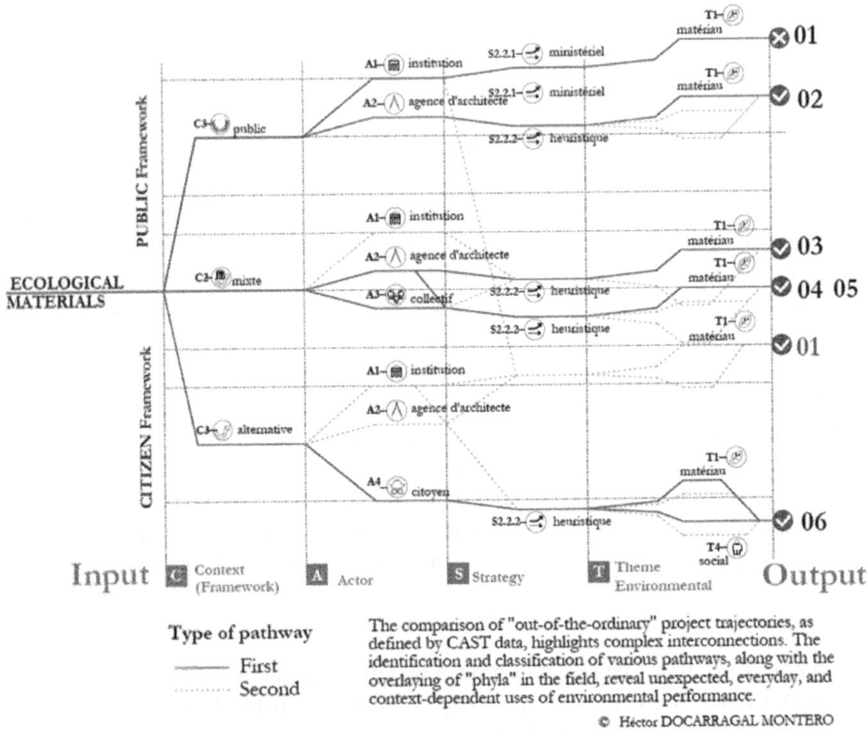

Figure 15.5. *Summary of the six INPUT/OUTPUT pathways based on the CAST framework. For a color version of this figure, see www.iste.co.uk/dobigny/rethinking.zip*

Across these diverse project trajectories, various experimental routines integrated a broad range of empirical resources – actors, regulations, materials and technical as well as normative knowledge – within the complex realities of their respective contexts. These elements, as outlined by key actors, were systematically identified, categorized and analyzed using our CAST framework. This third section aims to synthesize the empirical findings and demonstrate how these multidimensional project processes have revealed a more nuanced validation context for ecological materials than the regulatory framework established by RE 2020 and the ESSOC and LCAP laws. Below is a synthesized representation of the six interwoven trajectories (INPUT/OUTPUT).

An analysis of the comparative diagram of experimental pathways (Figure 15.5) reveals multiple transfers of performance strategies between project scenarios, following intersecting trajectories. Notably, the feedback and validation of the professional straw construction standards (CP, 2012) in the Chamarel participatory housing project (delivered in 2017) – a collective of residents organized as a Participatory Housing Association Chamarel-Les Barges *[C3, A4]* – later served to validate the strategy *[S.2.1]* for the high-rise wood-straw external thermal insulation (ITE) project on Rue de la Convention in Paris (delivered in 2021) *[S.2.2]*, led by the public social housing provider Paris Habitat *[C2, A1]*. For these projects, the CAST framework effectively integrates project trajectories within the construction sector, driving the transformation of locally sourced natural materials (wood-straw) *[T1.1.2]* into an asset for ecological and social action.

In addition, we observe other transfers of experimental resources through the CAST framework (see Figure 15.6). The renovation of the Halle de Colombelles (Project No. 03, delivered in 2019) involved the development of a protocol for characterizing and validating reclaimed materials *[T.1.2.3]*, known as "Lot Zéro" *[S.2.1]*. This protocol, validated by a diverse ecosystem of actors *[A3]* within the experimental setting of a third place *[C2]*, was subsequently transformed into technical knowledge capital for the renovation of the Maison des Canaux in Paris (Project No. 04, delivered in 2021). These findings reveal that the paralogical evolution of experimental trajectories in the field has generated multiple validation pathways, thereby expanding the range of shared potential project scenarios *[S.2.2]*.

The "fresque" above illustrates the diversity of daily work scenarios uncovered through our analysis of six selected projects. A key takeaway from this cross-analysis is the breakdown of heterogeneous experimental approaches into a multitude of coded data points, which can be classified and compared using the *[CAST]* framework (Experimental Context, Actors, Strategy for Challenging Norms, Environmental Theme). Each numerical label assigned during our ethnographic exploration represents a "micro-history". Collectively, these empirical and biographical – often spontaneous – data points, once gathered and analyzed, have become essential for understanding the experimental development trajectory of the projects.

More specifically, this approach has enabled the validation of an alternative, context-driven architectural framework that extends beyond basic regulatory objectives.

Figure 15.6. *Classification of experimental information transfers between projects (Project no. 01 © Arnaud Lombard; Project no. 02 © a+ samueldelmas architectes; Project no. 03 © Encore Heureux © WIP; Project no. 04 © Grand8, © Ville de Paris, © Bellastock; Project no. 05 © LandFabrik, © Edouard Vermès; Project no. 06 © Arketype Studio Architect). For a color version of this figure, see www.iste.co.uk/dobigny/rethinking.zip*

Based on these results, we can draw several scientific conclusions regarding our approach. First, the comparative analysis of these six heterogeneous projects, each involving complex implementation processes, has facilitated a deeper understanding of design complexities, the fragmentation of experimental trajectories, the multiplicity of choices and methodological cross-adaptations between projects in the field. Furthermore, our four-dimensional CAST analysis method has proven to be an effective framework for identifying, classifying and understanding a wide range of unexpected experimental resources within architectural project development.

Finally, we can conceptualize the essence of the experimental framework as a four-dimensional space defined by the coordinates C, A, S and T (Context[s], Actor[s], Strategy[ies], Environmental Theme[s]) (see Figure 15.7).

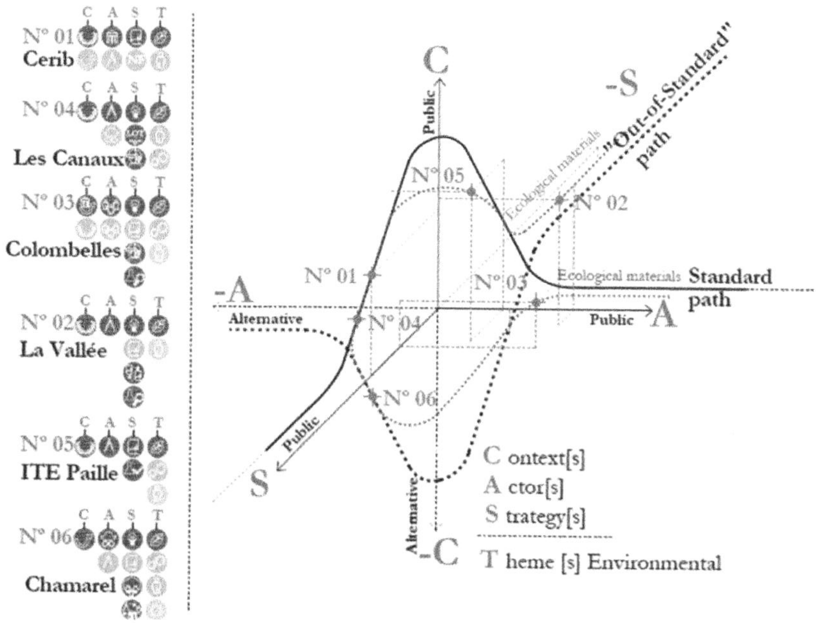

Figure 15.7. *CAST reference framework for the contextual analysis of projects*

This four-dimensional framework aims to reflect the diversity of perspectives through which, depending on circumstances, the work of designers is understood. From the standpoint of architectural design, the notion of "out-of-standard" takes on an "anthropogenic function of spatialization" tied to the understanding and assimilation of data from exploratory contexts, as expressed by Choay (2006, p. 10).

Ultimately, this design approach fosters the development of a complex ecological vision, deeply rooted in the everyday experiences of architects who navigate between transgression and the integration of evolving environmental requirements.

15.5. Toward an ecology of "out-of-standard" experimentation in architecture?

In conclusion, we outlined several avenues for further development to broaden and deepen our research on a continental scale. In the first part of our study, we showed that the harmonization of European standards has resulted in a highly complex bureaucratic system, particularly within the field of architectural design. The European Parliament and the European Commission issued a variety of legislative instruments – including regulations, directives and recommendations – such as the "Fit for 55" package, to advance the objectives set out in the EU treaties. However, in the construction sector, a binding regulatory framework is still lacking. Instead, non-mandatory directives on construction products – such as Directive 89/106/EEC of 1988 – have led to the development of the Eurocodes[11]. These codes aim to mitigate regulatory fragmentation across Europe by offering a unified set of technical standards for the design, dimensioning and verification of buildings and civil engineering structures, recognized as European benchmarks.

Confronted with a global context in which binding sustainability standards are imposed on local practices, our ethnographic approach – centered on "life stories" – sought to foreground the personal narratives of practitioners rather than institutional discourse. As this framework evolved, we began to critically examine the dominant, officially sanctioned eco-normative model that currently shapes project development. In doing so, the notion of the "out-of-standard" emerges to refer to an implicit world – one that extends beyond the predefined goals of normative, solution-oriented sustainable engineering that currently prevails in the construction sector. However, this experimental realm, which breaks from established norms, is inherently elusive. It relies on a sensitivity to the unexpected and a continuous openness to what has yet to emerge.

Within this context, the approach proposed in this chapter offers a shared foundation for better interpreting the fragmented nature of contemporary architectural practice. In our research, the project narrative is not treated as fiction, but rather as an exploratory tool – one that seeks to create points of convergence and generate resonance among diverse project situations, actors and strategies. Through this lens, we developed a comparative framework for analyzing the everyday realities of "out-of-standard" projects, particularly at the scale where the boundaries between local and global influences are increasingly blurred. Building on these premises and the methodological questions raised by our research, it is now essential to broaden our scientific framework by exploring the everyday realities of "out-of-standard" architectural projects – initially studied in the French context –

11 Eurocodes are standards developed under a mandate given to CEN by the European Commission within the framework of a European.

in other countries operating under the shared legislative framework of the European Union.

Examining the tangible, project-based dimensions of experimental architectural practice across Europe allows for a broader critique of the universal application of standardized environmental regulations. At the European level, initiatives such as the *Facilitating the Circulation of Reclaimed Building Elements in Northwestern Europe* program – funded by the European Commission from 2017 to 2021 – have provided a valuable overview of innovative material reuse practices in the Benelux region. This initiative highlighted the persistent administrative and regulatory obstacles that continue to impede the local implementation of experimental, "out-of-standard" reuse strategies, similar to those emerging in France.

In light of the urgent demands of the ecological transition, there is a growing need to adapt existing regulatory frameworks – such as RE 2020 – and institutional mechanisms for experimentation, including the ESSOC and LCAP laws, to better support innovative and experimental approaches in architecture. This shift calls for a deeper transformation in perspective: moving beyond the creation of ecological frameworks for individual projects toward cultivating *an ecology of the project* itself. In other words, can we begin to imagine a politically grounded, experimental ecological framework – one that acknowledges the complex interplay of human and non-human actors that bring a project into being? Such a framework would create space for diverse ecological imaginaries to take shape, spanning from the hyper-local to the global. This is the core ambition of this chapter. By proposing a broader structure for knowledge exchange, the research expands the representational scope of the CAST cartography of ecological imaginaries. It is grounded in the "out-of-standard" experimental practices that define the architectural projects examined here. Ultimately, the ethnographic approach seeks to establish a shared platform for dialogue and critical reflection – one attuned to the evolving, everyday environmental challenges highlighted in recent IPCC reports.

15.6. References

Bertaux, D. (1997). *Les récit de vie. Perspective ethnosociologique*. Nathan, Paris.

Bouchain, P. (2017). *Permis de faire. Leçon inaugural 2017 de l'École de Chaillot*. Éditions de la cité de l'architecture et du patrimoine. École de Chaillot, Paris.

Boutinet, J.-P. (2015). *Anthropologie du projet*. Presse Universitaires de France, Paris [Original published 1990].

Callon, M. (1996). Le travail de la conception en architecture. In *Situations. Les cahiers de la recherche architecturale*, Danièle, V. (ed.). Éditions Parenthèses, Marseille, 25–35.

Chesneau, I. (2021). Désobéir aux règles constructives deviendra-t-il la norme ? In *Profession architecte*, Chesneau, I. (ed.). Eyrolles, Paris, 287–290.

Choay, F. (2006). *Pour une anthropologie de l'espace*. Éditions du Seuil, Paris.

Descola, P. (2019). *Une écologie des relations*. CNRS Éditions, Paris.

Docarragal Montero, H. (2023). Transgression des normes. Un cadre à la création architecturale. In *Contre les normes ? Histoires d'architectes*, Carmen, P. and Julien, B. (eds). Publications du GRIEF, Rennes, 63–81.

Docarragal Montero, H. and Jeudy, O. (2021). Vers une écologie de l'expérimentation "hors norme" des ressources matérielles en architecture. In *Les cahiers de la recherche architecturale urbaine et paysagère*. OpenEdition Journals [Online]. Available at: www.journals.openedition.org [Accessed 30 April 2025].

Gauzin-Müller, D. (2001). *L'architecture écologique*. Groupe Moniteur, Paris.

Georges and Rollot, M. (eds) (2018). *L'hypothèse collaborative*. Éditions Hyperville, Paris.

Latour, B. and Woolgar, S. (1996). *La vie de laboratoire : la production des faits scientifiques*. La Découverte, Paris.

Maffesoli, M. (1979). *La conquête du présent. Pour une sociologie de la vie quotidienne*. Presse Universitaires de France, Paris.

Norberg-Schulz, C. (1981). *Genius Loci*. Mardaga, Brussels.

Patte, E. (ed.) (2017). *Hors la loi. Dépasser la loi.* Éditions ICEB, Montreuil.

Raynaud, D. (2004). Contrainte et liberté dans le travail de conception architecturale. *Revue française de Sociologie*, 45(2), 339–366.

Simonin, C. (2021). La fabrique de la norme et ses recompositions. In *Profession Architecte*, Chesneau, I. (ed.). Eyrolles, Paris, 276–285.

Violeau, J.-L. (2004). Un monde habitable. Le Team Ten et l'hypothèse du populisme. In *Les cahiers de la recherche architecturale et urbaine. Savant, Populaire*. Éditions du patrimoine, Paris, 147–166.

16

The Role of Social Media in the Climate Crisis: A Discourse Study of Environmental Conversations in France (2017–2022)

16.1. Climate change in France

Climate change has been stirring heated conversations across various social networks, yet the specifics vary depending on places and moments. In France, climate-related issues began to be regularly addressed during Emmanuel Macron's first presidential term (2017–2022). Notably, these issues were discussed due to the Citizens' Convention on Climate, held between 2019 and 2020, and again during the passage of the Climate and Resilience Law in August 2021. Since then, media outlets dedicated to climate issues have also been created (such as *Reporterre*, *Vert* and *Bon Pote*), and general media have likewise brought climate to the forefront of topics covered.

However, we also observe that social networks, through the devices they represent and the affordances they mobilize, have played a significant role in the formation and opposition of communities on this subject. Whether it involves discussions about potential solutions (such as electric cars), the rise of climate skepticism or the use of social networks by environmental activists, social networks are integral to societal consideration of all of these issues. Beyond climate marches, various activist actions and media events – including the occurrence of climate disasters in France and elsewhere – these five years have seen the emergence of

---

Chapter written by Albin WAGENER.

For a color version of all figures in this chapter, see www.iste.co.uk/dobigny/rethinking.zip.

distinct narratives encompassing a range of environmental and climate-related topics.

Whether concerning agriculture and food, energy or simply the beauty of nature, several narratives have gained strength and media impact, particularly (but not exclusively) on social media. The challenge here is to understand the stories and discourses circulating about climate and the environment – both positive and negative – and to identify the key questions that will need to be addressed by those seeking to support the climate cause.

16.2. Discourse analysis: a corpus study

16.2.1. *Discourse and corpus linguistics*

The goal of this study is to observe and understand the evolution of narratives related to these issues during Emmanuel Macron's presidency in France (2017–2022), in order to grasp the current relevance of these questions in contemporary France. Beyond these specifics, the aim is also to analyze how Emmanuel Macron's five-year term has influenced the evolution of these narratives (for instance, through the impact of the Citizens' Convention on Climate) and how society has responded to these topics.

By using methodologies specific to the study of social network conversations (Crawford 2009), I propose an in-depth analysis of the emblematic conversations that took place in France on social networks between 2017 and 2022. In this regard, the methodology used is based on netnography (Reid and Duffy 2018), which involves observing the most popular posts – i.e. those generating the highest levels of engagement. This high engagement rate is an important factor in understanding which topics – and, more importantly, which framing angles – generate the most discussion. For the purposes of this study, these posts were identified using CrowdTangle, a now-discontinued service formerly offered by Meta to conduct research across its various social media platforms. The next step is to extract the comments associated with these posts in order to analyze their content and obtain a snapshot of how discussions unfold online around the topics that elicit the greatest engagement.

This study is thus situated within the postdigital paradigm (Andersen et al. 2014), which posits that digital and nondigital worlds mutually influence one another, with significant social, cultural, economic and political implications. From this perspective, social media platforms are not separate from the broader social and political environment but rather constitute tools that offer complementary channels for communication and interaction. As such, the topics addressed – regardless of their nature or the platforms involved – unfold both online and offline through conversations that ebb and flow depending on the communication tools we use. In this sense, reality is expressed across both digital and nondigital environments, and the conversations within these two spheres constantly interpenetrate.

Our approach involves discourse analysis, combining corpus linguistics, ecolinguistics and lexicometric analysis of statements. These four epistemological and methodological standpoints are based on the following assumptions:

– Discourse is both the vehicle and the driver of our social and political imaginaries, including in environmental domains (Wagener 2019).

– In order to study these discourses, it is necessary to build corpora from large data sets to ensure sufficient representativeness of their strictly linguistic expression (Baker 2006).

– Language plays a crucial role in representing our relationship to the environment, nature, ecology, climate change and biodiversity (Stibbe 2015; Fløttum and Gjerstad 2017).

– A statistical lexicometric analysis of the corpus – allowing for the isolation of both verbal occurrences and co-occurrences, cross-referenced with their semantic contexts – provides a reliable representation of the narrative themes circulating within the corpus (Reiner 1990; Marchand 1998; Wagener 2021).

Our study highlights several central elements, namely: (1) the structuring of the most emblematic discourses on climate (Campion et al. 2015) in France during the given period; (2) the specifics related to the different networks studied, namely Facebook, Instagram, YouTube, Reddit and Twitter; and finally, (3) the importance of the structuring and evolution of online narratives. The aim of this study is therefore to understand how climate-related discourses evolve, interpenetrate and reveal underlying imaginaries, depending on the platforms studied and the comments published there.

In this sense, the present analysis is also grounded in the study of hypernarrativity (Rose 2012). For instance, the idea that online environments expose users to an intense variety of content through which we navigate based on our desires, as well as the constraints and affordances of digital platforms (notably, for example, social media and their algorithmic structures or the headlines of the news stories shared on these platforms).

This chapter also emphasizes the necessity of systemic thinking in networking narratives and conversations about the environment on social networks, in a logic where discourses intersect and collide (Garric and Longhi 2013), and where social networks and narrative networks contribute to the social and political legitimacy of certain statements (Campion 2016). In addition, our study aims to demonstrate the value of a multisemiotic analysis of discourses, capable of embracing the multiplicity of networks of meaning (Barbazan 2011) and the importance of understanding the multiplicity of discourse types concerning climate and environmental issues (Wagener 2023).

16.2.2. *Corpus extraction*

This narrative study is based on the extraction of corpora from the thirty publications with the highest engagement rate per social network, between May 7, 2017, and February 1, 2022. The social media platforms analyzed are Facebook, Instagram, Reddit, YouTube and Twitter (now known as X). For Facebook, Instagram and Reddit, Crowdtangle was used to isolate the posts with the highest engagement rate. For Twitter and YouTube, data extraction was conducted using either the API (for Twitter) or the search engine (for YouTube).

To extract the analyzed comments from Facebook, Reddit and YouTube posts, Facepager was used, allowing for the retrieval of a total of 70,000 Facebook comments, 17,000 YouTube comments and 1,200 Reddit comments; approximately 88,000 comments in total.

The narrative analysis first involves counting lexical co-occurrences, primarily using the Iramuteq software[1]. Word clouds are generated with Visual Tools, while other visualizations are produced using RAWGraph. Discourse analysis is conducted through an argumentative examination of the themes identified by Iramuteq, as well as a detailed study of the semantic implications of the terms used within the specific contexts of the comments. In other words, this is where the qualitative expertise of discourse analysis meets the quantitative representation of data.

16.3. Results

16.3.1. *Semantic network*

One of Iramuteq's most interesting features is to propose a semantic network, thus showing how words are interconnected in the corpus. At the center of Figure 16.1 lies the global issue of climate change. On the top of the network, the focus shifts to considerations regarding humanity's place in nature and our perception of the environment. In the bottom-right corner, attention is given to

1 See: https://www.iramuteq.org.

proposed technical solutions, such as electric cars (*voitures électriques*) and nuclear energy (*énergie nucléaire*). We can also note an emphasis on the psychological and reflective impact of climate change. In the bottom-left corner, climate change (*changement climatique*) is acknowledged as a phenomenon that must be taken into account.

Finally, in the central area, the necessity of dialogue and exchange in response to global warming is highlighted, underscoring its sheer social dimension.

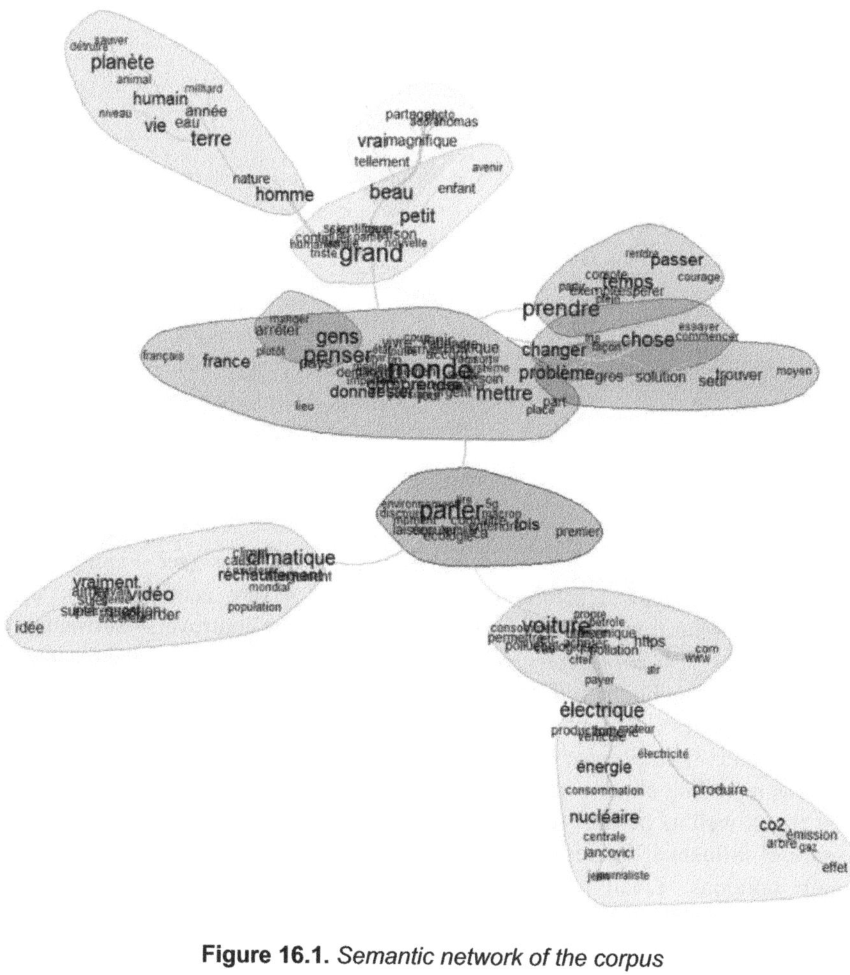

Figure 16.1. *Semantic network of the corpus*

16.3.2. *Narrative classes*

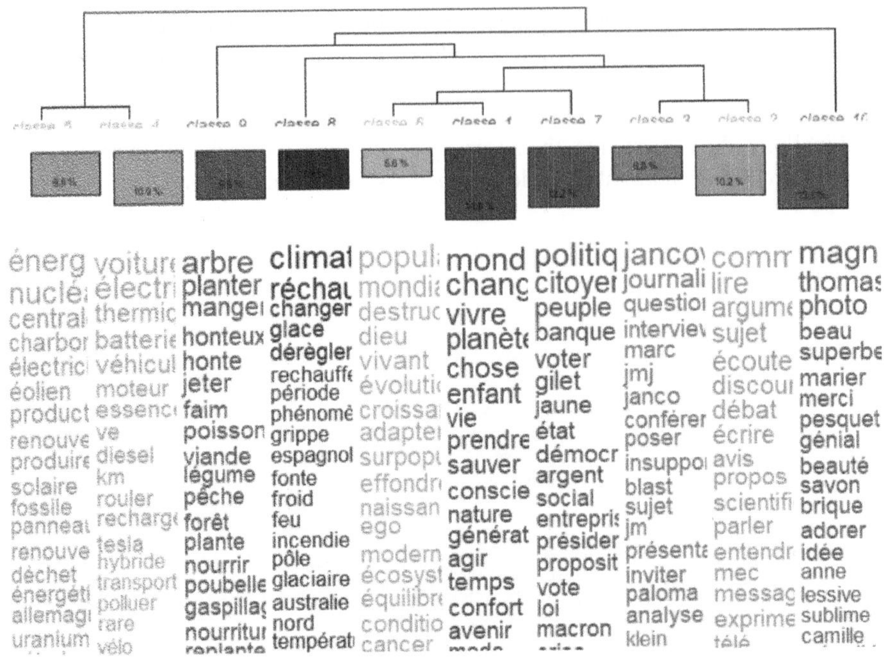

Figure 16.2. *Narrative classes represented within the corpus*

The thematic analysis reveals several key areas of focus (see Figure 16.2). Energy and technological solutions (Classes 5 and 4) are central topics, with discussions ranging from the broader energy debate to the controversy surrounding electric vehicles (*véhicules électriques*). Climate change effects and global responses (Classes 8 and 6) highlight both the perceived consequences of environmental shifts and how societies confront these challenges. Social and ethical dimensions (Classes 1 and 7) emphasize the need for political action and a growing ethical awareness regarding climate issues. Discussions also extend to food and vegetation (Class 9), as well as debates (*débats*) over potential solutions (Class 2), where the presence of influential figures such as Jean-Marc Jancovici (Class 3) plays a role in shaping opinions. Finally, positive reinforcements and incentives (Class 10) underline the importance of motivation and constructive engagement in the climate conversation.

The prominence of Classes 1 and 7 is no coincidence – it demonstrates that individuals directly associate the need for action with the necessity of implementing appropriate political change. In addition, the effects of climate change are clearly recognized, as is the demand for various solutions, including technical ones.

However, this technical aspect (represented by Classes 5, 4 and 2) does not overshadow the social dimension (Class 2) or the need for visible positive impact (Class 10).

16.3.3. *Narrative interactions*

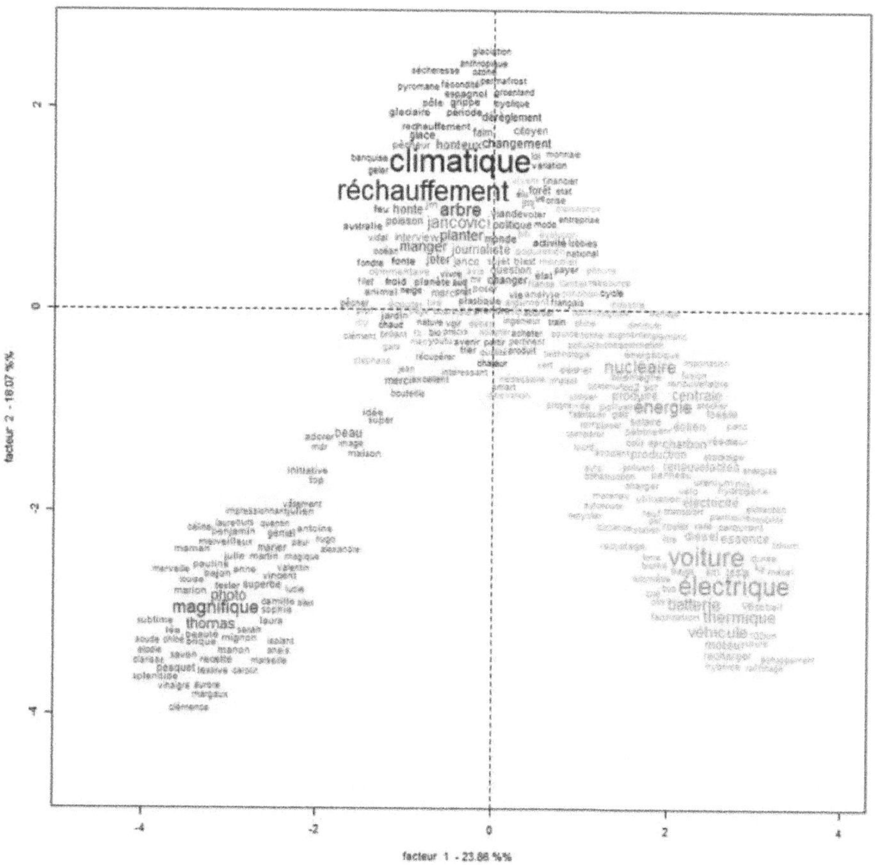

Figure 16.3. *Narrative interactions of the classes*

The ten narrative classes previously identified in the corpus interact within three main groups (see Figure 16.3). The first group (Classes 9, 8, 6, 1, 7 and 3) revolves around climate issues and their political, social and ecological implications. This cluster highlights the intersection of environmental awareness, ethical considerations and the need for political action while also reflecting the role of key figures in shaping public discourse. The second group (Classes 5 and 4) is strictly technical, encompassing engineering, energy and transportation. This set focuses on solutions driven by technological advancements, often detached from broader societal and political contexts. The third group (Class 10) pertains to an emotional and aesthetic connection to nature, emphasizing sensitivity to its beauty. However, this appreciation appears disconnected from both the urgency of climate action and the necessity of implementing solutions.

These distinctions reveal an underlying tension between technocratic discourse, which often disregards political and ethical concerns, and the broader societal implications of climate change.

Moreover, the capacity for wonder in the face of nature seems isolated from the pressing need to act. This absence of connection between these dimensions raises critical questions that warrant deeper reflection. In other words, this highlights that our societies carry imaginary and social representations that struggle to meet and connect, and that the difficulties in responding to climate change have deep anthropological and philosophical roots. It is our relationship with technological progress and with nature – which we are part of yet continue to define merely as an "environment", i.e. something external to us – that creates, among other things, a strong tension making it difficult to decide on appropriate courses of action.

16.3.4. *Corpus variables*

The analysis of various corpus variables – including the social network, page name and publication topic – provides a clearer understanding of how thematic classes are distributed (see Figure 16.4). Notably, Facebook appears to lean towards more positive narratives (Class 10), whereas YouTube focuses primarily on the strong presence of Jean-Marc Jancovici in the discourse (Class 3). In addition, the technical dimension emerges through debates surrounding the production of electric vehicles, particularly concerning their environmental and humanitarian costs. These distinctions highlight the crucial role that social media and their topical framing play in shaping the development of climate-related narratives.

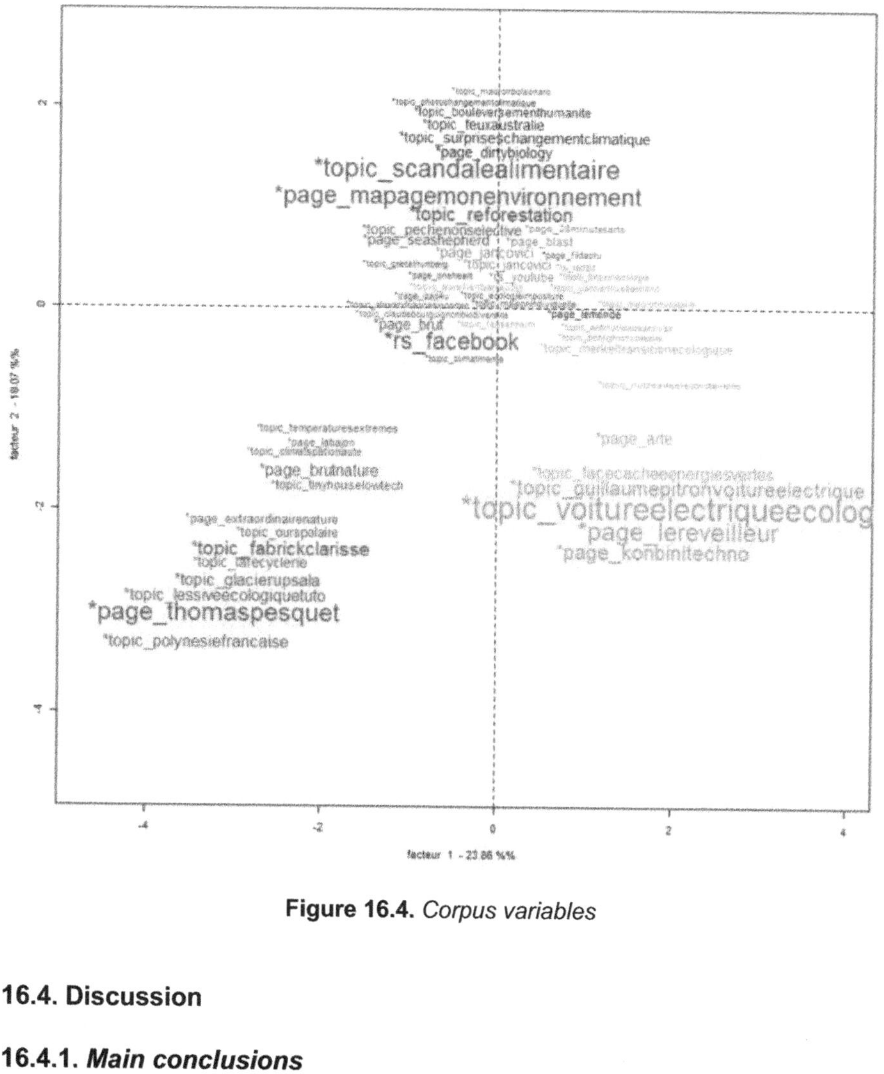

Figure 16.4. *Corpus variables*

16.4. Discussion

16.4.1. *Main conclusions*

Through the analysis of tens of thousands of social media comments, we gain a faithful representation of the conversations that have shaped public and media discourse between 2017 and 2022.

This analysis allows us to isolate the main narratives surrounding climate change, providing insight into the topics and arguments that resonate with the public. These narratives shed light on the key issues that drive public reactions when discussing global warming and environmental concerns. One prominent theme is the growing awareness of the reality of global warming. This has become a significant focus in public conversations, highlighting the urgent need for action. Alongside this, there is an increasing emphasis on the importance of positive stimuli, which encourages proactive engagement and solutions to climate challenges.

Political action plays a crucial role in the discourse, with many calling for stronger commitment to tackle climate change. At the same time, the debate has highlighted the scandal surrounding the production of electric vehicles, exposing the issues of false solutions and greenwashing that undermine genuine efforts to address environmental concerns. The value of discussions and debates is another central theme, as these are seen as essential for understanding the complexity of climate issues and for finding the most appropriate solutions. In addition, the impact of agricultural and food models is increasingly recognized as a key area that needs to be addressed in the fight against climate change, particularly in France, where food is part of national culture and often offers vigorous debates.

There is also considerable attention on energy production models and energy consumption, with debates focusing on how to transition to more sustainable practices. The tangible, real-world effects of climate change, such as extreme weather events and rising sea levels, are another critical issue discussed in the public sphere.

Prominent figures, such as Jean-Marc Jancovici, have gained significant influence in the discourse, shaping public understanding of the urgency of climate action. Finally, there is a philosophical dimension to the discussion, with growing awareness of humanity's place on the planet and the need to reconsider our role in the environment to ensure a sustainable future.

More broadly, the discourse analysis conducted on this corpus highlights the value of a lexicometric approach, which enables us to depict the diversity of imaginaries and social representations circulating in response to the climate crisis, while also emphasizing their complexity and our collective difficulty in interconnecting highly complex issues. In this sense, our interdiscursive approach reveals the links – or the lack thereof – between certain narrative themes, while also allowing us to grasp their positioning within online conversations.

In other words, it is precisely the complexity of the subject – both from a postdigital and ecolinguistic perspective – that is clearly exposed here. The circulating discourses are highly diverse, often competing and make the systemic actions required increasingly difficult to implement.

16.4.2. *Perspectives*

As the first presidential mandate of Emmanuel Macron unfolded, it was marked by many significant events concerning the climate crisis and environmental debates, such as climate marches, Intergovernmental Panel on Climate Change (IPCC) reports, climate disasters and the Citizens' Convention for the Climate. These events highlight how these issues are at a tipping point. Divided between the divergent interests of groups and opinion leaders, and the narratives circulating on the ground, the entire public, media and political space must now be restructured in a way that allows more unifying voices to emerge.

The presence of researchers and engineers seemed essential to communicate effectively on climate change issues, and Jean-Marc Jancovici serves as proof that this is possible. His approach shows how experts can play a vital role in informing and shaping public opinion on such critical topics. Alongside this, there is a significant level of skepticism and criticism towards green solutions and greenwashing. This skepticism, in turn, can foster a broader distrust of climate change science and the industrial technologies promoted as solutions to the crisis. There are also real political battles underway to propose solutions to the climate crisis. These debates are centered around three main ecological approaches: liberal ecology, which seeks to work within the current system; reformist ecology, which advocates for changing certain aspects of the existing system; and radical ecology, which calls for a complete overhaul of the system to address climate emergency.

Moreover, the weight of everyday issues such as energy bills and food consumption plays a significant role in shaping concrete narratives about ecological and environmental concerns. These topics are directly linked to individuals' daily lives, making the environmental crisis more tangible and relatable to the public. Furthermore, there is also a growing emergence of more ethical and philosophical questions, particularly regarding humanity's place on the planet and our relationship with other species, including animals. These discussions open up a broader reflection on the role humans should play in ensuring the future sustainability of the planet.

For all of this to be possible, it remains crucial to propose positive narratives that motivate individuals to anticipate upcoming changes and adapt their behaviors in ways that are not experienced as burdens but as chosen horizons. This issue is not only social, economic and political but also philosophical and anthropological. That said, this questioning cannot avoid the need for truly courageous political and economic choices; choices that must be made based on the scientific data available to us in the face of the climate crisis, even though these dimensions remain relatively absent from online conversations, like an elephant in a burning room. Thus, the stimulation offered by positive narratives must also be accompanied by a deep reflection on the neoliberal, productivist system that generates and accelerates the climate crisis or else risk being reduced to a socially acceptable form of greenwashing.

16.5. References

Andersen, C.U., Cox, G., Papadopoulos, G. (2014). Postdigital research. *A Peer-Reviewed Journal About Everything is a Matter of Distance*, 3(1), 4–7.

Baker, P. (2006). *Using Corpora in Discourse Analysis*. Continuum, London.

Barbazan, M. (2011). Énonciation ou représentation du monde ? Le jeu dynamique de la construction du sens dans les échanges verbaux. *Cahiers de praxématique*, 56, 117–166.

Campion, B. (2016). Dynamiques de construction et instrumentalisation de la légitimité dans les débats en ligne relatifs au réchauffement climatique. *Questions de communication*, 30, 159–179. doi: 10.4000/questionsdecommunication.10752.

Campion, B., Tessier, L., Bourgatte, M. (2015). Interactions en ligne sur les changements climatiques. Dynamiques d'échanges et affordances des dispositifs. *Hermès*, 73(3), 181–190.

Crawford, K. (2009). Following you: Disciplines of listening in social media. *Journal of Media & Cultural Studies*, 23(4), 525–535.

Fløttum, K. and Gjerstad, Ø. (2017). Narratives in climate change discourse. *WIREs Climate Change*, 3(1), e429, 1–15.

Garric, N. (2012). Construire et maîtriser l'hétérogénéité par la variation des données, des corpus et des méthodes. *Langages*, 187(3), 73–92.

Herring, S. (2013). Discourse in web 2.0: Familiar, reconfigured, and emergent. In *Discourse 2.0. Language and New Media*, Tannen, D. and Trester, A.M. (eds). Georgetown University Press, Washington, D.C.

Marchand, P. (1998). *L'analyse du discours assistée par ordinateur : concepts, méthodes, outils*. Armand Colin, Paris.

Pepperell, R. and Punt, M. (2000). *The Postdigital Membrane: Imagination, Technology and Desire*. Intellect, Bristol.

Reid, E. and Duffy, K. (2018). A netnographic sensibility: Developing the netnographic/ social listening boundaries. *Journal of Marketing Management*, 34(3–4), 263–286.

Reinert, M. (1990). ALCESTE, une méthodologie d'analyse des données textuelles et une application: Aurélia de Gérard de Nerval. *Bulletin de méthodologie sociologique*, 26, 24–54.

Rose, E. (2012). Hyper attention and the rise of the antinarrative: Reconsidering the future of narrativity. *Narrative Works: Issues, Investigations & Interventions*, 2(2), 92–102.

Stibbe, A. (2015). *Ecolinguistics: Language, Ecology and the Stories We Live By*. Routledge, London.

Wagener, A. (2019). *Discours et système*. PIE Peter Lang, Brussels.

Wagener, A. (2021). "Que ferait-on des prairies, puisque l'homme ne sait pas digérer l'herbe ?" : les discours climatosceptiques à l'assaut des discours véganes. *Mots. Les langages du politique*, 127, 45–63.

Wagener, A. (2022). Climat et environnement en France (2017–2022). Study, Komons and Open Society Foundations [Online]. Avaliable at: https://sysdiscours.hypotheses.org/1038 [Accessed 21 March 2025].

Wagener, A. (2023). *Blablabla : en finir avec le bavardage climatique*. Le Robert, Paris.

17

Transitioning to Sustainability: An Emotional Odyssey

17.1. Introduction

The transition to a sustainable lifestyle is not merely a behavioral shift but an emotionally complex journey influenced by personal, societal and psychological factors. While rational incentives for sustainability exist, emotional barriers – such as anxiety, guilt and resistance to change – often dominate decision-making. This chapter situates these emotional challenges within established psychological and behavioral theories, demonstrating how emotions shape sustainable choices. By bridging emotional and rational perspectives, we will offer a nuanced analysis of the sustainability transition. Unlike prior studies that focus primarily on policy or economic incentives, this chapter highlights the deeply personal struggles individuals face. Through this approach, it fills a critical gap in sustainability discourse, emphasizing that long-term change requires addressing both external structures and internal emotional responses. Future research should explore how targeted interventions – such as emotional resilience training – can facilitate smoother lifestyle transitions.

17.2. The challenges of embracing the consumer-citizen role: a cognitive and rational decision-making perspective

All stakeholders, across varying scales, have been called to adapt their strategies and practices to facilitate the shift from the dominant social paradigm – responsible for the perpetuation of social inequalities and the depletion of environmental resources – to a paradigm grounded in the fundamental principles of sustainability

Chapter written by Cristina LONGO and Ishan JALAN.

(Kilbourne et al. 1997; Schaefer and Crane 2005). This dominant paradigm is fundamentally rooted in the economic imperative of continuous growth, which underpins profitability and the accumulation of capital, resulting in an uneven distribution of resources and wealth (Kilbourne et al. 1997). The transition to sustainability, therefore, necessitates a cultural shift that involves profound transformations in values, norms and practices, ultimately aiming to preserve and respect planetary resources (Fuchs and Lorek 2005). Achieving this transformation requires the adoption of an ecosystemic perspective, reflecting the interconnectedness of micro, meso and macro levels (Dolan 2002; Ganglmair-Wooliscroft and Wooliscroft 2022; Pham et al. 2024). The move toward a sustainability-oriented paradigm is a macro-level phenomenon that extends beyond individual behavioral change to encompass broader community-level transformations and adaptations (Pham et al. 2024).

However, an increasing focus is placed on the role of consumers as citizens in shaping the transition to sustainable ethos and practices, especially as individual consumption accounts for nearly two-thirds of global greenhouse gas emissions (Williamson et al. 2018). This shift necessitates a societal transformation that requires individuals to critically examine and revise ingrained consumption habits, aiming to transition to more sustainable lifestyles (Brosch and Sauter 2023). Sustainable consumption, however, often presents a paradox, as it challenges the deeply entrenched consumerist behaviors that glorify materialism and sustain a throwaway culture, while attempting to foster more environmentally and socially responsible practices (Schaefer and Crane 2005; Gössling and Humpe 2020).

Eco-citizenship, as an emerging framework, involves fulfilling our needs within the consumption cycle by considering its environmental and societal impacts (Ginsburger 2020). Sustainable living can be pursued either by maintaining current consumption levels or reducing them (McDonagh et al. 2012). Mainstream approaches to sustainable consumption tend to emphasize efficiency – using fewer resources – rather than a broader reduction in consumption. This form of "weak sustainability" is reflected in alternative purchasing behaviors, such as acquiring information about responsible products and then purchasing them. However, it maintains overconsumption patterns, limiting the environmental benefits (Fuchs and Lorek 2005; Gössling and Humpe 2020). Much of the research focuses on individual actions, such as the purchase of eco-friendly products, but neglects to address systemic barriers, resulting in consumer-driven sustainability efforts that often lead to minimal societal change (Gough 2017). In contrast, achieving sustainability through the reduction of consumption is less widespread, yet it involves efforts to decrease ecological footprints. Downshifting, which entails scaling back consumption and reducing working hours to improve quality of life, is seen as a voluntary, long-term lifestyle change that contributes to a redefined personal identity (Soper 2007, 2008, 2013).

The formation of an "ecological habitus" (Ginsburger 2020) represents a significant challenge for consumer-citizens. Information dissemination has been a central strategy used to raise awareness about environmental, social and economic challenges (Spangenberg and Lorek 2019). The underlying assumption in much of this discourse is that consumers are rational, utility-maximizing decision-makers who engage in sustainable consumption practices through a process of information processing, which unfolds in a linear, phase-based manner (Carrington et al. 2010). The decision-making framework, particularly as outlined in the Theory of Reasoned Action (Fishbein and Ajzen 1975) and the Theory of Planned Behavior (Ajzen 1991, 2002), has been extensively applied to investigate the cognitive processes underlying sustainable consumption, with particular attention to how perceptions of consumer effectiveness shape decision-making (Antonetti and Maklan 2014).

However, this rational consumer paradigm is increasingly being challenged. It tends to overlook the tacit knowledge involved in sustainable consumption and production practices, as well as the complex and nonlinear nature of behavioral change (Hobson 2003; Shove 2003; Ozanne et al. 2022). This perspective fails to account for the nuanced emotional and experiential dimensions of living sustainably and engaging with other stakeholders to create a more virtuous value chain (Dolan 2002; Schaefer and Crane 2005). By prioritizing cognitive factors such as awareness and knowledge, it neglects the emotional complexities that influence consumer choices, thus limiting our understanding of why individuals adopt or resist sustainable practices. Therefore, a comprehensive approach to understanding the transition to sustainable lifestyles must consider both cognitive and emotional processes (Xie et al. 2019).

17.3. Investigating the emotional dimensions of sustainable consumption

To address the limitations inherent in the rational decision-making perspective, a growing body of research has begun to explore the emotional dimensions of sustainability, seeking to better understand the complexities and inconsistencies that underlie the actualization of sustainability principles in consumer behaviors. Emotions can play a pivotal role in influencing sustainable consumption, as they are integral in driving behavioral change (Amatulli et al. 2019; Castro-González et al. 2019).

Moreover, the emotive qualities of communication can impact the effectiveness of environmental policies by influencing behavior change (Pham et al. 2024).

Castro-González et al. (2019, p. 847) define emotions as:

> A form of affection that involves visceral responses associated with a specific referent, which give rise to action [...] made up of critical factors that stimulate the way consumers perceive, feel, and behave toward brands as well as toward organizations.

More broadly, emotions are conceptualized as "psychological and physiological states related to a variety of feelings, thoughts, and internal (physical) or external (social) behaviors" (Shao et al. 2022, p. 3). Despite this growing attention to the emotional components of sustainability, much of the existing literature falls short of adopting a holistic perspective on the emotional experiences of different stakeholders grappling with sustainability challenges. Current studies primarily focus on how corporate communication and messages can foster pro-environmental attitudes and behaviors (Di Russo and Myrick 2021; Shao et al. 2022), often with an emphasis on negative emotions such as guilt, shame and anger (Baek and Yoon 2017; Amatulli et al. 2019). However, there is limited focus on positive emotions such as gratitude, pride and admiration (Romani et al. 2013; Antonetti and Maklan 2014; Castro-González et al. 2019).

This chapter seeks to broaden the scope of understanding by highlighting the emotional complexities involved in engaging with sustainability. In doing so, it responds to Kurth and Pihkala's (2022) call to recognize that different eco-emotions can have divergent functions. The following sections will examine both the "negative" and "positive" emotions that underpin sustainable consumption practices, discuss the limitations of current conceptualizations and propose a research agenda aimed at deepening and extending our understanding of the emotional nature of the transition toward more sustainable lifestyles.

17.3.1. *Negative emotions*

Previous research has often framed sustainable living as a path of sacrifice and restriction (Valor 2008), with much of the literature concentrating on negative emotions (Brosch and Sauter 2023).

For example, Newholm (2005) underscores the ethical dilemmas faced by consumers, particularly the guilt they experience when inconsistencies arise between their sustainable intentions and their actual behaviors, while some individuals find these inconsistencies enriching. Similarly, consumers may encounter various negative emotions (e.g. shame and anger) when they experience a disconnect between their values and behaviors. This disconnect can arise when they struggle to adopt more sustainable consumption practices or become aware of the deterioration of their environment.

In addition, negative emotions may be elicited by marketing messages designed to encourage the purchase of environmentally friendly products or support for societal causes. Di Russo and Myrick (2021) investigate how fear-based messages, which highlight potential threats, can induce defensiveness, while those that conclude with a sense of efficacy may more effectively stimulate sustainable behavior. Their study also explores how emotions such as hope, fear and anger can influence memory and persuasion, shedding light on the psychological implications of such emotional appeals in sustainability communications.

Similarly, Amatulli et al. (2019) explore how negatively framed messages can evoke anticipated shame in consumers, prompting them to engage in pro-environmental behaviors as a means of regaining a positive self-image. Shame is often described as a self-threatening emotion, which can motivate individuals to adopt sustainable behaviors as a coping mechanism to restore their self-perception (ibid). Furthermore, Antonetti and Maklan (2014) argue that both guilt and pride influence consumers' perceptions of personal responsibility in relation to the outcomes of their actions, thus affecting their perceived consumer effectiveness and shaping their engagement in sustainability efforts. Understanding these psychological dynamics is crucial for the design of interventions that leverage negative emotions to foster positive behavioral change in sustainability efforts.

17.3.2. *Positive emotions*

Sustainability is often viewed not just as a series of obstacles, but as a potential source of joy and satisfaction. In response to calls for a deeper investigation into the pleasurable aspects of sustainability (McDonagh et al. 2012; Xie et al. 2019), this chapter aligns with literature that recognizes the hedonistic benefits of sustainable practices (Soper 2007, 2008, 2013; Autio et al. 2009). The concept of "alternative hedonism" encapsulates this perspective, suggesting that sustainable living can provide not only self-fulfillment but also contribute to communal well-being (Soper 2007, 2008, 2013).

Alternative hedonism posits that consumers derive pleasure from sustainable practices that involve time investment, rather than resource depletion. Practices such as cooking from scratch or reducing work hours to enjoy more leisure activities serve as prime examples (Soper 2008, 2013). This shift in consumption habits leads to a deeper sense of satisfaction and contributes to the broader communal well-being. The redistribution of prosperity, through a more equitable allocation of time and resources, is a fundamental component of this alternative hedonism (Soper 2007, 2008, 2013). This approach encourages a revaluation of everyday activities, fostering both individual satisfaction and environmental benefits (Soper 2008).

Furthermore, eco-friendly practices can yield joy and satisfaction, as evidenced in case studies where individuals derive pleasure from sustainable living, thus forging deeper emotional connections to their choices (Autio et al. 2009; McDonagh et al. 2012).

However, the complexity of the emotional spectrum has yet to be fully explored, particularly the role of positive emotions in sustainability discourse. As noted earlier, existing literature predominantly focuses on the negative emotions that consumer-citizens experience – emotions of guilt or shame arising from perceived moral failings, or frustration directed toward corporate irresponsibility. Few studies, however, have examined the emotional intricacies of consumer advocacy in response to virtuous corporate strategies.

Positive emotions, such as admiration for socially responsible corporate actions, can foster a sense of collective responsibility towards sustainability, increasing the consumer's sense of well-being (Xie et al. 2019). Admiration, as generated by consumer perceptions of socially responsible corporate strategies, encourages consumer advocacy (Castro-González et al. 2019). Likewise, consumer-citizens may experience gratitude toward companies and public institutions when they perceive corporate and institutional efforts as enhancing collective well-being, thus increasing trust and commitment to the organization (Romani et al. 2013; Xie et al. 2019; Shao et al. 2022). Consumers are likely to experience both admiration and gratitude when they feel that responsible corporate strategies align with their moral values and foster a sense of being part of a collective movement toward sustainability (Romani et al. 2013; Xie et al. 2019).

For example, beyond message framing, negative emotions can also emerge from personal decision-making processes in sustainable consumption. For instance, Wang and Wu (2016) found that consumers often experience anxiety and regret when faced with complex trade-offs, such as choosing between cost and energy efficiency in household appliances. This decision-making stress highlights how sustainable choices are not only shaped by external messaging but also by internal conflict and cognitive overload.

Similarly, frustration may arise when consumers encounter structural barriers – like limited access to sustainable alternatives – despite a strong personal commitment to sustainability. These everyday emotional tensions underscore the broader psychological costs of navigating sustainable consumption and suggest that emotional responses are embedded in both interpersonal and systemic contexts. Including these dimensions enriches our understanding of the emotional landscape beyond institutional or brand messaging, reinforcing the need to address emotional ambivalence and decision fatigue within sustainability interventions.

17.4. Unpacking the emotional complexity of sustainable consumption and fostering deeper sustainable practices: next steps

The study of environmental and societal transitions through the lens of emotions is crucial for understanding the complex networks of various stakeholders involved. Existing literature tends to focus more on negative emotions than on the full spectrum of emotions that emerge when individuals engage in responsible practices. Thus, the potential for strengthening sustainable consumption practices and the adoption of an alternative hedonism lifestyle is not fully realized if the focus is solely on stimulating guilt and shame. A more nuanced analysis of positive emotions could open more powerful avenues for fostering behavioral change, both in personal consumption habits and in workplace environments.

Public policies can effectively target emotions by framing sustainability as a collective and positive experience. Campaigns such as "Keep America Beautiful" or India's "Swachh Bharat Abhiyan" evoke pride and a sense of community spirit, while initiatives such as the UK's "Love Food, Hate Waste" or France's "Tu fais ça, toi?" rely on guilt and empathy to reduce food waste. These forms of emotional engagement help cultivate a sense of responsibility and connection, thus encouraging individuals to adopt sustainable consumption practices. Moreover, developing social–emotional learning programs to nurture collective emotional intelligence and community engagement could significantly aid the transition to more sustainable lifestyles and foster lasting behavioral change. Future studies could examine emotions such as joy and empathy, both of which have received relatively little attention. Positive emotions such as empathy enhance individuals' awareness of environmental issues, motivating them to adopt sustainable practices for the well-being of future generations and the planet. Furthermore, collective joy from participating in community-based sustainability initiatives creates a sense of belonging, reinforcing pro-environmental behaviors and encouraging collaboration among individuals (Soper 2007, 2008, 2013). In addition, hedonistic pleasure derived from sustainable practices can extend to social acknowledgment, with consumers gaining status and showcasing their environmental ethos through their sustainable behaviors (Hartmann and Apaolaza Ibáñez 2006; Miller et al. 2009). Awareness-raising initiatives and community interaction can further amplify these emotional benefits. For example, participants in community sustainability initiatives experience synergy and social recognition, often symbolized by tangible tokens of their commitment (Miller et al. 2009).

Current analyses of emotions are often oversimplified, focusing on the dichotomy between negative and positive emotions, rather than exploring their more nuanced dimensions. While pride and hope are often considered positive emotions that sustain the realization of sustainable consumption and production practices, they can also reveal an underexplored darker side. When pride becomes too predominant, it may blind consumers and employees to the practical feasibility of corporate social responsibility (CSR) strategies, undermining the pursuit of a balanced triple bottom line. Similarly, hope, if relied upon alone, can result in passive actions that shift responsibility to others rather than motivating individuals to take ownership. It would be valuable to further investigate how negative emotions can lead to positive ones and vice versa.

Additionally, exploring how a combination of negative and positive emotional message framing could stimulate greater consumer awareness of sustainability and its implications for both individual and collective well-being would be insightful (Di Russo and Myrick 2021). Such an exploration could illuminate how these emotions, when effectively harnessed, can enhance sustainable behaviors and their broader societal impact.

A particularly growing area of investigation is that of “eco-anxiety”. Eco-anxiety is a complex emotion that has received limited attention despite its relevance to sustainability, as it is inherently linked to the temporal and intergenerational aspects of ecological responsibility. Eco-anxiety reflects the fear of present and future ecological threats that could disrupt current and future generations’ opportunities and memories (Cunsolo et al. 2020; Kurth and Pihkala 2022). The concept of eco-anxiety is multifaceted, encompassing emotional responses to uncertain ecological dangers, self-assessments of personal responsibility for environmental harm and the mourning of environmental losses (Kurth and Pihkala 2022).

While eco-anxiety can be paralyzing, it also plays a critical role in raising awareness and motivating individuals to engage in sustainable actions. It fosters ecological conscientiousness and encourages consumers to act in ways that promote individual and collective well-being (Cunsolo and Ellis 2018; Cunsolo et al. 2020; Kurth and Pihkala 2022). Similarly, anger has been identified as an emotion that can catalyze consumer action, particularly in the context of addressing perceived ecological threats (Di Russo and Myrick 2021). Public institutions and corporations could collaborate to normalize and destigmatize negative emotions such as eco-anxiety and eco-grief, re-framing them as proactive emotions that can foster behavioral change (Brosch and Sauter 2023).

Except for the work of Demasi and Voegtlin (2023), the analysis of the link between emotions and the implementation of sustainable practices has been somewhat fragmented, with some studies focusing on the emotional experiences of individual consumers, and others examining the emotional landscape of CSR in the workplace. These emotional spheres often overlap, yet they are not typically studied as interconnected, even though engaging in sustainability should be a common thread across both private and public domains. The study of these emotional boundaries could uncover barriers and unresolved dilemmas that inhibit the development of more integrated, sustainable practices.

17.5. Concluding thoughts

This chapter explores the emotional complexities involved in transitioning to sustainable lifestyles, emphasizing the interplay between cognitive and emotional factors in shaping consumer behavior. It critiques traditional rational decision-making models, highlighting how emotions such as guilt, anxiety, pride and hope influence sustainability choices.

Advocating for a more holistic approach, the chapter recognizes both negative and positive emotions in driving sustainable actions. In addition, it suggests that future research could examine emotions within various geographical, cultural and socio-economic contexts (Schneider and van der Linden 2023). For example, to enrich the socio-cultural perspective on emotions in sustainable consumption, future research could benefit from exploring specific, real-world contexts where emotional dynamics play a pivotal role. For example, studies might investigate how feelings of pride and community belonging emerge in grassroots movements such as urban gardening or local zero-waste initiatives, particularly in cities where sustainability is tied to identity and collective action.

Conversely, research could examine how emotions such as guilt and frustration surface in rural or economically marginalized regions where structural limitations constrain sustainable choices, such as lack of access to public transport or affordable eco-friendly goods. Emotional responses to sustainability messaging could also vary based on cultural narratives, for example, fatalism versus empowerment, which influence whether consumers respond with hope or apathy. Another promising avenue involves integrating emotional resilience frameworks into sustainability education, helping individuals cope with eco-anxiety or climate grief, particularly among younger populations. These examples illustrate how a nuanced understanding of emotion – grounded in socio-cultural contexts – can inform more targeted, empathetic and effective sustainability interventions across diverse global settings. Developing this complexity will not only advance theoretical insights but also offer practical strategies for fostering enduring behavioral change.

Thus, while the transition to sustainability is a global goal, different countries experience environmental and societal threats in distinct ways. Exploring cultural nuances within the emotional web of sustainability could provide valuable insights into cross-cultural variability in emotional responses to sustainability challenges.

This chapter further proposes research into emotional resilience training and the emotional dynamics of sustainability across diverse contexts, emphasizing that fostering positive emotions could be key to achieving more effective and lasting behavioral change.

17.6. References

Ajzen, I. (1991). The theory of planned behavior. *Organizational Behavior and Human Decision Processes*, 50(2), 179–211.

Ajzen, I. (2002). Perceived behavioural control, self-efficacy, locus of control, and the theory of planned behaviour. *Journal of Applied Social Psychology*, 32(4), 665–683.

Amatulli, C., De Angelis, M., Peluso, A.M., Soscia, I., Guido, G. (2019). The effect of negative message framing on green consumption: An investigation of the role of shame. *Journal of Business Ethics*, 157(4), 1111–1132.

Antonetti, P. and Maklan S. (2014). Feelings that make a difference: How guilt and pride convince consumers of the effectiveness of sustainable consumption choices. *Journal of Business Ethics*, 124(1), 117–134.

Autio, M., Heiskanen, E., Heinonen, V. (2009). Narratives of 'green' consumers – the antihero, the environmental hero and the anarchist. *Journal of Consumer Behaviour*, 8(1), 40–53.

Baek, T.H. and Yoon, S. (2017). Guilt and shame: Environmental message framing effects. *Journal of Advertising*, 46(3), 440–453.

Brosch, T. and Sauter, D. (2023). Emotions and the climate crisis: A research agenda for an affective sustainability science. *Emotion Review*, 15(4), 253–257.

Carrington, M.J., Neville, B.A., Whitwell, G.J. (2010). Why ethical consumers don't walk their talk: Towards a framework for understanding the gap between the ethical purchase intentions and actual buying behaviour of ethically minded consumers. *Journal of Business Ethics*, 9(1), 139–158.

Castro-González, S., Bande, B., Fernández-Ferrín, P., Kimura, T. (2019). Corporate social responsibility and consumer advocacy behaviors: The importance of emotions and moral virtues. *Journal of Cleaner Production*, 231, 846–855.

Cunsolo, A. and Ellis, N. (2018). Ecological grief as a mental health response to climate change-related loss. *Nature Climate Change*, 8(4), 275–281.

Cunsolo, A., Harper, S.L., Minor, K., Hayes, K., Williams, K.G., Howard, C. (2020). Ecological grief and anxiety: The start of a healthy response to climate change? *The Lancet Planetary Health*, 4(7), 261–263.

Demasi, R. and Voegtlin, C. (2023). When the private and the public self don't align: The role of discrepant moral identity dimensions in processing inconsistent CSR information. *Journal of Business Ethics*, 187(1), 73–96.

Di Russo, C. and Myrick, J.G. (2021). Sustainability in CSR messages on social media: How emotional framing and efficacy affect emotional response, memory and persuasion. *Environmental Communication*, 15(8), 1045–1060.

Dolan, P. (2002). The sustainability of "sustainable consumption". *Journal of Macromarketing*, 22(2), 170–181.

Fishbein, M. and Ajzen, I. (1975). *Belief, Attitude, Intention and Behaviour: An Introduction to Theory and Research*. Addison-Wesley, Reading, MA.

Fuchs, D.A. and Lorek, S. (2005). Sustainable consumption governance: A history of promises and failures. *Journal of Consumer Policy*, 28(3), 261–288.

Ganglmair-Wooliscroft, A. and Wooliscroft, B. (2022). An investigation of sustainable consumption behavior systems – Exploring personal and socio-structural characteristics in different national contexts. *Journal of Business Research*, 148, 161–173.

Ginsburger, M. (2020). De la norme à la pratique écocitoyenne. Position sociale, contraintes matérielles et diversité des rapports à l'écocitoyenneté. *Revue française de sociologie*, 1(61), 43–78.

Gössling, S. and Humpe, A. (2020). The global scale, distribution and growth of aviation: Implications for climate change. *Global Environmental Change*, 65, 102194.

Gough, I. (2017). *Heat, Greed and Human Need: Climate Change, Capitalism and Sustainable Wellbeing*. Edward Elgar Publishing, Cheltenham.

Hartmann, P. and Apaolaza Ibáñez, V. (2006). Green value added. *Marketing Intelligence & Planning*, 24(7), 673–680.

Hobson, K. (2003). Thinking habits into action: The role of knowledge and process in questioning household consumption practices. *Local Environment*, 8(1), 95–112.

Kilbourne, W.E., McDonagh, P., Prothero, A. (1997). Sustainable consumption and the quality of life: A macromarketing challenge to the dominant social paradigm. *Journal of Macromarketing*, 17(1), 4–24.

Kurth, C. and Pihkala, P. (2022). Eco-anxiety: What it is and why it matters. *Frontiers in Psychology*, 13, 981814.

McDonagh, P., Dobscha, S., Prothero, A. (2012). Sustainable consumption and production: Challenges from transformative consumer research. In *Transformative Consumer Research for Personal and Collective Well-Being*, Mick, D.G., Pettigrew, S., Pechmann, C., Ozanne, J.L. (eds). Routledge, New York, 267–282.

Miller, E., Buys, L., Bell, L. (2009). Living smart homes: A pilot Australian sustainability education programme. *Journal of Education for Sustainable Development*, 3(2), 159–170.

Newholm, T. (2005). Case studying ethical consumers' projects and strategies. In *The Ethical Consumer*, Harrison, R., Newholm, T., Shaw, D. (eds). Sage Publishing, London, 107–124.

Ozanne, J.L., Davis, B., Ekpo, A.E. (2022). Research pathways for societal impact: A typology of relational engagements for consumer psychology research. *Journal of Consumer Psychology*, 32(1), 127–144.

Pham, H.N., Nguyen, T.T., Troy, W.H., Reynolds, N. (2024). Environmental policies and the promotion of pro-environmental consumer behavior: A systematic literature review. *Journal of Macromarketing*, 44(1), 30–58.

Romani, S., Grappi, S., Bagozzi, R.P. (2013). Explaining consumer reactions to corporate social responsibility: The role of gratitude and altruistic values. *Journal of Business Ethics*, 114(2), 193–206.

Schaefer, A. and Crane, A. (2005). Addressing sustainability and consumption. *Journal of Macromarketing*, 25(1), 76–91.

Schneider, C.R. and van der Linden, S. (2023). An emotional road to sustainability: How affective science can support pro-climate action. *Emotion Review*, 15(4), 284–288.

Shao, J., Zhang, T., Wang, H., Tian, Y. (2022). Corporate social responsibility and consumer emotional marketing in big data era: A mini literature review. *Frontiers in Psychology*, 13, 919601.

Shove, E. (2003). Converging conventions of comfort, cleanliness and convenience. *Journal of Consumer Policy*, 26(4), 395–418.

Soper, K. (2007). Re-thinking the 'good life': The citizenship dimension of consumer disaffection with consumerism. *Journal of Consumer Culture*, 7(2), 205–229.

Soper, K. (2008). Alternative hedonism, cultural theory and the role of aesthetic revisioning. *Cultural Studies*, 22(5), 567–587.

Soper, K. (2013). The dialectics of progress: Irish 'belatedness' and the politics of prosperity. *Ephemera*, 13(2), 249–267.

Spangenberg, J. and Lorek, S. (2019). Sufficiency and consumer behaviour: From theory to policy. *Energy Policy*, 129, 1070–1079.

Valor, C. (2008). Can consumers buy responsibly? Analysis and solutions for market failures. *Journal of Consumer Policy*, 31(3), 315–326.

Wang, J. and Wu, L. (2016). The impact of emotions on the intention of sustainable consumption choices: Evidence from a big city in an emerging country. *Journal of Cleaner Production*, 126, 325–336.

Williamson, K., Satre-Meloy, A., Velasco, K., Green, K. (2018). *Climate Change Needs Behavior Change: Making the Case for Behavioral Solutions to Reduce Global Warming.* Rare, Arlington, VA.

Xie, C., Bagozzi, R.P., Grønhaug. K. (2019). The impact of corporate social responsibility on consumer brand advocacy: The role of moral emotions, attitudes, and individual differences. *Journal of Business Research*, 95, 514–530.

PART 4

Networks of Universities in Transition

18

Schools in Synergy for Sustainability: Experiences and Insights from Networks of Universities in Socioecological Transitions

18.1. Sustainability and universities

While the Brundtland Commission (1987) defined sustainable development almost 40 years ago, the recent years have witnessed a stronger push for socioecological transitions throughout the globe. Various sectors have played a role in overcoming inertia, keeping momentum and driving forward in ensuring that the socioecological transitions are carried out in various contexts.

The global community, through the United Nations, came together to adopt the 17 Sustainable Development Goals in 2015. This monumental document created a renewed sense of commitment to advancing environmental stewardship, social responsibility and economic prosperity in the next 15 years. This holistic approach to transition and planning was later on cascaded to other documents such as the Paris Agreement on Climate Change, the Sendai Framework for Disaster Risk Reduction and the Addis Ababa Action Agenda on Financing for Sustainable Development (United Nations 2023).

Aside from governments, other sectors played a key role in advancing sustainable development. Individual businesses and conglomerates have been developing transitions to promote a triple-bottom line (Elkington 1994). Recently, Crona et al. (2024) released "Doing business within planetary boundaries" to

---

Chapter written by Emmanuel D. DELOCADO, Philip Arnold P. TUAÑO and Ma. Margarita Christina A. LACDAO-UMALI.

promote corporate sustainability post-pandemic. Meanwhile, the Catholic Church, driven by the encyclical *Laudato Si'* (Pope Francis 2015), has been advocating for climate justice and environmental protection with social equity and protection of the marginalized as goals.

In spite of this, universities or higher education institutions (HEIs) play a key role in knowledge generation through research and knowledge dissemination through teaching. Numerous literature and toolkits have been written on education for sustainable development (ESD), which are focused on the deliberate design of curricula and activities that promote knowledge, skills, values and attitudes on sustainability (Leicht et al. 2018; Kestin et al. 2020).

However, beyond these, universities are deemed to play an integral role in promoting socioecological transitions as part of their responsible operations and duty to drive social change to the greater community. Clear evidence of the burgeoning movement among universities includes the growing number of external grants on campus sustainability and growing participation in university rankings on sustainability.

While many universities have sufficient resources to mobilize socioecological transitions in their operations, the looming reality points to the necessity for collaborative effort and synergistic activities in order to advance the SDGs on national and global scales. Thus, networks of universities have been formed to leverage vast resources to accelerate sustainable solutions and to upskill universities that are in the infancy of their sustainability journey.

In a field as dynamic as sustainable development, the role of networks is especially crucial because the work is as daunting as it is complex. Thus, undeniably, collaboration and synergy among universities is relevant and is a practical way of proceeding. The work in sustainable development requires expertise in multi-disciplinary fields, which no university holds a monopoly of, and so this necessitates constant collaboration and partnerships. These allow for the sharing of knowledge and resources that can help in expanding capacity and building collective action.

18.2. Institutionalization of sustainability in Philippine HEIs

The progression of sustainability strategies within Philippine HEIs has been marked by significant policy developments and institutional initiatives (Bernardino 2000; Galang 2010).

18.2.1. *Policy framework and institutionalization*

The integration of sustainability into education in the Philippines commenced in the mid-1970s with Presidential Decrees 1121 and 1152, mandating the then Department of Education, Culture and Sports to incorporate environmental education across all educational levels. This directive was influenced by the 1977 United Nations Economic, Social and Cultural Organization (UNESCO), United Nations Environment Program (UNEP) and Intergovernmental Conference on Environmental Education, which emphasized the global importance of environmental awareness.

In 1987, Executive Order 192 under President Corazon Aquino reinforced this commitment by directing the Environmental Management Bureau, an agency of the government's Department of Environment and Natural Resources, to actively promote environmental education. Subsequently, the 1989 Philippine Strategy for Sustainable Development, formulated through multi-sectoral consultations, positioned environmental education as a central component, highlighting its critical role in national development.

The early 1990s saw the development of the National Environmental Education Action Plan (NEEAP) for the 1992–2002 period, aiming to enhance environmental education nationwide. This plan encouraged greater involvement of tertiary institutions, particularly their education departments, in curriculum integration, sustainable campus practices and community outreach. During this period, universities began offering specialized undergraduate and graduate programs focused on environmental studies.

Two national networks emerged to support these initiatives: the Environmental Education Network of the Philippines (EENP), which concentrated on developing sustainability curricula and a voluntary accreditation system for schools, and the Philippine Association of Tertiary Institutions for Environmental Protection and Management (PATLEPAM), which provided faculty training workshops. In 2003, a consultation workshop assessed the implementation of sustainability education in line with national and international frameworks, including the Philippine Agenda 21.

Balanay and Halog (2016) identified several factors crucial for embedding environmental sustainability in the Philippine education sector:

– Legal frameworks and policies: legislation such as Republic Act No. 9512 (Environmental Awareness and Education Act of 2008) and Republic Act No. 9163 (National Service Training Program Act of 2002) mandate the inclusion of environmental education at all educational levels.

– Initiatives by the Commission on Higher Education (CHED), the regulatory and development body overseeing the Philippine higher education sector: CHED's Strategic Plan for 2011–2016 outlines a roadmap for HEIs to align with sustainable development goals. Memorandum orders require HEIs to implement programs such as the National Service Training Program (NSTP) and encourage affiliation with environmental education associations such as PATLEPAM.

– NSTP: serving as a formal platform for ESD in tertiary education, NSTP focuses on enhancing civic consciousness, defense preparedness and environmental protection among youth.

– Environmental education associations: organizations such as PATLEPAM and EENP play significant roles in streamlining environmental education and promoting sustainable development within university curricula and campus management.

– Curricular development: the introduction of new programs in environmental studies, guided by CHED's Memorandum Order 35 (2005), provides standards for offering environmental science as a baccalaureate degree, embedding essential training policies.

– Practical and theoretical interventions: HEIs are encouraged to engage in practical and theoretical ESD interventions, including fieldwork, research collaborations and community extension activities to boost environmental awareness and disaster preparedness.

– NEEAP 2005–2014: this plan outlines strategies for ESD, such as establishing centers of excellence in environmental education, offering incentives for professionals specializing in environmental science and mobilizing youth in environmental projects.

– Social marketing and community engagement: HEIs participate in social marketing and community programs to instill environmental consciousness and promote disaster preparedness.

18.2.2. *Institutional initiatives*

Beyond these national drivers for embedding sustainability in university education, certain HEIs have identified sustainability as a key priority area in their mandate. An example of the evolution of sustainability strategies in HEIs is contained in Cuyegkeng and Favis (2018) which features Ateneo de Manila University, a private university based in the Philippine capital region Metro Manila. In 2009, the university established the Ateneo Environmental Management Coalition (AEMC), comprising faculty and students, to address environmental concerns on campus. The AEMC operated through subgroups focused on waste audits, research and analysis, advocacy and logistics. By 2011, the university's strategic planning included sustainability as a key focus, leading to the implementation of a new strategic plan in 2012. This plan aims to maintain sustainable development, reduce disaster risks associated with climate change and geohazards and promote a sustainable campus environment. To coordinate these efforts, the Ateneo Institute of Sustainability was established in 2013, serving as the university's hub for environment and development initiatives.

Similarly, the tertiary education neighbor of Ateneo de Manila, Miriam College, has integrated environmental education into its programs, reflecting its mission to cultivate genuine "stewards of creation". The college's efforts align with national strategies and benefit from the support of networks such as EENP and PATLEPAM, which advocate for the incorporation of sustainable development principles into school curricula and campus operations.

18.3. Case studies on networks of universities

In this section, we will look at networks of universities on socioecological transitions that Ateneo de Manila University is part of in various capacities.

18.3.1. *Sustainable Development Solutions Network Philippines*

Established in August 2015 through the leadership of then Leadership Council Chair Prof. Emmanuel de Dios of the University of the Philippines School of Economics (UPSE), the Sustainable Development Solutions Network Philippines (SDSN PH) is a network of mainly universities and research institutions collaborating together in localizing the SDGs and developing solutions to sustainability issues in the Philippines. The network highlights the invaluable role of universities in generating and propagating knowledge and in being models in their operations and community leadership (Kestin et al. 2020). In 2025, with 32 member institutions, the network is jointly hosted by the Ateneo de Manila University and De La Salle University.

Similar to other SDSN networks globally, SDSN PH aims to mainstream and localize the SDGs deemed crucial to the context of the Philippines by leveraging the financial and human resources of the members and by developing suitable approaches to advance ESD (SDSN 2024). In pursuit of this aim, SDSN PH partners with local and national governments in order to translate the SDG expertise of the members into policy formulation ranging from recommendations on city-wide initiatives and national development plan. Moreover, SDSN PH partners with nongovernmental organizations (NGOs), civil society organizations (CSOs) and the private sector to provide support in undertaking deeper diagnoses of the sustainability challenges in the Philippines and in the development of solution initiatives. In the process of doing so, in 2021, SDSN organized a series of workshops stemming from the manual SDSN PH developed called "Strengthening Academe-Business-Government Partnerships to Address SDGs in the Philippines", which is funded by Deutsche Gesellschaft für Internationale Zusammenarbeit (GIZ).

In addition, highlighting that sustainability is a shared responsibility and tapping into the strengths of the members, SDSN PH has Network Projects or thematic working groups, which are composed of like-minded network members who have a shared specialization and passion for particular sustainability issues. Through these network projects, SDSN PH intends to promote socioecological transitions by upskilling the network members and cascading it to the wider community. To date, there are three Network Projects under SDSN PH, namely:

– *SDG2 Zero Hunger Network Project*: stemming from the UN Food Systems Summit 2021, the SDG2 Zero Hunger has been organizing webinars and developing compendia on initiatives leading to reduction of hunger and social inequality. In addition, the SDG2 Network Project aims to communicate how multifaceted the issue of food security is by featuring different sectors, including academia, the government, the business sectors, CSOs, NGOs and young people. To date, the group has organized 28 webinars, which are all archived and can be accessed by anyone.

– *ESD Network Project*: the ESD Network Project was born from the frustration of members on the disconnect between awareness and urgency on the SDGs and its integration in the national curricula. Thus, the ESD Network Project intends to make a compilation of best sustainability education practices and to make a caravan-style workshop for educators from different parts of the Philippines. By "teaching the teachers", the working group hopes to introduce innovative ways of integrating the SDGs into the practiced curricula by not necessarily adding subjects or modules to the overloaded national curriculum.

– *Carbon Neutrality Network Project*: in its infancy, the Carbon Neutrality Network Project was established as a continuation and further localization of the ASEAN Green Deal Project, a regional strategy for transition to net zero emissions as part of honoring the commitment to the Paris Agreement. Now, the Carbon Neutrality Network Project is exploring how to mainstream net zero on Philippines campuses by learning the best practices and leveraging resources to make it a reality.

Beyond the Philippines, SDSN PH has been engaging with other national and regional SDSN networks in order to leverage knowledge on socioecological transitions across the region and beyond. Since the pandemic years, inter-network webinars have been organized on a variety of topics ranging from progress on SDGs to promoting campus sustainability, from biodiversity conservation to ESD.

18.3.2. *ASEAN University Network on Ecological Education and Culture*

The ASEAN University Network (AUN) on Ecological Education and Culture (AUN-EEC) is one of the 19 thematic networks of AUN, the regional mother organization working with ASEAN. Established in 2018 with the Ateneo Institute of Sustainability in Metro Manila as the secretariat, AUN-EEC identifies itself in a unique position of promoting not only knowledge dissemination but also behavioral and mindset shifts among Southeast Asian youth. This vision is deemed pertinent, especially in the region, as it faces insurmountable environmental threats set against the backdrop of colorful cultural and sociopolitical realities.

Grounded in this vision, AUN-EEC has a three-fold objective, namely, to foster cooperation among members on curriculum, research and outreach on building a sustainability culture; to further the formal and informal education for sustainable education programs in the region; and to generate contributions on ecological education solutions, such as policy formulation, information campaigns and sustainability solutions.

To date, AUN-EEC has 42 member institutions believing in this three-fold objective. The membership of AUN-EEC is composed of universities, university-based student organizations, government offices and NGOs. While AUN and its thematic networks have been working with these different sectors in pursuit of its mission, AUN-EEC differs from most of the other thematic networks by having student organizations as part of its membership base. As a university network, AUN has a set of 30 core university members and other HEIs as adjunct members. However, AUN-EEC has youth groups as members because the network believes that socioecological transitions can be furthered by developing the competences and mindset of the youth, who can pass it on to their peers and the future generations.

In order to ensure engagement with its growing membership base, AUN-EEC identifies a banner theme each year, ensuring that the different areas of expertise of the network are covered. In recent years, AUN-EEC identified pandemic recovery (2022), reconciling with nature (2023), urban sustainability (2024) and climate change education (2025) as its banner themes. Branching from these themes, AUN-EEC organizes a series of events, including:

– a *webinar* that tackles the problems and current realities attached to the banner theme;

– a *conference* exploring a framework on how the banner theme can be best addressed against the complexity of diverse cultures in the region;

– a *workshop* that seeks to promote individual embodiment of the framework.

Beyond these "template" of activities, AUN-EEC partners with groups that can further its commitment to ESD and sustainability culture in designing various projects and programs. Hoping to learn from the best practices of the ASEAN community, as well as other regions, AUN-EEC, in partnership with the University of Turku, developed a university course called "Integral Ecology for Socio-ecological Problem Solving" and a massive open online course called "MOOC for Sustainability: Empowering Global Campuses" (Posio and Delocado 2025).

18.3.3. *Promotion of Sustainability in Postgraduate Education and Research (ProSPER.Net)*

Promotion of Sustainability in Postgraduate Education and Research (ProSPER.Net) is a network of universities and academic institutions in the Asia-Pacific region that focuses on integrating sustainable development into postgraduate curricula. One of the main objectives of the network is to respond to the present environmental, social and economic realities and challenges by creating multi-disciplinary approaches.

Founded in 2008, ProSPER.Net is currently composed of 55 member universities from the Asia-Pacific region, three of which are in the Philippines: Ateneo de Manila University, University of the Philippines Diliman and De La Salle University. Through education and research programs, the network desires to inspire a new generation of sustainability leaders whose vision builds on innovating areas in governance, education, research and outreach. These are mainly done through workshops, training programs and research collaboration, as well as conferences that serve as a platform for sharing knowledge and best practices.

Depending on their expertise, members of ProSPER.Net are invited to join the Expert Advisory Panel (EAP) which advances particular focus areas of the network through funded and collaborative projects. To date, ProSPER.Net has four EAPs, namely EAP on Climate Change, EAP on Circular Economy, EAP on Equity and Inclusion, and EAP on Mobility Programme.

Recent activities of ProSPER.Net include the ProSPER.Net 2024 Sustainability Forum hosted by Ateneo de Manila University, which focused on the role of the circular economy in higher education, and the 2024 ProSPER.Net Young Researchers' School by the University of Peradeniya and the United Nations University Institute for the Advanced Study of Sustainability (UNU-IAS). The YRS focused on integrating nature-based solutions as a response to climate change.

18.3.4. *Philippine Human Development Network*

The Philippine Human Development Network (HDN) is a nonstock, nonprofit organization that was founded in 1992 through the initiative of Prof. Solita Collas-Monsod, UPSE Professor and former Socioeconomic Planning Secretary of the country, with the cooperation of Kevin McGrath, then the United Nations Development Program Resident Representative in the Philippines. The HDN's mission is to propagate and mainstream the concept of sustainable human development through research and advocacy.

A group of academics and civil society advocates met in a series of "brainstorming" sessions to discuss how best to apply the major findings and conclusions of the global Human Development Report in the Philippine setting.

From an informal group in 1992, the HDN became a registered organization in 1997. It has grown to more than 140 members in the academe and the civil society, although more recently it has restructured its membership.

Over the past three decades, HDN has released several country reports called Philippine Human Development Reports (PHDRs), each focusing on a specific theme:

– 1994: *Human Development* introduced the concept of human development and computed the Human Development Index (HDI) for Philippine regions.

– 1997: *Women and Gender in Development* focused on enhancing women's capabilities and opportunities and highlighted gains in education, employment and political participation.

– 2000: *Quality, Relevance and Access in Basic Education* addressed the decline in the quality of Philippine education and proposed a framework for improvement.

– 2002: *Work and Well-Being* examined unemployment, the profile of the unemployed and employment quality, as well as recommended strategies for job creation.

– 2005: *Peace, Human Security and Human Development in the Philippines* analyzed the causes and costs of ideology-based armed conflicts and assessed government policies related to peace efforts.

– 2008/2009: *Institutions, Politics and Human Development* investigated the role of institutions and politics in human development, with a focus on agencies such as the Department of Education, Civil Service Commission and Department of Budget and Management.

– 2012/2013: *Geography and Human Development in the Philippines* explored how geographic factors influence human development outcomes across different regions.

– 2020/2021: *Socioeconomic Mobility and Human Development* provided insights into factors enabling individuals and families to improve their socioeconomic status over time and sustain these improvements across generations.

While the earlier reports of the network centered on topics of education, employment and institution reforms, the 2012/2013 PHDR had increasingly recognized the intricate relationship between geography and human development by highlighting how physical and environmental factors influence development outcomes across the Philippines' diverse provinces. This focus on geography inherently ties into sustainability, as it underscores the necessity of considering environmental contexts, as viewed through the lens of issues in agriculture and health, in crafting long-term development strategies.

By examining how geographic features affect access to resources, economic opportunities and vulnerability to natural hazards, the report emphasizes that sustainable development must be tailored to the unique environmental conditions of each region. This perspective advocates for policies that are environmentally attuned and resilient by ensuring that development efforts are both effective and enduring. While these areas are crucial, the emphasis on geography provides a foundational understanding of how environmental factors underpin and influence all aspects of human development. This approach offers a more holistic view, integrating sustainability into the core of development discourse and policy formulation. The network continues to take this approach in an upcoming volume on stock-taking on the progress of development outcomes at the sub-national level.

18.3.5. *Philippine Society for Public Administration*

The Philippine Society for Public Administration (PSPA) is a nonprofit professional organization established in 1977 dedicated to enhancing public service education in the Philippines. Its primary objectives are to promote professionalism in public administration and disseminate research findings in the field.

Over the years, the PSPA has evolved to address contemporary challenges, placing a significant emphasis on sustainability in development, public administration and policy. This evolution is evident in its initiatives and thematic focus areas, which include:

– *Principles of Effective Governance for Sustainable Development*: the PSPA advocates for governance frameworks that integrate sustainability into public administration practices. In 2019, for example, its international conference had the theme “Governance 4.0 and the Sustainable Development Goals”. Then, in 2024, the theme of the conference was “New Governance Approaches for Sustainability”, which punctuates the following themes in response to the changing environment:

– *Institutions, Climate Action and Environment*: the organization emphasizes the role of institutions in driving climate action and environmental stewardship, as well as recognizes their impact on sustainable development.

– *Digital Government*: the PSPA explores the adoption of digital technologies to enhance public sector efficiency and service delivery, as aligned with the SDGs.

– *Public Sector Workforce*: focusing on capacity building, the PSPA addresses the need for a skilled and responsive public sector workforce to implement sustainable policies effectively.

The PSPA’s commitment to these areas is further demonstrated through its conferences and workshops. For instance, in November 2022, the PSPA organized a hybrid conference titled “A Public Sector Reform Agenda for the New Administration”. This event underscored the organization’s proactive approach in shaping public administration reforms that prioritize sustainability and effective governance. At the same time, the PSPA has strengthened coordination with the regional offices of the Department of Interior and Local Government (DILG) in strengthening sustainability throughout provincial, city and municipal governments in the country.

Through these efforts, the PSPA continues to play a pivotal role in advancing sustainable development within the realm of public administration and policy in the Philippines.

18.4. Challenges in networks of universities

In the previous section, some case studies of networks of universities were presented. In this section, the focus shifts to examining these case studies for common challenges among and across networks of universities in socioecological transitions.

18.4.1. *Finding commonalities and synergies*

While networks of universities intend to leverage their resources and synergize their existing initiatives, the reality is that each member university has its own priority areas reflective of its vision–mission that it seeks to promote. Therefore, one challenge among universities is finding commonalities and harmonizing initiatives.

One potential threat in harmonizing synergies is the drive of many universities to highlight their individual identity, considering the growing popularity of university rankings, especially those devoted to sustainability. Case in point, the Philippines witnessed a growing number of HEIs partaking in the Times Higher Education (THE) Impact Rankings, which assesses universities' performance vis-à-vis the 17 UN SDGs. There has been an increase from 1 HEI in 2019 to 15 HEIs in 2022, and from 29 HEIs in 2023 to now 113 participating HEIs in 2025 (Times Higher Education 2025).

The work of the networks featured in this chapter has shown, however, that this threat also presents an opportunity to continue to engage HEIs and encourage their contribution in leveraging resources for the networks. This can be done by focusing on the strengths and specializations of the universities and providing them with a platform to showcase these through the network events. This opens opportunities for networking among universities working in the same area.

18.4.2. *Transitioning within universities*

Socioecological transitions for networks of universities have been challenging in itself. However, jumpstarting the sustainability journey of a university is an arduous task as it calls for a shift in their operations including, but not limited to, managing their electricity, water, waste, physical structures, food sourcing, compensation packages and governance structures. Many equate socioecological transitions to be an expensive process, and there is truth to it. University administrators and their managing/administrative boards are thus tasked to assess the financial viability of investing in such sustainability measures.

In addition, socioecological transitions in universities include embedding sustainability in the conduct of the three-fold identity of universities, namely teaching, research and outreach. This is often met with resistance among university faculty and staff, as it requires a drastic shift in the way of proceeding compared to what they have been accustomed to. Sustainability transitions may be regarded as a disruption to the practice and as a challenge to convenience.

18.4.3. *Sustaining the work of networks*

When networks are up and running, another question must be asked: how do we guarantee impactful and sustainable operations? The involvement of HEIs in networks of universities in the Philippines is influenced by several key factors, particularly financial stability and organizational resilience.

– *Financial sustainability*: Philippine HEIs face significant financial challenges due to outdated regulations, political disruptions and escalating operational costs. Many private HEIs are facing difficulties in terms of enrollment, especially those based outside of major urban areas, because of a law passed in the late 2010s that provides free university tuition in state universities and colleges. Thus, private institutions have difficulties in participating in network activities that are beyond the scope of their core activities.

To address these issues, HEIs are adopting strategies such as fiscal prudence, green initiatives and income diversification. Some institutions have implemented income-generating projects to supplement their budget by focusing on market demand and community engagement to enhance profitability and sustainability. This has allowed them to use these resources to be able to better participate in networks and associations.

– *Organizational sustainability*: effective leadership and a culture of collaboration are vital for the resilience of networks of universities. However, given that sustainability is a journey, HEIs are in different stages of the transition with some being advanced and ready to influence the larger community while others are still exploring the necessary structures and resources. Several universities have prepared and published sustainability reports documenting the university's economic, social and environmental initiatives. Other institutions engaged in collaborative research and community extension programs, such as the establishment of marine protected areas, which demonstrate a commitment to environmental sustainability and community involvement.

In summary, the sustainability of networks of universities in the Philippines depends on proactive financial management and robust organizational practices. By embracing diverse funding strategies, fostering collaborative leadership and integrating sustainability into their core operations, these institutions can navigate challenges and continue to fulfill their educational missions.

18.4.4. *Maintaining the relevance of networks*

Maintaining relevance requires a network of universities to be attuned to the shifting interests and requirements of their members. This involves regularly assessing and updating program offerings, research focuses and collaborative projects. For example, the transition from traditional face-to-face learning to online and hybrid models has been accelerated by the need to accommodate diverse student populations and those with various commitments. While this shift offers flexibility, it also presents challenges in ensuring engagement and maintaining educational standards. Networks must balance innovation with quality to ensure that new approaches meet the high expectations of their members.

18.4.5. *Finding and generating resources*

Networks enable partners to ideate and come up with innovative and relevant ways to further the understanding and appreciation of sustainable development. Inasmuch as there is a wealth of knowledge and information resources, the financial mechanisms to get these projects and activities running can be quite challenging to secure.

One reason for this is that there are limited resources available across the board – and as a result, the requirements for particular funding opportunities are ultra-specific and nuanced – and therefore, difficult to fulfill. Another challenge related to this is these proposed projects must then be unique and instantly standout in order to even be considered for support at the onset. This therefore further limits the options for potential funding for some otherwise very promising projects and endeavors.

18.5. Best practices in advancing networks

Working in and with networks of universities in socioecological transitions, however, is not just marred with challenges. The wealth of experience from the five case studies presented in this chapter led us to identify numerous best practices for like-minded universities seeking to further sustainability.

18.5.1. *Mapping of strengths*

Given that the networks are established to leverage resources of the member universities, their expertise must be capitalized for the benefit of the network. Thus, one practice that benefits universities is an inventory of the strengths of the universities in the network.

For one, SDSN PH developed its SDG Specialization Database, which contains 22 sustainability experts and 297 SDG-specific experts from 24 member institutions. This guided the network in setting up SDG2 and Carbon Neutrality as its Network Projects. Aside from the identification of experts for each SDG, the network identified experts on SDG trends, sustainability reporting, sustainability leadership, and systems and future thinking among others.

Mapping the strengths of the member universities of a network allows for the identification of key thrust and programs of the network. In addition, it allows the network secretariat to create a more seamless and data-driven way of proposing avenues for collaboration among the members. Moreover, developing the inventory of specialization provides insights into the competence gaps within the network and can thus provide evidence for targeted recruitment of universities as members or even tapping into other sectors to complement the strengths of the universities.

18.5.2. *Collaboration in special working groups*

Taking off from the practices of SDSN PH on its Network Projects and of ProSPER.Net on its EAP, another best practice in advancing socioecological transitions through networks is to synergize efforts in well-defined or focused areas or topics. Recognizing that there is strength in numbers, collaborating in subgroups within a network enables groups to share knowledge and resources, as well as make meaningful contributions towards a common goal.

Special working groups allow for a more focused and purposeful way to proceed. This means that more depth can also be created in collaborative work and partnerships that will allow for more intentional discourse and response to determined thematic areas. There is merit in working with like-minded individuals to further drive towards a common message and capture various audiences.

18.5.3. *Network crossovers*

AUN-EEC and SDSN PH are privileged to be part of a bigger mother network, namely AUN and SDSN, respectively. Because of this, there are parallel networks that have similar visions situated against a different field or in a different geographical region. For example, SDSN PH has been partnering with the SDSN networks of Southeast Asia (i.e. Southeast Asian countries with no national SDSN network), Thailand, Malaysia and Hong Kong SAR for a variety of webinars. The partnership started when the network managers met in a regional SDSN event.

Meanwhile, as one of the 19 thematic networks of AUN, AUN-EEC developed partnership with the ASEAN University Network on Sustainable Cities and Urban Development (AUN-SCUD) housed at the Universitas Indonesia for a series of activities on urban sustainability. The two networks jointly organized a conference and a city planning workshop cum competition. The presentations submitted to the conference also formed a special issue in the *Journal of Infrastructure Development*.

18.5.4. *Partnerships beyond university network members*

Even though networks of universities have been shown to accomplish remarkable feats in socioecological transitions, seeing the SDGs to fruition requires a whole-of-society approach. Therefore, all networks presented in this chapter, namely AUN-EEC, SDSN PH, ProSPER.Net, HDN and PSPA, have been working closely with governments, civil society and businesses in order to access greater reach.

Government engagement is truly important in translating the knowledge generated by the universities to policy and programs related to socioecological transitions. The PSPA worked with the DILG in capacity-building projects for public sector reform; SDSN PH worked with the National Economic Development Authority on the voluntary report on the SDGs and on communicating the SDGs to the public; and AUN-EEC provided recommendations to the Philippine Congress as part of the consultation process for various bills on ecological education and ecosystem services.

The leadership councils of SDSN PH and HDN also include representatives outside of academia who can provide a more nuanced take on mobilizing projects given the perspectives of the civil society and the business community. Such partnerships also yielded financial support to some projects of the networks of universities.

18.6. Concluding remarks

Any work in the field of sustainability can be overwhelming when considering all that needs to be done (i.e. the problems that need to be solved, the issues that must be responded to, and the obstacles that should be overcome). The work is vast and perhaps intimidating, as it is complex, but there is good merit in proceeding with partnerships towards socioecological transitions. Universities, as centers of knowledge generation and propagation, play an integral role in this pursuit through the formation of networks which mobilize and amplify their work.

As in the experiences elaborated on and shared in this chapter, the secret lies in synergizing efforts through the networks of universities and forging meaningful connections to create meaningful and lasting impact as universities journey in contributing to the achievement of the Sustainable Development Goals.

18.7. Acknowledgments

The authors would like to thank Dr. Ma. Assunta Cuyegkeng, Dr. Charlotte Kendra Gotangco Gonzales, Ms. Jean Meir Jardeleza-Mijares and Ms. Abigail Marie Favis (Ateneo de Manila University) for their contributions in setting up and managing the networks presented here. Also, the authors would like to thank Dr. Lizan Perante-Calina (Graduate School of Public and Development Management, Development Academy of the Philippines) for information on the

Philippine Society for Public Administration (PSPA), in which she served as president and executive director.

18.8. References

Balanay, R.M. and Halog, A. (2016). Teaching education for sustainable development at university level: A case study from the Philippines. In *Teaching Education for Sustainable Development at University Level*, Leal Filho, W. and Pace, P. (eds). Springer, Cham.

Bernardino, C.S. (2000). Exploring education for sustainable development: Its theory and practice in Philippine higher education institutions. PhD Dissertation, University of Alberta, Edmonton.

Brundtland Commission (1987). Our common future: Report of the world commission on environment and development. Report, WCED, Oslo.

Crona, B., Wassénius E., Parlato, G., Kashyap, S. (2024). Doing business within planetary boundaries. Research Brief, Stockholm Resilience Centre (Stockholm University) and the Beijer Institute of Ecological Economics, Royal Swedish Academy of Sciences, Stockholm.

Cuyegkeng, M.A.C. and Favis, A.M.T. (2019). Sustainability practices in higher education: Stakeholder engagement in promoting campus sustainability. *Journal of Management for Global Sustainability*, 7(2), 101–117.

Elkington, J. (1994). Towards the sustainable corporation: Win-win-win business strategies for sustainable development. *California Management Review*, 36(2), 90–100.

Galang, A.P. (2010). Environmental education for sustainability in higher education institutions in the Philippines. *International Journal of Sustainability in Higher Education*, 11(2), 173–183.

Kestin, T., Lumbreras, J., Puch, M.C. (2020). Accelerating education for the SDGs in universities: A guide for universities, colleges, and tertiary and higher education institutions. Report, Sustainable Development Solutions Network, New York.

Leicht, A., Heiss, J., Byun, W.J. (eds) (2018). *Issues and Trends in Education for Sustainable Development*. UNESCO, Paris.

Pope Francis. (2015). *Laudato Si'. Encyclical letter*. Vatican Press, Vatican City.

Posio, P. and Delocado, E.D. (2025). MOOC for sustainability: Cultivating sustainability leadership in campus management through open online education. *Journal of Management for Global Sustainability*, 13(1), 1013103.

SDSN (2024). SDSN networks in action 2024. Report, Sustainable Development Solutions Network, Paris, New York, Kuala Lumpur.

Times Higher Education (2025). Impact Rankings 2025: Results announced [Online]. Available at: https://www.timeshighereducation.com/depth/impact-rankings-2025-results-announced [Accessed: 24 August 2025].

United Nations (2023). 2023: Times of crisis, times of change: Science for accelerating transformations to sustainable development. Global Sustainable Development Report (GSDR), United Nations, New York.

19

Better Together? Exploring What Can Be Achieved through Real-world Examples of Innovative Knowledge Exchange Partnerships between SMEs and Universities

19.1. Introduction

Where does responsibility lie for tackling climate change? Between the two extremes of "big" business and multi-national corporations and individuals are two organizational sectors that offer potential: small- and medium-sized enterprises (SMEs) and universities. This chapter presents the findings from a knowledge exchange (KE) project between Nottingham Trent University (NTU) and SMEs in the Nottingham region through a whole range of business support, student-led consultancy and financial support offered to local SMEs to help them on their journey to net zero.

It is becoming well established that SMEs have a crucial role as an economic and social force in driving the changes required in the UK's transition to a net-zero economy (Baranova and Conway 2017). Their environmental footprint is significant with smaller businesses collectively accounting for approximately half (43–53%) of greenhouse gas emissions by UK businesses (British Business Bank 2021). The picture is much the same in Nottingham itself, albeit with a higher proportion of manufacturing companies alongside wholesale and retail trade and automotive organizations. Thus, SMEs have significant potential to influence and bring change

Chapter written by Richard BULL, Ana Rita DOMINGUES, Muhammad MAZHAR and Gamze YAKAR-PRITCHARD.

toward the low-carbon economy collectively. They also have an essential role in delivering UK and EU carbon emissions targets (net zero by 2050). The capacity to implement and monitor environmental and carbon aspects differs significantly compared to larger businesses (Fawcett and Hampton 2020). The specific characteristics of SMEs, often characterized by restricted resources (such as human and financial), explain in part their slower transition to implementing systems that enable them to measure, manage and plan the transition to a low-carbon economy.

Universities play a significant role in their local economies, with a substantial impact on the environment – both positive and negative. Much has been written on their impact in terms of their curriculum, carbon management and research (Tilbury 2014; Mahzar et al. 2017; Rodríguez Aboytes and Barth 2020). Less work has been done on the potential within the emerging field of "KE" to support and work with local businesses to enable the transition to net zero. This chapter proceeds as follows: first, the concept of KE is defined and situated within wider understandings of partnership working and collaboration, and then the research findings and case studies are discussed. Finally, conclusions and recommendations are offered.

19.2. Research themes: KE, sustainability and partnerships

KE has risen rapidly up the agenda in universities, partly helped by the UK government taking steps to promote participation as a tool for KE (Johnson 2022). The Knowledge Exchange Framework (KEF) was introduced by Research England (RE) in 2020 as a new assessment tool (UKRI 2022) and the Office for Students (OfS) and RE launched a student involvement program worth £10 million in 2020 to develop effective practices for student participation in KE activities outside of their curriculum. A recent report published by the Office for Students (2022) shows that approximately 16,700 students and 2,300 partners have been involved in KE projects. KE is about universities engaging in activities beyond the traditional teaching and research agenda. KE is a concept that contributes to developing innovative learning strategies by providing a better understanding of the outcomes of participatory learning methods, for example, problem-based learning, real-life consultancy courses or experiential learning activities. It implies a two- or multi-directional knowledge-sharing process with mutual benefit and multi-learning between researchers, decision-makers, practitioners, businesses and other beneficiaries (Fazey et al. 2013, 2014). Instead of the traditional approach where academics are considered the sole producers of knowledge, these approaches involve multiple interactions, such as engaging students and decision-makers in the co-production of knowledge (Fazey et al. 2014; Karcher et al. 2022). A key principle of KE practice is focused on engaging Higher Education Institutions (HEIs) with stakeholders and broader society to bring change. Specifically, KE aims to measure the academic, economic and societal impact, change or benefit of learning, teaching or other activities in HEIs (Johnson 2022).

This has particular relevance when considering education for sustainable development and the aspiration of universities to prepare students for future careers that will address the challenge of net zero. Student KE projects offer a functional and innovative model for real-world learning opportunities. This approach, which integrates teaching and learning, enables students to go beyond mere knowledge acquisition. It helps them grow into value-driven graduates capable of strategic and systematic thinking on sustainability issues, especially on topics such as net zero (Brundiers et al. 2010; Konrad et al. 2021; Birdman et al. 2022). Such projects enable students to (1) address real-world sustainability issues in collaboration with businesses, local communities and civil society; (2) develop skills such as problem-solving and collaborative working (Wiek et al. 2014; Kricsfalusy et al. 2018); and (3) help students develop the necessary confidence and leadership skills to address sustainability issues in their future work lives (Brundiers et al. 2010; Heiskanen et al. 2016; McPherson et al. 2016).

With increasing expectations on the universities to impact their local and regional environments beyond "merely" providing education, there are questions over the efficacy of this. Gohari et al. (2019) have explored the complex relationship and governance arrangements that universities operate within. NTU has a civic partnership arrangement with its other local university (Universities for Nottingham) to work together on shared initiatives. NTU also sits on various boards within the city and the region. Parallels are observed with debates over recent years around the changing nature of governance over the last few decades.

In the UK, in the late 90s, democratic renewal was seen to be as a result of greater participation and partnerships in local and regional government and changes, for example, in planning regulation (Newman 2001). Williams (2004) noted that New Labour policy drew from radical democratic discourse and saw public engagement and partnerships as a vital component of state governance. Davoudi and Evans (2005) drew attention to the effects of the restructuring of society and the role of the state in mediating power through devolving responsibilities to regional agencies and partnerships. The notion of networks and partnerships was fiercely debated between Rhodes (1997) and Davies (2001). Rhodes (1997, p. 53) asserted that "governance refers to self-organizing, inter-organization networks", i.e. a series of different actors, for example, private sector companies, local community groups and local authorities coming together to accomplish a task.

The task in a city like Nottingham is carbon reduction, with Nottingham having set an ambitious target of net zero by 2028. As key actors in cities, universities have a vital role in contributing to the economy, employment and the ambitions of the cities and regions in which they are situated. This chapter presents one example in which NTU, a university of 40,000 students with a strategic priority to "embrace sustainability", worked with SMEs to support Nottingham's ambition to be net zero by 2028 by embarking on a KE project, "Sustainability in Enterprise".

19.3. Research method and context

This research was conducted in the context of the Sustainability in Enterprise (SiE) project, which worked with SMEs in the Greater Nottingham area to progress toward net zero by providing free, practical support from specialists in sustainable business operations, building management, product design and employee engagement. SMEs signed up to seek support, provided they met the criteria.

The project helped businesses understand their current carbon footprint and identify effective carbon management (repeated phrase) measures. Financial support was available to help companies to implement efforts to reduce carbon emissions and carbon management (repeated phrase) workshops were offered to SME managers to build their in-house capacity and know-how to implement carbon management (repeated phrase); also consultancy from academics, practitioners and students from multi-disciplinary teams at NTU. Student sustainability consultancy was also provided to SMEs by students from Nottingham Business School (NBS) and the School of Architecture, Design and the Built Environment (ADBE), which was a unique part of the project in terms of KE. The project has similar elements to the concept of Learning Energy Efficiency Networks (LEEN) which was originally developed in Switzerland in 1987 (Köwener et al. 2014).

Since then, the network concept was transferred and adapted to the German, Swedish and Austrian context. LEEN are designed to accelerate the take-up of energy efficiency measures by companies, and networks are managed by local change agents such as business groups, trade bodies and banks with the central government playing a key role (Mallaburn 2018).

The research adopted a mixed methods project evaluation approach including an online survey questionnaire sent to all enrolled SMEs, pre- and post-intervention surveys and case studies. The number of participants who responded to the pre-intervention survey was 215, and the number of participants who had responded to the post-intervention survey was 124. Only SMEs who responded to both pre- and post-intervention surveys were included in this study to measure the impact on SMEs. After the elimination of the participants who had responded to neither of the surveys and those who had not given consent to use their data for the research, the final study sample was 101.

19.4. Findings and discussion

19.4.1. *SiE project*

The SiE project unlocked NTU expertise to support and partner with SMEs. Part-funded by the European Regional Development Fund, the £4 million SiE program supported over 200 SMEs to understand and reduce their carbon footprint and drive low-carbon growth. Eligible businesses were located in a geographic area, comprising Nottingham City Council and Broxtowe Borough Council, parts of Rushcliffe and Gedling Borough Councils and the Hucknall area of Ashfield District Council. Running between 2021 and 2023, the project provided support from sustainability specialists across the university, enabling local businesses to benefit from free, practical support and interdisciplinary expertise delivered through one holistic project. SMEs received at least 12 hours of support or a grant. However, the support was often much more (please see Table 19.1 for a full overview of the SiE business support offer). The product design strand averaged over 100 hours of consultancy to each business they worked with.

SiE program strand	Details
Full Environmental and Carbon Audits	This support offer for businesses was led by NTU's award winning sustainability team. 56 businesses were supported and provided with baseline carbon footprint and detailed, practical carbon reduction recommendations with carbon emission reduction, cost savings and payback periods calculated. This was a 15-hour consultancy package, including a site visit to the business premises.
Premises Consultancy	ADBE led this strand of the SiE programmed. 27 businesses seeking to move or refurbish their premises supported with retrofit survey and recommendations by the SiE sustainability consultant architect directly or by student teams supervised by the consultant architect. Over 220 undergraduate students participated. Each consultancy included a minimum of 15 hours of consultant architect time including a site visit.
Product Design Consultancy	Led by the School of Art and Design, 14 bespoke product design consultancies were provided by commercial product design consultants to reduce carbon, including product/packaging life cycle analysis, practical design concepts, material recommendations and data for prototype production. These were 50–200-hour consultancy packages depending on the brief.
Student Sustainability Consultancy	Nottingham Business School dedicated an elective final-year undergraduate module to providing carbon consultancy to 43 businesses. 280 students participated, supervised by a small specialist team of academics.

In-house capacity building programs: Carbon Management Program (CMP)	Nottingham Business School delivered a skills development program on carbon management to 66 businesses. The NTU employability department delivered a peer support program for early career staff to 77 individuals in 68 businesses – to develop a strategic sustainability project within their business.

Table 19.1. *Business support offers in the SiE project*

19.4.2. *Insights and impacts*

While the project reports to an overall carbon reduction target of 750 tCO2 e, a wider lens is required to understand the full impact of the support. All SiE beneficiaries completed a sustainability survey before and after receiving support. The survey was developed based on the key elements of the carbon management (repeated phrase) process within businesses. Analysis of the quantitative data from participating SMEs shows carbon management (repeated phrase) and sustainability practices significantly progressed. This is likely to result in an overall carbon emission-reduction impact greater than that we can confidently calculate. An interdisciplinary research team from Nottingham Business School and the School of ADBE at NTU supported the SiE impact evaluation and data analysis as part of the SUSchange project.

19.4.3. *Physical actions*

The SiE program was action oriented with an aim to support and encourage SMEs to take actions to reduce energy and carbon emissions of their operations and value chain. The research found that SMEs had planned to take actions in a range of areas of carbon impact. SMEs report planned energy and carbon reduction actions in building energy (e.g. electricity and gas consumption) (70%), waste (52%), business travel (50%), procurement and supply chain (50%), staff commute (40%), own transport fleet (27%) and water consumption (23%).

As a result of the university support, 62% of businesses are now effectively managing their carbon emissions and successfully implementing carbon reduction projects as planned, compared with 9% before the SiE support. As a result, in total, the SiE project has managed to reduce 1000 tCO_2e based on the actions reported by participating businesses (please see Table 19.2 for an overview of the physical actions taken by SMEs).

Physical actions taken by SMEs
Almost 1000 tCO_2e reduced. Carbon calculations were only completed for businesses receiving SiE consultancy.
SMEs report planned energy and carbon reduction action in key areas such as building energy (70%), waste (52%), business travel (50%), procurement and supply chain (50%), staff commute (40%), own transport fleet (27%) and water (23%).
62% of businesses are effectively managing their carbon emissions and successfully implementing carbon reduction projects as planned, compared with 9% before support.

Table 19.2. *Physical actions taken by SMEs for carbon management*

19.4.4. *Awareness, engagement and strategy*

Awareness, engagement and strategy development are key to delivering net zero ambitions. Comparative pre- and post-business support data analysis from businesses shows that the SiE project improved sustainability and carbon management (repeated phrase) performance of SMEs through real actions such as awareness raising, engagement and strategy. 72% of the businesses have developed an environmental or sustainability policy and 57% of the businesses now have a carbon management (repeated phrase) strategy and plan (10% before the project support). 70% now have, or are progressing toward, setting, carbon reduction targets (25% before the project support). 56% see carbon management (repeated phrase) as central to our business activities and decision-making process. Table 19.3 provides detailed results against the key indicators.

Awareness, engagement and strategy	Before	After
We have developed an environmental or sustainability policy for the business	57%	72%
We have a dedicated carbon management plan or strategy	10%	57%
Carbon management is central to our business activities and decision-making	20%	56%
We have, or are progressing toward, setting carbon reduction targets	25%	70%
We are able to manage the conflicts between carbon management and core business growth	23%	59%
We have carbon reduction behavioral change and awareness raising programs for staff	12%	59%
We communicate our carbon reduction targets, strategies and performance to all relevant stakeholders	12%	53%

Table 19.3. *Awareness, engagement and strategy of SMEs*

19.4.5. *Knowledge and capability*

The SiE project had a significant impact on improving knowledge and capability of SMEs. The majority of the SMEs have now either measured their carbon footprint to establish baseline or it is in progress and SMEs are of the view that they now have the required knowledge and skills to calculate the carbon footprint for business. Furthermore, SMEs are now tracking and reviewing progress toward their carbon management (repeated phrase) target to report progress and communicate to their stakeholders.

Table 19.4 details the "Before" and "After" figures for comparison and impact assessment. A lot of this is due to the university partnership with SMEs and support via consultancy and carbon management (repeated phrase) training. NTU was able to act as a knowledge hub and was enabled with resources and expertise to support SMEs in knowledge and capability development. The project resulted in three-way learning: learning for SMEs, university and the wider project team who learned from real-world industry challenges.

Knowledge and capability	Before	After
Does your enterprise measure its carbon emissions/footprint? Yes/in progress	18%	71%
We have the required knowledge and skills to calculate the carbon footprint of our business	12%	73%
We are effectively tracking and reviewing progress toward our carbon management target	8%	58%

Table 19.4. *Knowledge and capability of SMEs in carbon management*

19.4.6. *Monitoring resource use*

SiE support led to a marked improvement in business resource monitoring which was an area SMEs struggled with. Capacity to collect resource data was a significant barrier for some businesses accessing the program due to various reasons such as lack of access or control of the metering or billing, residing in rented buildings and lack of systematic data management procedures within the business.

Many SMEs received considerable SiE help to identify and collect data in the correct format prior to formal support commencing. SiE improved the monitoring of resource use for effective carbon management (repeated phrase) including electricity, gas, water, water and fuel consumption, as presented in Table 19.5.

Monitoring resource use	Before	After
Electricity	52%	83%
Gas	37%	59%
Water	34%	46%
Waste	30%	60%
Fuel – owned fleet vehicle	8%	31%
None	40%	9%

Table 19.5. *Monitoring resource use for carbon management*

19.4.7. *Addressing barriers to decarbonization*

Before the SiE project commenced, SiE businesses participated in a baseline survey and reported the lack of knowledge and expertise and lack of time and resources were the biggest barriers to decarbonization. Exit surveys showed SiE-style support significantly reduces these barriers to change (see Table 19.6).

SiE businesses told us that access to capital funding and factors outside their control were the biggest barriers going forward. Being in leased premises and the challenges associated with that were referenced prominently by exiting businesses. Barriers such as lack of knowledge and expertise, lack of time and resources, lack of data collection, monitoring or management systems and lack of policy or strategy were significantly addressed through the support strands of SiE.

Barriers	Before	After
Lack of knowledge and expertise	79%	2%
Lack of time and resources	58%	15%
Lack of capital funding	46%	48%
Lack of data collection, monitoring or management systems	44%	8%
Lack of policy or strategy	42%	2%
Not a priority	7%	2%
Staff resistance to behavioral change	6%	7%
Factors out of control (e.g. rented premises/accessing suppliers)	Not asked	40%

Table 19.6. *Barriers to decarbonization*

19.5. Case studies

19.5.1. *Case study 1: Murphy & Son Ltd*

Basford-based manufacturing business, Murphy & Son Ltd, has been supporting the world's leading breweries for over 130 years, providing technical support and supplies for brewers, winemakers and manufacturing chemists. Their small team, operating from a large Victorian brewery, were at an early stage of improving their sustainability but keen to develop. Time and knowledge were core barriers to progress with limited in-house company knowledge around sustainability. The NTU Sustainability team led the support, providing the company's first full carbon audit, with a roadmap to reducing their emissions by over 200 tCO2e. A project carbon grant of £4,000 helped them take forward the priority carbon reduction measures. Subsequently, the company hired their first sustainability specialist role, with support from the NTU Employability team (support to source talent, 20% salary grant and a sustainability peer-support program). After a few months in post, the new staff member also attended the Nottingham Business School Carbon Management Program, learning how to measure and reduce the company's carbon emissions. Overall, Murphy & Son received 46 hours of support and £7,000 of grant. Within a year of receiving the support, they had achieved a reduction of 96 tCO_2e and had developed vital in-house skills and know-how to continue this progress.

> I thought it would just be looking for one or two quick fixes and that's it. But what has been great is the scope of the sustainability consultancy is so broad. It has helped to bring everything together and has helped inform us how to coordinate it better (Head of Technical, Murphy & Son Ltd).

19.5.2. *Case Study 2: Alpkit*

Alpkit are an award-winning outdoor and bike brand born out of a passion for the outdoors. The Derbyshire-based company, founded in 2004, has retail stores around the UK and sells online to approximately 70 countries worldwide. Alpkit were the first outdoor brand to publish a comprehensive sustainability report and have been certified by B-Corp since 2021. Alpkit wanted to review two of their best-selling products, already designed to be low-impact, repairable and recyclable. Fifty-four BSc (Hons) Product Design students from the NTU School of ADBE and 17 international exchange students calculated the carbon footprint of the Alpkit Soloist tent, making design recommendations to make it more sustainable.

For a second product, the Bru-Kit stove, the NTU commercial product design consultancy "Design Matter" from the School of Art & Design conducted a similar process, assessing the carbon footprint and making several design recommendations focused on carbon reduction, as well as look and usability. Finally, the NTU Sustainability team (Professional Services) provided a full carbon audit of their business operations, providing recommendations and a grant to help the company implement changes. Alpkit received 90 hours of professional consultancy support overall, a much greater number of student consultancy hours and £7,000 of grant. Recommendations provided have led to a reduction of 53 tCO2e, of which 14.8 tCO_2e were adopted by the company in the same year. Teaching and research academics in two schools as well as 71 undergraduate students gained valuable experience of applied life cycle analysis and sustainable design principles:

> Focusing on our Soloist tent enabled us to explore the data-driven redesign of one of our best-selling products ... At the end of the project, we had 21 innovative ideas to choose from. Using aspects from every student project, we're exploring switching to recycled materials, eliminating waste in production, redesigning the pole hubs to reduce weight and material use, tweaking the tent's dimensions and structure, improving our packaging, and prolonging the Soloist's usable life with improved care instructions and repair kits (Senior Product Developer, Alpkit).

19.6. Conclusions and recommendations

At a time of intense scrutiny over the role of universities alongside increasing climate anxiety and concern over rising global temperatures, this example of partnership working between NTU and SMEs in Nottingham offers an example of what can be achieved. Funding and leadership are fundamental to this. Thankfully, the UK is now again part of Horizon funding, as the ERDF investment is critical to this project's success. Strong leadership and an ambitious vision of a university to engage in KE activity were also key. Increasingly, universities may need to play an important neutral "brokering" and leadership role within city landscapes, especially when funding constraints are severely impacting the role of municipalities to lead in this space, with multiple UK municipalities declaring bankruptcy being the most notable challenge as the state is increasingly hollowed out. For SMEs, future research needs to address the organizational change process that led SMEs to behave differently, as they are all under the same regulatory framework. The industry could play a role as multiple initiatives are developed in different sectors, which might influence the implementation of environmental actions and effective carbon management. Cross-sector SME analysis can be carried out through segmentation to draw upon similarities and differences as well as lessons learnt. At the same time, cultural identity and image might also play a key role, as key internal stakeholders might align their values with the business organization.

19.7. References

Baranova, P. and Conway, E. (2017). Transition to a low-carbon economy: An SME perspective. In *The Low Carbon Economy*, Baranova, P., Conway, E., Lynch, N., Paterson, F. (eds). Palgrave Macmillan, Cham, 87–115.

Birdman, J., Wiek, A., Lang, D.J. (2022). Developing key competencies in sustainability through project-based learning in graduate sustainability programs. *International Journal of Sustainability in Higher Education*, 23(5), 1139–1157. doi: 10.1108/IJSHE-12-2020-0506.

British Business Bank (2021). Smaller businesses and the transition to net zero [Online]. Available at: https://british-business-bank.co.uk/wp-content/uploads/2021/10/J0026_Net_Zero_Report_AW.pdf [Accessed 20 October 2021].

Brundiers, K., Wiek, A., Redman, C.L. (2010). Real-world learning opportunities in sustainability: From classroom into the real world. *International Journal of Sustainability in Higher Education*, 11(4), 308–324. doi: 10.1108/14676371011077540.

Davies, J. (2001). *Partnerships and Regimes: The Politics of Urban Regeneration in the UK*. Taylor Francis, Aldershot.

Davoudi, S. and N. Evans (2005). The challenge of governance in regional waste planning. *Environment and Planning C*, 23, 493–517.

Fawcett, T. and Hampton, S. (2020). Why & how energy efficiency policy should address SMEs. *Energy Policy*, 140, 111337. doi: 10.1016/j.enpol.2020.111337.

Fazey, I., Evely, A.C., Reed, M.S., Stringer, L.C., Kruijsen, J., White, P.C.L., Newsham, A., Jin, L., Cortazzi, M., Phillipson, J. et al. (2013). Knowledge exchange: A review and research agenda for environmental management. *Environment Conservation*, 40(1), 19–36. doi: 10.1017/S037689291200029X.

Gohari, S., Medalen, T., Aranya, R. (2019). Exploring the impact of complex multi-level governance structures on the societal contribution of universities to knowledge-based urban development. *Social Sciences*, 8(10), 279.

Fazey, I., Bunse, L., Msika, J., Pinke, M., Preedy, K., Evely, A.C., Lambert, E., Hastings, E., Morris, S., Reed, M.S. (2014). Evaluating knowledge exchange in interdisciplinary and multi-stakeholder research. *Global Environmental Change*, 25(1), 204–220. doi: 10.1016/j.gloenvcha.2013.12.012.

Heiskanen, E., Thidell, Å., Rodhe, H. (2016). Educating sustainability change agents: The importance of practical skills and experience. *Journal of Cleaner Production*, 123, 218–226. doi: 10.1016/j.jclepro.2015.11.063.

Johnson, M.T. (2022). The knowledge exchange framework: Understanding parameters and the capacity for transformative engagement. *Studies in Higher Education,* 47(1), 194–211. doi: 10.1080/03075079.2020.1735333.

Karcher, D.B., Cvitanovic, C., van Putten, I.E., Colvin, R.M., Armitage, D., Aswani, S., Ballesteros, M., Ban, N.C., Barrag´an-Paladines, M.J., Bednarek, A. et al. (2022). Lessons from bright-spots for advancing knowledge exchange at the interface of marine science and policy. *Journal of Environmental Management*, 314, 114994. doi: 10.1016/j.jenvman.2022.114994.

Konrad, T., Wiek, A., Barth, M. (2021). Learning to collaborate from diverse interactions in project-based sustainability courses. *Sustainability*, 13(17), 9884. doi: 10.3390/su13179884.

Köwener, D., Nabitz, L., Mielicke, U., Idrissova, F. (2014). Learning energy efficiency networks for companies–saving potentials, realization and dissemination. In *Proceedings of the European Council for an Energy Efficient Economy (ECEEE) Industrial Summer Study*, Arnhem, 91–100.

Kricsfalusy, V., George, C., Reed, M.G. (2018). Integrating problem- and project-based learning opportunities: Assessing outcomes of a field course in environment and sustainability. *Environmental Education Research*, 24(4), 593–610. doi: 10.1080/13504622.2016.1269874.

Mallaburn, P. (2018). Principles of successful non-residential energy efficiency policy. In *Proceedings of the European Council for an Energy Efficient Economy (ECEEE) Industrial Summer Study*, Berlin, 15–22.

Mazhar, M., Bull, R., Lemon, M. (2017) Critical success factors for embedding carbon management in organisations: Lessons from the UK higher education sector. *Carbon Management*. 8(5–6), 379–392. doi: 10.1080/17583004.2017.1386533.

McPherson, S., Anid, N.M., Ashton, W.S., Hurtado-Martín, M., Khalili, N., Panero, M. (2016). Pathways to cleaner production in the Americas II: Application of a competency model to experiential learning for sustainability education. *Journal of Cleaner Production*, 135, 907–918. doi: 10.1016/j.jclepro.2016.06.138.

Newman, J. (2001). *Modernising Governance: New Labour, Politics and Society*. Sage Publishing, London.

Office for Students (2022). Student engagement in knowledge exchange competition. Student Engagement in KE formative report 3 [Online]. Available at: www.officeforstudents.org.uk [Accessed 24 November 2022].

Rhodes, R. (1997). *Understanding Governance: Policy Networks, Governance, Reflexivity and Accountability*. Open University Press, Buckingham.

Rodríguez Aboytes, J.G. and Barth, M. (2020). Transformative learning in the field of sustainability: A systematic literature review (1999–2019). *International Journal of Sustainability in Higher Education*, 21(5), 993–1013. doi: 10.1108/IJSHE-05-2019-0168.

Tilbury, D. (2014). Higher education for sustainable development. Commissioned Evaluation of the Impact of the UN Decade of Education for Sustainable Development. UNESCO, Paris, 80.

UKRI (2022). Review of the first iteration of the Knowledge Exchange Framework. Report [Online]. Available at: https://www.ukri.org/wp-content/uploads/2022/02/RE-030222-KEFReviewReport.pdf [Accessed 24 November 2022].

Wiek, A., Xiong, A., Brundiers, K., van der Leeuw, S. (2014). Integrating problem and project-based learning into sustainability programs: A case study on the school of sustainability at Arizona state university. *International Journal of Sustainability in Higher Education*, 15(4), 431–449. doi: 10.1108/IJSHE-02-2013-0013.

Williams, M. (2004). Discursive democracy and new labour: Five ways in which decision-makers manage citizen agendas in public participation initiatives. *Sociological Research Online*, 9(3), 1–17. doi: 10.5153/sro.949.

20

Research for Energetic and Societal Transition: Experience of the Live TREE Program

20.1. Introduction

Universities are ideal places for experimentation at scale and in real conditions of different axes of energy and societal transition. This is due to the support of interdisciplinary research, and also thanks to information, awareness and involvement of students and staff, as well as the populations within the territories in which they are anchored, through training and education missions.

Since 2013, in the Hauts-de-France region in the wake of the Third Industrial Revolution, the Catholic University of Lille has been engaged in the "Live TREE" program (Lille Vauban-Esquermes in energy, ecological and economic transition). This program aims to support university establishments in achieving carbon neutrality and to develop a living laboratory on a real scale (called a living lab), as well as to address other challenges. This is achieved via experimental buildings in a sustainable and desirable district, which enable us to experiment and implement green mobility strategies, bring nature back to the city, develop the student experience by encouraging the active involvement of students in the transition and develop transdisciplinary research between human and social sciences and engineering sciences, etc.

After briefly presenting the Catholic University of Lille and the University Convention for Climate (that it organized in the fall of 2022, which was a democratic approach to identify actions to act in favor of the climate and the planet,

---

Chapter written by Benoît ROBYNS.

For a color version of all figures in this chapter, see www.iste.co.uk/dobigny/rethinking.zip.

and also an object of sociological research), the university's demonstrator development approach is described. These demonstrators force researchers to leave the laboratory and confront reality, inducing essential interdisciplinary research to find solutions to successfully carry out the energy and social transition.

Some interdisciplinary research projects conducted by the university lead to the observation that this research is essentially high-tech with a significant and still poorly understood climatic and environmental impact, and an appropriation by everyone that is more than uncertain, even taking into account the sociological profiles of the different actors.

So, the question "should we continue scientific research?" is discussed. It emerges that research must be increasingly interdisciplinary, that an analysis of the contributions of the results of each research work to the UN Sustainable Development Goals (SDGs) is recommended, as well as the development of low-tech approaches in all areas where this is possible, by encouraging research work on these approaches, which are also a source of innovation but also of change in lifestyle.

20.2. University: the ideal place for research and implementation of energy and societal transition

20.2.1. *Introduction*

In November 2013, the Catholic University of Lille[1] decided to commit itself resolutely to the Third Industrial Revolution theorized by the American economist Jeremy Rifkin (2012) and thus to contribute to the essential energy, ecological, economic and social transition of the university and the Hauts-de-France region by developing an orientation plan for the "Live TREE" program (Lille Vauban-Esquermes in energy, ecological and economic transition).

The Live TREE program aimed to set up a global approach which concerns all aspects of university life and is connected to its environment: students, staff, inhabitants of the Vauban district, partner companies, communities, associations, etc. Through Live TREE, the university wished to strengthen its role of providing

1 The Catholic University of Lille is a multidisciplinary federation that brings together faculties, colleges, engineering and business schools, and research centers on the one hand, and a hospital group, on the other hand. At the start of the 2024 academic year, the Catholic University of Lille had welcomed 42,000 students in more than 300 training courses and 7,000 staff members.

"global education" for students as well as its social responsibility as a university and to engage its students, once they have been launched in their adult and professional life, to take action and build new models of a society which is more respectful of the environment and of the individual human, which is more just and equitable. Live TREE is therefore a global program, which crosses the various functions of the university and which implies a transformation of the modes of governance and management used by the university (Robyns et al. 2023).

Based on these objectives and the geographical context of the university, the Live TREE program is organized around several themes:

– carbon neutrality of the university;

– student experience: student training and involvement, learning campus;

– living lab: experimental buildings and living spaces in a sustainable and desirable district, openness to the neighborhood and third places, green mobility, climate and nature in the city;

– transdisciplinary research between human and social sciences and engineering sciences.

These themes make it possible to meet a large number of the United Nations' (UN) sustainable development goals.

20.2.2. *University convention for the climate*

Through its mission of training and education, the university has a special responsibility and opportunity to raise awareness among its students and staff so that they contribute to reducing the university's carbon footprint, and also to reduce their personal carbon footprint in their daily activities in the process. This interaction is aimed at informing, raising awareness and mobilizing university players so that they divide their carbon footprint by 5 to 6, which amounts to going from ten tons of CO_2/year in 2020 to approximately two tons of CO_2/year in 2050. An important step in this mobilization is the organization in the fall of 2022 of a university convention for the climate, the aim of which was to allow students and staff to propose actions that contribute to the development of the university's carbon trajectory which aims for neutrality in 2050, involving actions and visions that transform habits and behaviors, as well as ways of seeing and being in the world. Then, during the fall of 2022, 150 citizens – students, high school students, university staff and residents of the Vauban-Esquermes district – took part in an unprecedented training program and collective intelligence workshops; they proposed solutions with a view to reducing the university's greenhouse gas emissions and climate action.

This convention was also a subject of research in sociology. As observers, sociologists followed and analyzed the process of ideas to offer feedback on the course of the event and recommendations during the convention. Another challenge is the implementation of the convention's 55 proposals in 2023 and beyond by measuring their climatic and sociological impacts. The sociologists will analyze by means of surveys the appropriation of the proposals of the convention by the various actors: establishments of the university, students, staff, etc., and also inhabitants of the district (see Chapter 21). The university will collaborate on these subjects with the region, the European Metropolis of Lille and the city of Lille, because the university's carbon emissions are also dependent on the policies of the territorial authorities (e.g. for the development of public transport, heating networks, for urban planning reducing the effects of climate change and to bring nature back to the city, etc.).

20.3. Demonstrators

Implementing the energy and societal transition will require full-scale experimentation with a wide range of technological, economic and societal solutions while accepting trial, error and adjustment. It is a question of validating the viability of technologies and their effectiveness and impact on the environment, social acceptance (which is by no means automatic), finding new economic models, devising new legal frameworks, etc. The success of the project also depends on effective persuasion. Persuasion plays a role in the degree of resistance to change.

Science and technology are generally developed in laboratories where testing conditions are under control. This generally leads to the development of prototypes for innovations. The complexity and speed of the transition mean that emerging solutions need to be tested earlier and more systematically in real-life conditions, where not everything is under control, particularly climatic conditions and human behavior. For example, in the case of smart buildings, technologies will have to be tested and, ideally, designed with the users of a building in normal operation.

Confronting the reality on the ground, i.e. with the real world, requires us to engage with a wide range of disciplines: electrical and communication techniques for smart grids, for example, and also sociology, to assess the acceptability and involvement of stakeholders' economics, because new viable models need to be found, law, ethical issues, etc. Interdisciplinarity, and even transdisciplinarity, will be essential to the success of the transition (Robyns et al. 2024).

As a part of the Live TREE demonstrators, the university renovates its buildings in order to limit its carbon footprint and to make it a successful model in the acquisition of new technologies such as smart buildings interacting with smart grids (Robyns et al. 2023, 2024).

The Live TREE program encourages experimenting with soft mobility and carries out work on electric mobility. University establishments are pursuing proactive policies in favor of alternative mobility, including mobility management, travel planning, rechargeable electric and hybrid service vehicles, provision of electric scooters or bicycles, bicycle repair stations, reimbursement of the bicycle travel allowance, remote working, etc.

Bringing nature back into the city is also a subject of real-scale experimentation inside and outside buildings. Sustainable gardening, restoration of social bonds, the study of economic models, etc., at the university and nature in the city respond to diverse needs, dynamics and practices.

These different demonstrators converge toward a sustainable and desirable demonstrator neighborhood, as shown in Figure 20.1 (Robyns et al. 2024).

Figure 20.1. *Demonstrators of the ecological and societal transition toward a sustainable and desirable neighborhood (source: UCL and Willy Pulse d'Atelier 9.81)*

An energy control center, as well as a smart building control center, has been developed to collect, manage and store data retrieved from electrical equipment and buildings, to process consumption and production predictions and to control photovoltaic sources, storage and controllable loads such as electric vehicle loads. Figure 20.2 shows part of the control center equipped with a SCADA (Supervisory Control And Data Acquisition) system controlling the first block of demonstration buildings (the historic block of the university) equipped with photovoltaic power plants, storage systems and electric vehicle charging stations, as seen from the SCADA system (Robyns et al. 2024).

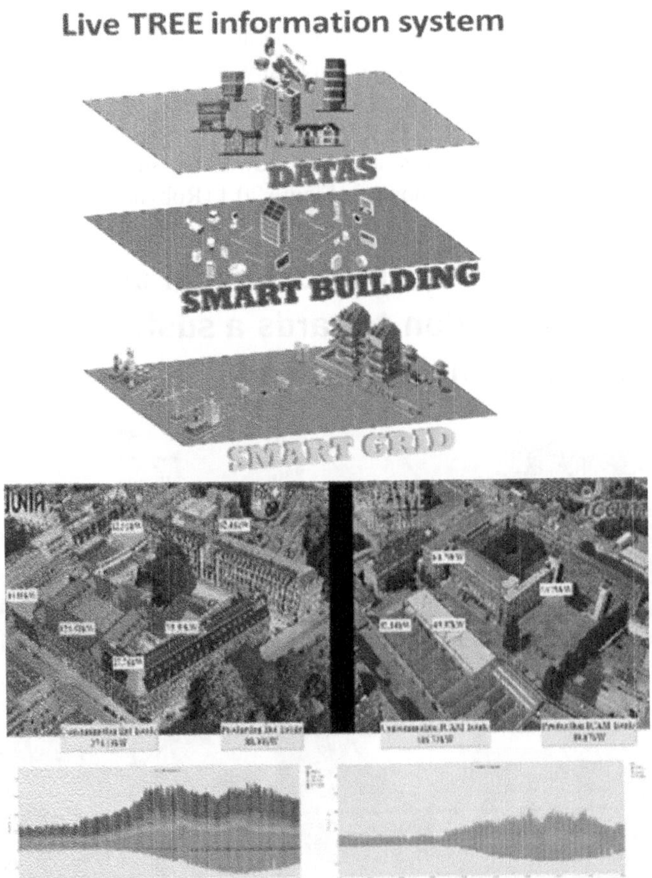

Figure 20.2. *Live TREE information system and part of the control center equipped with an SCADA (Supervisory Control And Data Acquisition) system (Robyns et al. 2024)*

20.4. Interdisciplinary research

20.4.1. *Necessary interdisciplinary research*

To make sure that energy and societal transition is a success, we need to get the various scientific and humanities disciplines to work together, in order to develop the interdisciplinary approaches between the humanities, social sciences and engineering sciences, which are essential to a successful transition that belongs to everyone. Each field of science has developed its own methods and language, which are shared by an often-restricted community. This makes dialogue between the sciences difficult. Researchers need to simplify their discourse to make it more accessible to non-experts in their discipline. Exchanges between disciplines on vocabulary and methods are essential to identify more precisely the possibilities for dialogue. Interdisciplinary research, which crosses disciplines, and transdisciplinary research, which makes equal use of the expertise of at least two disciplines, must be encouraged and recognized by the authorities who guide and evaluate research, because this effort to cross disciplines is an adventure that takes time to be productive; up to now, evaluations of research activities have been essentially mono-disciplinary. Time is needed to build trust and respect for different disciplines on an ethical basis.

20.4.2. *Example of research projects*

As part of the Live TREE program, the Catholic University of Lille has developed several interdisciplinary projects contributing to the transition, for example:

– Working with the European Metropolis of Lille as part of the "So MEL, So Connected"[2] project on self-consumption of photovoltaic production in intelligent interaction with the distribution electrical grid manager (Robyns et al. 2024).

– The university managed the European project MaPerEn Life[3] "Energy Performance Management driving a new governance", in which the city of Lille and a social housing organization (LMH) are partners (Robyns et al. 2023).

– The university was a partner as a demonstrator site of the H2020 ebalance+[4] which aims to increase the electric grid flexibility with an energy balancing (management) platform integrating smart energy technologies (store, generation and consumption) (Robyns et al. 2024).

2 See: https://www.cre.fr/documents/fiches-demonstrateurs-smartgrids/so-mel-so-connected.html.

3 See: https://www.maperen.eu/.

4 See: https://www.ebalanceplus.eu/.

– Junia managed the chair "Smart Buildings as nodes of Smart Grids"[5] (Robyns et al. 2024).

– PhD thesis work on technological, societal, economic and environmental approach of renewable energy communities by interdisciplinary approaches taking into account the sociological profiles of the different actors (Durillon et al. 2020, 2021; Bossu et al. 2024a, 2024b; Robyns et al. 2024).

A part of these works is presented in Chapters 10 and 11).

– The INCLUNIV project works on mobilization and inclusion in the ecological transition of universities presented in Chapter 22.

However, all of this research develops mainly high-tech solutions with a significant and still poorly understood climatic and environmental impact and an appropriation by everyone that is more than uncertain, even taking into account the sociological profiles of the different actors.

20.4.3. *Should we continue scientific research?*

> Innovation, we tell you. Research and development, green growth, 'high tech' products and therefore necessarily clean and resource-efficient, here is the answer to all our ills, those of the bloodless planet, as well as those of our societies in search of new life. Sleep soundly, good people, because in ultramodern laboratories and cutting-edge companies, serious researchers and engineers in white coats work and invent for you. We are looking, and since we are looking, we will find – we have always found, haven't we? (Bihoux 2014).

Then, should we continue scientific research?

This questioning was carried out by Alexandre Grothendieck, an eminent researcher in mathematics who won the Fields Medal (equivalent to the Nobel Prize) in 1966, who, noting the state of the world and social relations, asked himself the question "Are we going to continue scientific research?" during a conference at the European Council for Scientific Research (CERN) in Geneva in 1972.

Starting from the observation that scientific research leads to the technoscience that is primarily responsible for the current climate and environmental situation, that technology could certainly provide answers to limit the damage, or even rebuild ecosystems, Grothendieck pointed out that scientific research, especially of a

5 See: https://rev3-energie.fr/project/chaire-smart-buildings-as-nodes-of-smart-grids-sbnodessg/.

fundamental nature, does not ensure that it will have a positive contribution to meeting global challenges.

After 20 years of fruitful research, Grothendieck stopped all research activity for the following reason: “it was because I realized that there were such urgent problems to solve the crisis of survival that it seemed madness to me to waste energy on pure scientific research”. He analyzed that:

> Within the usual system of reference, let’s call it Western or industrial civilization, there is no possible solution; the interweaving of economic, political, ideological and scientific problems is such that there are no possible solutions. At the beginning, we thought that with scientific knowledge, by making it available to enough people, we would be able to better find a solution to the problems that arise. We have come back from this illusion. We now think that the solution will not come from an addition of scientific knowledge, from an addition of technology, but that it will come from a change of civilization. The truly profound change that will take place is a change in mentalities and human relations (Jouventin 2023).

Grothendieck encourages us to ask ourselves the question: “Is current science in general, or my research in particular, useful, neutral or harmful to all people?”

This is too rarely asked, but fortunately, this is slowly changing, as some researchers are realizing a branching of their research themes.

20.4.4. *Societal impact of the research*

While it is important to assess the environmental impact of research (laboratories, publications, conferences, travel, etc.), it is becoming necessary to assess the societal impact of the results of a research activity. This assessment can be based on the 17 UN’s SDGs and its 169 subgoals. It is no longer just a question of scientific evaluation based on the impact of scientific publications in specialized journals but of assessing whether the results and innovations of the theme are accessible, appropriable and have concrete spin-offs for society (i.e. a wide audience) through popularization actions (conferences, publications, mass media, social networks, etc.), training, dissemination of technical or software applications or services or new economic models (circular economies), etc.

Open Science, which makes scientific publications freely available on the Internet, is a step in the direction of accelerating the transition, but these publications are generally not understandable and appropriate for a wide audience.

The Junia Graduate Engineering School has developed a methodology based on the 169 subobjectives of the UN SDGs to identify the contributions of research themes to these subobjectives, requiring the writing of a report and validation by experts in sustainable development and SDGs (Robyns et al. 2024).

20.4.5. *From high-tech to low-tech...*

Rather than stopping scientific research, it is rather a question of reorienting it, because knowledge and research will always be needed but oriented toward different ends.

We will have to do with less and more sustainably: fewer materials in quantity, and also in diversity in a product and of more local origin, less energy, more robust, appropriable by users (whoever they are) and having the capacity to repair themselves.

Low-tech therefore appears to be a serious way to reduce environmental impact while satisfying our needs.

According to Bihoux (2014), the principles of low-tech are to “question needs, produce and design sustainably, direct knowledge towards saving resources, seek a balance between performance and user-friendliness, relocate without losing the good effects of scale, demechanize[6] services and know how to remain modest”.

ADEME (2022) offers a more precise definition:

> The term low-tech applies to an approach and not to its result. Thus, an object is not low-tech in absolute terms; it is more (or less) low-tech than an alternative solution that meets the initial need. The low-tech approach, sometimes called frugal innovation, is an innovative and inventive approach to the design and development of products, services, processes or systems that aims to maximize their social utility, and whose environmental impact does not exceed local and global limits. The low-tech approach involves questioning the need in order to keep only the essential, reducing technological complexity, maintaining what exists rather than replacing it. The low-tech approach also allows as many people as possible to access the answers it produces and to control their content.

6 Demachinize: replace resource- and energy-consuming machines with human or animal labor.

However, if low-tech appears more natural in countries in need of development than in highly developed or rapidly developing countries, low-tech approaches appear essential for leading more sober and frugal lifestyles that preserve the planet, its climate and its biodiversity. Research that is mainly focused on high-tech should be encouraged to focus on the development of low-tech solutions that also call on creativity and innovation, as illustrated in this section.

It should be noted that low-tech is not suitable for all technologies. No one will complain about medical engineering developing high-tech devices that can specify diagnoses, closely monitor patients and help them improve their comfort of life and lifespan.

20.5. Conclusion

As scientific research, especially of a fundamental nature, does not ensure that it will have a positive contribution to meeting global challenges induced by energetic and societal transition, we have to ask us the question "should we continue scientific research?" It emerges that research must be increasingly interdisciplinary, that an analysis of the contributions of the results of each research work to the UN SDGs is recommended, as well as the development of low-tech approaches in all areas where this is possible, by encouraging research work on these approaches, which are also a source of innovation and also of change in lifestyle.

20.6. Acknowledgments

The development of the Live TREE demonstration sites has benefited from funding from the Hauts-de-France region, several European funds from Feder (European Regional Development Fund) and H2020 (e-balanceplus project) and funding from the Université Catholique de Lille foundation. It has received support and funding from the European Metropolis of Lille (So Mel So Connected project funded by ADEME and financing of Junia's *Smart Buildings as nodes of Smart Grids* chair) and from la Ville de Lille. It has also received funding under the State-Region plan via the Sunrise project. Several companies are supporting the development of Live TREE. The INCLUNIV project was financed by ADEME.

Numerous actors are contributing to the development of these demonstration sites; the authors would like to thank them via this non-exhaustive list: Benoit Bourel, Yohann Rogez, Grégoire Destombes, Flovic Gosselin, Grégory Vangreveninge, Khaled Almaksour, Anthony Aouad, Arnaud Davigny, Christophe Saudemont, Dhaker Abbes, Benoit Durillon, Youssef Lyousfi, Kahina Hassam, Nicolas Gouvy, Serkan Surkulu and Jad Nassar.

20.7. References

ADEME (2022). Procedures. Low-tech [Online]. Available at: https://librairie.ADEME.fr/.

Bihouix, P. (2014). *L'âge des low-tech. Vers une civilisation techniquement soutenable*. Editions du Seuil, Paris.

Bossu, A., Durillon, B., Davigny, A., Barry, H., Robyns, B., Belaïd, F., Saudemont, C. (2024a). Coalitional game-based gain generation and distribution for collective self-consumption in an energy community. *Mathematics and Computers in Simulation*, 225, 129–147.

Bossu, A., Durillon, B., Davigny, A., Barry, H., Robyns, B., Belaïd, F., Saudemont, C. (2024b). Energy management based on coalitionnal game subdivision applied to energy communities. *Sustainable Cities and Society*, 117, 105911. doi: 10.1016/j.scs.2024.105911.

Durillon, B., Davigny, A., Kazmierzcak, S., Barry, H., Saudemont, C., Robyns, B. (2020). Decentralized demand response considering residential profiles for load smoothing. *Sustainable Cities and Society*, 63, 102464. doi: 10.1016/j.scs.2020.102464.

Durillon, B., Salomez, F., Davigny, A., Kazmierczak, S., Barry, H., Saudemont, C., Robyns, B. (2021). Consumers' sensitivities and preferences modelling and integration in a decentralized two levels energy supervisor. *Mathematics and Computers in Simulation*, 183, 142–157.

Jouventin, P. (2023). *Qui était Alexandre Grothendieck ?* Libre & Solidaire, Paris.

Rifkin, J. (2012). *La troisième révolution industrielle*. Les Liens qui Libèrent, Paris.

Robyns, B., Lenglet, C., Barry, H., Bozzo-Rey, M. (2023). *Smart Users for Energy and Societal Transition*. ISTE Ltd, London, and John Wiley & Sons, New York.

Robyns, B., Dobigny, L., Abbes, D., Durillon, B., Barry, H., Saudemont, C. (2024). *Smart Grids and Buildings for Energy and Societal Transition*. ISTE Ltd, London, and John Wiley & Sons, New York.

21

Socioecological Analysis of the Impact of the University Climate Agreement: What Prospects for Transition of the Catholic University of Lille

21.1. Socioecological Analysis of the Impact of the University Climate Convention: What Transition Perspectives for the Université Catholique de Lille (UCL)?

This presentation, entitled *SocioEcological Analysis of the Impacts of the University Climate Convention (UCC): What Prospects for UCL's Transition?*, offers a sociological analysis of the convention's impact on the actors and entities within the Université Catholique de Lille (UCL) (including higher education institutions and the university hospital). The objective is to identify the socioeconomic and structural dynamics that either support or hinder the socioecological and energy transition.

The analysis draws on empirical data collected primarily from staff and student delegates involved in the convention. These data were complemented by interviews with members of UCL's central administration and affiliated institutions, as well as with those responsible for the campus decarbonization process and the coordination of the convention itself. This data collection made it possible to understand the context in which the convention was launched, participants' views on its implementation, and both the expected and already observable outcomes.

Chapter written by Kokou Kouzouahin SOMABE and Hervé BARRY.

For a color version of all figures in this chapter, see www.iste.co.uk/dobigny/rethinking.zip.

Following a presentation of the structure and ambitions of the UCC, the article will explore its impacts on participants, highlighting several key aspects: the evolution of values and the generation of action proposals for UCL's governance bodies. The final section will summarize the main findings from the survey data collected at the end of each phase of the convention.

21.2. Presentation of the UCC

The UCC was organized by the UCL, with the involvement of the Métropole Européenne de Lille (MEL) and ADEME (the French Agency for Ecological Transition). It forms part of the implementation of the *Grenoble Agreement*, which commits UCL's institutions to accelerating their socioecological transition. The convention is also embedded within the broader framework of the *Live TREE* program: UCL's leadership initiative for ecological transition. In addition, the UCC was inspired by the *French Citizens' Convention for Climate*, held between October 2019 and June 2021 at the initiative of Prime Minister Édouard Philippe.

21.2.1. *A tactical step toward the university's transition*

The UCC aligns with the objectives of the *Live TREE* program, which seeks to foster a deep transformation in education and organizational practices, oriented towards a just and sustainable society. The program's primary goal is to catalyze the decarbonization of UCL's various components. To achieve this, it involves the transformation of both the university's infrastructure and its operational models, supported by academic research (Robyns et al. 2023). Remaining faithful to its academic mission, UCL also aims to support students and staff in revising their frames of reference and adopting more sustainable practices.

Through the *Live TREE* program, the university seeks to enhance its role in the general education of students and to reaffirm its social responsibility. As future professionals and citizens, students are expected to contribute to building new societal models; ones that are more environmentally respectful, socially equitable and aligned with the dignity of the human person (Robyns et al. 2023, p. 92).

21.2.2. *A process of discussion and learning*

The CUC is a space for learning, debate and democratic deliberation on climate issues, bringing together approximately 100 student delegates, 50 staff representatives and the leadership of all institutions and schools within UCL. The aim is to define proposals for action that support the reduction of UCL's climate impact and promote "transformative visions of habits and behaviors, of ways of seeing and being in the world" (Robyns et al. 2023, p. 127). It was an eco-citizenship training process for students and staff, carried out through seven three-hour sessions on transition-related themes, spread over three months. Discussions and joint work focused on the concrete aspects of how the university's entities operate, as well as the most everyday activities of students that have a carbon impact on the environment, namely: commuting, consuming, working/playing, connecting, eating, living and self-care.

Figure 21.1. *Photo of a plenary session of the university climate convention (source: Médialab ICL)*

At the individual level, the CUC represents a learning and training journey for participating students, focused on ecological issues and the assessment of environmental impact (Renouard 2021, 2022). It is also a democratic opportunity and experience to collectively define the values needed to move beyond the current status quo: openness to others, sobriety, solidarity and sacrifice. As one student delegate put it:

> The University Climate Convention allowed me to learn a great deal about the environment and climate change through readings, presentations, lectures, and group workshops. The diversity of participants enriched the debate, bringing new perspectives and challenges, all of which elevated the discussions. This experience helped me grow and reflect on the importance of collective thinking when it comes to tackling societal issues such as climate change (student delegate, participant in the University Climate Convention).

The individual impacts of the CUC become more visible as they unfold across institutional and collective dimensions. For example, through his involvement, Sébastien Carcelle contributes to the visibility and implementation of the ecological education objectives within the Humanities curriculum at ICL:

> One of my first assignments at ICL is coordinating the CUC. My commitment aims both to engage students in the project and to recruit them for courses in literature and the humanities. This course provides skills in ethics, humanism, and service learning. My participation in the CUC has allowed the humanities course to turn the CUC into an alternative form of civic education focused on ecology, climate action, and related themes (Sébastien Carcelle, Literature and Humanities team, ICL).

21.2.3. *A process for developing actions*

The CUC invited delegates to work in groups to reflect on and formulate action proposals aimed at strengthening the campus transition project. A total of 55 proposals were developed. These were presented and discussed with a panel of experts, academics and local, municipal and regional authorities during the CUC's closing session, held in the MEL Council Chamber in December 2022. The proposals were then submitted to UCL's governing bodies, which ultimately selected seven priority actions, with some potential content adjustments.

The seven projects, presented in the table below, serve as concrete indicators of the institution's commitment to managing its various pledges, unlike many institutions that struggle to commit or implement actions to protect our planet (Pope Francis 2015, 2023). This contributes to revitalizing the *Live TREE* program, in line with institutional transition policies. Finally, the seven key and priority projects revolve around three main objectives:

1) to develop the operational conditions of the campus in order to live, work and study in new ways;

2) to adopt a lifestyle that combines greater comfort with reduced carbon emissions;

3) to create a campus that is better connected to its surrounding neighborhood.

In this sense, the CUC is part of an eco-campus approach. By eco-campus, we refer to a set of ecological and societal transition initiatives led and embodied by a university: across its curricula, governance, infrastructure, operations and the lifestyles of its stakeholders.

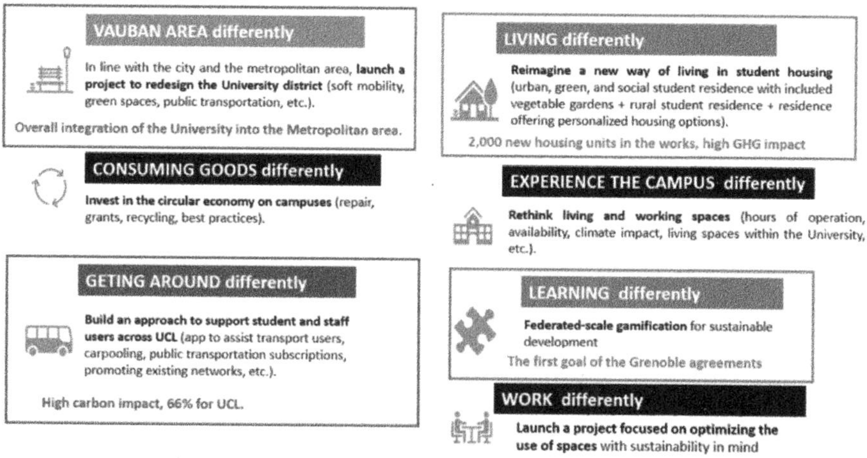

Figure 21.2. *The seven actions formulated by the UCL Governance, following proposals from the CUC (European Parliament 2024)*

To support the transition towards carbon neutrality, complementary measures are being implemented in the area of commuting, through the project "Getting Around Differently". This initiative supports the use of various modes of transport by introducing mechanisms that encourage and promote sustainable mobility through awareness-raising, guidance, financial incentives, car-sharing, the installation of charging stations for electric vehicles and more. The project "Working Differently" focuses on optimizing the use of university facilities through a partial remote work scheme for staff. As an indicator of the eco-campus project's impact, the implementation of partial remote working at UCL resulted in an 11.7% reduction in CO_2 emissions from commuting between 2019 and 2022, decreasing from 8,200 to 7,200 tons of CO_2. Through the "Living Differently" project, UCL plans to introduce new lifestyles within student residences: urban green student housing featuring community gardens and vegetable plots, rural student residences and housing offering personalized accommodation.

Similarly, the project "Experiencing the Campus Differently" promotes a new way of living together and a healthy, ecological environment: green spaces, green roofs, green walls and green rooms on campus. UCL's urban transformation project, part of the "ZEN Campus" (Zero Net Emissions) initiative, is being developed in collaboration with local authorities to promote several climate-conscious construction projects on campus, thereby offering new opportunities for innovation.

In this context, structural transformations are accompanied by the strengthening of the technical team, equipped with both socioenvironmental and technical expertise, to enhance the quality of life at UCL. This reality is expressed in the following statement by the Director of Real Estate Projects at the Institut Catholique de Lille:

> In our line of work, we are increasingly involved in sustainable use, particularly through building renovation and energy consumption optimization based on clear guidelines. We have consolidated our themes and practices in support of the transition. This has led to structural and operational changes in the Transition Chair (Research). We created a new department dedicated to CSR, led by Benoît Bourel at ICL, which translates our transition commitments into concrete CSR initiatives. Our current project, in collaboration with the CUC, is to enhance the spaces available to students in order to sustainably improve the living environment at UCL (including student dining areas and shared workspaces) (Etienne de Boisredon, Director of Real Estate Projects, Institut Catholique de Lille).

The project "Consuming Differently" supports circular economy principles through virtuous socioecological practices in the daily operations of campus facilities: solidarity initiatives, student aid programs, recycling and frugality, among others.

The project "Learning Differently" focuses on experimenting with gamification at a federative scale for sustainable development (via a MOOC) and on launching a training platform addressing ecological transition issues for sustainable development across the UCL. The ODYSSEE MOOC is a pragmatic approach aiming to "create a compulsory common course for all students […] devoted to socio-ecological questions […]" and to integrate a "training module on socio-ecological issues" into the initial training of all teaching staff (Grenoble Agreement 2021, Articles 2.2 and 2.4). In this context of pedagogical transition, the President for Humanities and Student Life at UCL explains:

> Our ambition is to energize the logic of our transition – a transition based on voluntary sobriety – and to enhance the performance of our training and professions by integrating socio-ecological challenges. The ODYSSEE MOOC is both a decision and a concrete realization of a federated training program on ecological issues and the Sustainable Development Goals, led by Junia. Junia's experimental farm is a demonstrative urban farm – a strong choice in favor of the transition. Other examples include the EDHEC Sustainable Finance program, the Sustainable Development Master's at FGES, and more

> (Thierry Magnin, President Delegate for Humanities and Student Life).

This initiative provides a shared training framework for students, focused on issues of transition and sustainable development, with 2,600 users within the first year. Forestry and agronomy students from ISA-Junia share their experience:

> After two years at ISA, we've seen that our school raises awareness of environmental issues, but doesn't actually integrate these topics into our curricula. [...] Thanks to our actions and negotiations with student associations, we succeeded this year in convincing the school administration to enter into a training partnership with ODYSSEE. This partnership will allow us to organize workshops and seminars on ecological issues, CSR (Corporate Social Responsibility), and sustainable development, which will be credited with ECTS points for all students. [...] This is a real step forward in mobilizing, raising awareness, and involving students in the fight against today's ecological crises (ARGO Student Association, Junia, UCL).

In this perspective, several researchers argue that teaching ecological transition in higher education represents a significant contribution towards entering the era of the *Great Transition* (Jouzel 2020; Renouard et al. 2020, 2022; Jouzel et al. 2022; Petit et al. 2024). To this end, Petit et al. (2024) assert that the renewal of higher education should aim to:

> Transform all higher education programs [...] so that they integrate a common foundation of knowledge and skills based on five transversal competencies: understanding the balances and limits of our world through a systemic approach; grasping scales and uncertainties through forward-looking analysis; co-constructing diagnoses and solutions; using tools to operationalize transitions; and acting responsibly (Petit et al. 2024, p. 2).

This co-constructive approach to developing the eco-campus has impacts that reach beyond the academic sphere. It involves a set of actions integrated into a broader neighborhood project, designed in collaboration with the city and the metropolitan area. In the context of eco-campus, the goal is to extend the dynamic transition of UCL's various entities to surrounding neighborhoods. This initiative is entitled "Rethinking the Vauban Neighborhood". This transition initiative is thus embedded in an inclusive, cooperative sociological approach rooted in the local territory. It reflects a sociological configuration of networked actors (*acteurs-réseau*) (Callon 1986; Callon and Latour 1991; Latour 2004; Walsh and Renaud 2010), centered on ecological transition. This logic of transition entails the interweaving of multiple actors at the local level – including university institutions, external stakeholders and territorial public officials – working collectively to tackle climate change (Castellanos 2016). From the perspective of *pragmatic sociology* (Boltanski and Thévenot 1991; Thévenot 2001, 2006), the eco-campus emerges as a shared framework for co-constructing the common good at the intersection of multiple *orders of worth*. Climate is understood by actors in this network as a *common good*, a higher principle for which they commit themselves and coordinate actions, transcending individual interests and divergent values.

21.2.4. *Diversifying actions to achieve better results*

The implementation of the seven priorities will enable UCL to continue reducing its environmental footprint, fostering changes in the lifestyles and work habits of students and staff both on campus and in the surrounding areas.

Efforts and initiatives already undertaken have allowed UCL to reduce its overall carbon footprint from 81,569 to 76,342 tons between 2019 and 2022, representing a 6.41% decrease over four years (UCL Carbon Trajectory Committee, June 2024). Regarding other indicators, UCL's approach has been recognized multiple times. Internationally, IESEG School of Management was honored by the *Financial Times* for its commitment to sustainable development and corporate social responsibility. Ranking at the top in the "Teaching" and "Best Overall School" categories at the *Responsible Business Education* laureate's forum (IÉSEG 2024) underscores the importance of ecological education promoted by UCL's dynamic eco-campus approach. More specifically, in the "Responsible Teaching Resources" category, IESEG received a *Highly Commended* mention for its Transition 2026 training program, followed by all administrative and teaching staff. In the "Best Overall School" category, the school was also among the top five laureates. UCL has also received other national distinctions, such as the EuraClimat label (2023, 2024). Beyond its educational component, the *eco-campus* project led by the Live TREE program supports structural transformation projects at the university (building renovations and eco-friendly construction) as well as interdisciplinary research initiatives, through partnerships with private and public sector actors, all serving the common good.

Although "organizations responsible for higher education and research have been slow to address the issues of ecological and social transition" (Petit et al. 2024, p. 2), a growing momentum for campus transitions is emerging within French universities such as UCL (Renouard 2022; Robyns 2023). From this perspective, UCL can be considered among research institutions and universities actively engaged in "developing pioneering approaches that demonstrate their commitment to sustainable development" (Petit et al. 2024, p. 2). However, this new eco-campus dynamic needs to be further strengthened in light of the progress made in university transitions abroad, particularly within Anglo-Saxon academic and research contexts (Robyns 2023).

21.3. Appreciation and impacts of the CUC according to delegates

The presence of students and employee delegates during the two phases of the Congress enabled rapid online surveys to be conducted, addressing the current state and potential evolution of knowledge, attitudes and practices regarding sustainable development, as well as the appreciation of the proposed proceedings.

21.3.1. *Survey methodology*

Invitations to participate in each survey were sent by email to all students and staff delegates present at each of the two days of the convention.

21.3.1.1. *Respondent sample*

The order of contributions (see Table 21.1) follows the participation of the two profiles. Student delegates, more numerous on the first day, were also more likely to respond (n = 39). Since many students from December 2022 had left the University by January 2024, staff delegates became more represented in the second survey (n = 20).

Respondent sample	Survey 1	Survey 2
Student delegates	39	4
Staff delegates	12	20
Total	**51**	**24**

Table 21.1. *Profile of respondents to both surveys*

These variations in delegates present and responding explain that the survey samples have somewhat limited representativeness, due to sample size and self-selection bias (Berthier 2023), as delegates present chose whether or not to respond.

21.3.1.2. *The value of engaging young delegates*

Conducting this survey with volunteer student delegates was particularly valuable because the 20–25-year-old generation is identified as rather skeptical (even anxious) about the ability of decision-makers – politicians, international organizations and large corporations – to agree on sufficiently effective solutions to combat climate change (Lachance and Mane 2022). This generation is also known to be more aware of environmental issues. Compared to their elders, this manifests less through voting, due to a loss of trust in institutions, and more through specific forms of engagement: public protests, occasional mobilizations, abstention, online actions and so forth (Loncle 2022). They also exhibit a stronger tendency to adopt lifestyle choices with a lower carbon footprint (such as eating less meat, supporting local food systems, limiting air travel, etc.). Finally, as a general characteristic of the student population, it is important to note that an increasing number is taking into account the environmental impact of companies when making career choices (Harris Interactive Toluna Institute 2022).

21.3.2. *Delegates experiencing eco-anxiety*

A large majority of respondents (91%) consider themselves well-informed about climate issues. This is nearly 30 points higher than what most opinion polls report across the general population (Abel et al. 2024). Beyond the possible effect of incidental or deliberate personal information seeking, this result can also be explained by the university context: awareness modules within students' academic curricula, campus activities, recognition of the campus decarbonization initiative, etc.

This clear awareness of the climate problem is associated with a low level of confidence in the future (75%). By comparison, this score is 10 points higher than that of a 2025 survey conducted on a national population sample (ParlonsClimat 2025). It is precisely this lack of confidence combined with the perceived certainty of being at an impasse that causes worry, which can escalate to eco-anxiety, which is defined as "mental distress (discomfort) in response to environmental issues". According to a recent survey (ADEME 2025), eco-anxiety affects 11% of the population, and even more intensely among young people (aged 15–34), who are more "strongly eco-anxious" (9% vs 5% on average).

It is interesting to note that this lack of confidence is accompanied by a low level of formal engagement in favor of sustainable development (15%), through membership in a political party or an association working for the environment (ADEME 2023). The survey does not provide further insight into the origin of this low level of engagement: distrust towards intermediary organizations? Priority is given to building one's future, etc. What is possible, however, is that the growing diversification of forms of engagement outside of voting or formal membership limits the visibility of actual commitments.

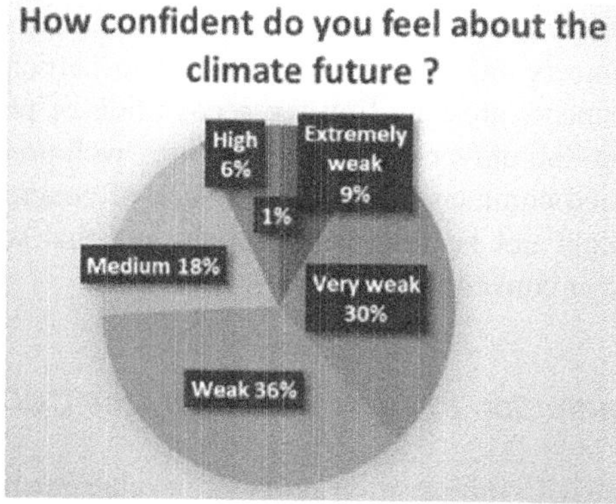

Figure 21.3. *Perceived trust in climate prospects among CUC delegates (source: CUC survey)*

21.3.3. *A majority of present delegates are engaged*

Engagement, understood here as the willingness to participate in the CUC process in order to influence future university governance decisions, emerged as the main motivation shared by both groups. This response was more common among staff members (67% vs 51%), who wished to contribute to a participatory management approach. On the contrary, students, still in their academic path, were more focused on acquiring new knowledge (44% vs 25%), which supports the known observation (Lachance and Mane 2022) that this age group's environmental concern is more future-oriented (what kind of organization?) than focused on repair (what corrections?). Curiosity represented only 5% to 8% of reasons for participation.

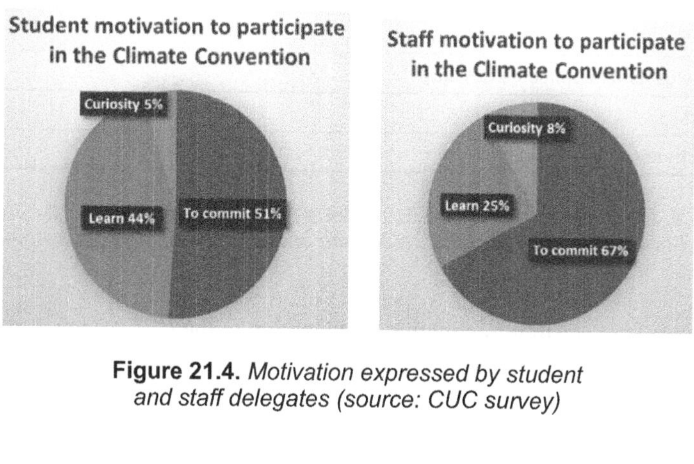

Figure 21.4. *Motivation expressed by student and staff delegates (source: CUC survey)*

Despite differences in motivation and expectations, the satisfaction rate reached 70% among all volunteers and 95% among those who participated on both days. These positive assessments are based on the recognition of past achievements, as well as the feeling that the university's initiatives are "well underway". Among the critiques, some nuanced comments referred to a lack of concrete actions, concerns that commitments might not be followed through, and that some proposals were considered unrealistic or conventional.

21.3.4. *Increased attention to sustainability by the CUC*

All aspects of the CUC (information provided, exchanges between participants, group work, expert input, etc.) had a notable impact on the attention paid to the carbon effects of adopted lifestyles, which can be credited to the convention. This shift was verified through the submission of a series of attitudes and consumption behaviors regarding transportation (less use of cars and planes), product and service purchases (production location, materials, reparability, etc.) and energy sobriety (general attitude). However, it should be noted that changes were more modest in more personal and intimate areas of practice, such as leisure, health and connected devices.

Would you say that the convention contributed to (further) changing your criteria or practices in terms of ...?	%
Purchasing products and services	77
Energy self-sufficiency	65
Transport	58
Product repair	54
Food	50
Company selection	46
Purchasing or using connected devices	46
Healthcare	35
Leisure activities	31

Table 21.2. *Evolution of social practice criteria by suggested domains (source: CUC survey). For a color version of this table, see www.iste.co.uk/dobigny/rethinking.zip*

Finally, it is interesting to note that participation in the CUC changed the criteria for choosing companies for 46% of respondents (new or increased criterion). Recent data (Elabe 2024) show that this score aligns more closely with that of employees. It would likely have been higher within a student cohort (between 70% and 85%, depending on the wording).

21.3.5. *Delegates remain optimistic*

At the end of the convention, a majority of respondents (26 out of 42) expressed optimism about the prospects for solving climate challenges. This optimism was particularly strong among employees (22 out of 34). They were encouraged by the active participation of students, whom they considered essential for the future. Moreover, the priorities set by the university leadership provide employees with clear guidance, reassuring them about the intentional and motivating nature of governance.

State of mind	Students	Students	Students
Optimistic or rather optimistic	4	22	26
Mixed		4	4
Pessimistic or rather pessimistic	4	8	12
Total	**8**	**34**	**42**

Table 21.3. *State of mind of student and staff delegates regarding the effective management of climate change (source: CUC survey)*

Finally, it is noted that the expressed pessimism largely rests on three key points. The first concerns the speed of change, considered insufficiently rapid. The responsibilities highlighted primarily involve the political sphere in two ways: public officials, more specifically the government, and secondly, the French political climate, marked by an attitude described as "populist", which makes it difficult to take environmental concerns into account in the daily management of public affairs. Finally, a lack of ambition is pointed out. While the will to act is present, it is neither fast nor massive enough to meet the environmental challenges facing the contemporary world.

21.4. Conclusion

The CUC is a collective program of training and workshops that has enabled participants to better understand global climate and energy challenges, the university's carbon footprint and its various emission sources, and to identify possible solutions for the transition. From a sociological perspective, this CUC can be seen as a mechanism reinforcing eco-citizen and democratic values, namely: ecological and social associative engagement; strong mobilization of students and staff in local and international socioecological events (observer status granted to the university at COP Climate); and initiatives promoting eco-friendly practices to raise awareness among individuals and communities about sobriety and solidarity.

Beyond behavioral aspects, structural and functional transformations are also taking place. Within UCL, there is an evolution in pedagogical approaches and teaching and research programs (renewal of the training offer, e.g. Uschool, Faculty of Medicine, Junia, FGES, etc.). Work and lifestyle modes on campus and in the neighborhood are increasingly inspired by sustainability concerns. This highlights the fundamental role and responsibility of educational, training and scientific research institutions in this transition process. This transition dynamic, unfolding through the CUC, is part of a structural and pedagogical transformation of higher education that can be called an "eco-campus" (Pope Francis 2015; Dardelet et al. 2021).

Finally, the collaborative dynamic between institutions proves effective, with cross-pollination effects that support ecological transition. In this regard, the Live TREE program successfully fulfills its role as a catalyst and diffuser of progress.

21.5. References

Abel, J.D., Boucherand, S., Goguet, P. (2024). Transition écologique : croissance vs décroissance, de quoi parle-t-on. Study, CESE, Paris.

ADEME (2023). Enquête : les 15–25 ans face aux enjeux environnementaux. Survey, ADEME, Montrouge.

ADEME (2025). L'éco-anxiété en France. Final Report, ADEME, Montrouge.

Berthier, N. (2023). *Les techniques d'enquête en sciences sociales*. Armand Colin, Paris.

Callon, M. (1986). Éléments pour une sociologie de la traduction. La domestication des coquilles Saint-Jacques et des marins-pêcheurs dans la baie de Saint-Brieuc. *L'année sociologique*, 36, 169–208.

Callon, M. and Latour, B. (1991). Réseaux technico-économiques et irréversibilités. In *Les figures de l'irréversibilité en économie*, Boyer, R., Chavanche, B., Godard, O. (eds). Éditions de l'EHESS, Paris.

Castellanos, I.J.C. (2016). Enjeux environnementaux transnationaux. Politiques et acteurs sociaux. *Émulations*, 20, 113.

Convention pour la Transition des Établissements du Supérieur (CTES) (2021). Accord de Grenoble. Document, Grenoble [Online]. Available at: https://la-ctes.org/presentation-accord-de-grenoble.

Dardelet, C., Dambrine, L.-M., Majou de la Debutrie, G., Crouzet, S., Balbo-Bonneval, V., Chycki, P., Foulon, H., Chauvin, F. (2022). *Transformation des campus. Petits manuels de la grande transition*. Les Liens qui Libèrent, Paris.

Econoïa (2025). L'impact de l'éco-anxiété sur la santé mentale des Français (15 à 64 ans) [Online]. Available at: https://librairie.ademe.fr/actualites/quel-est-l-impact-de-leco-anxiete-sur-la-sante-mentale-des-francais--n16.

Elabe (2024). Le travail en transitions. Survey, Elabe, Paris.

European Parliament (2024). Compte rendu in extenso des débats du 26 novembre 2024. *Journal officiel FR de l'Union européenne*, C/2025/5034 [Online]. Available at: https://eur-lex.europa.eu/legal-content/FR/TXT/PDF/?uri=OJ:C_202505034.

Harris Interactive Toluna for Challenges (2022). Échantillon de 10 020 personnes représentatif de la population française âgée de 18 ans et plus. Survey, Harris Interactive Toluna for Challenges.

IÉSEG (2024). Financial Times Responsible Business Education Award: IÉSEG wins the 'Best Schools' Award! IÉSEG [Online]. Available at: https://www.ieseg.fr/en/news/responsible-business-education-awards-2024/.

Jouzel, J. (2022). Sensibiliser et former aux enjeux de la transition écologique et du développement durable dans l'enseignement supérieur. Report, Ministère de l'enseignement supérieur, de la recherche et de l'innovation, Paris [Online]. Available at: https://www.enseignementsup-recherche.gouv.fr/sites/default/files/2022-02/sensibiliser-et-former-aux-enjeux-de-la-transition-ecologique-dans-l-enseignement-sup-rieur-16808.pdf.

Jouzel, J. and Abbadie, L. (2020). Enseigner la transition écologique dans le supérieur. Report, Ministère de l'enseignement supérieur, de la recherche et de l'innovation, Paris [Online]. Available at: https://www.enseignementsup-recherche.gouv.fr/fr/enseigner-la-transition-ecologique-dans-le-superieur-51505.

Lachance, J. and Mane, I. (2022). Le rapport des jeunes aux enjeux environnementaux en France. Les15-24 ans sont-ils différents de l'ensemble de la population ? Report, Université de Pau et des Pays de l'Adour, TRansitions Energétiques et Environnementales (TREE), Pau.

Latour, B. (1991). *Nous n'avons jamais été modernes. Essai d'anthropologie symétrique*. La Découverte, Paris.

Latour, B. (1995). Moderniser ou écologiser. A la recherche de la Septième Cité. *Écologie Politique*, 13, 5–27.

Latour, B. (2004). *Politiques de la nature. Comment faire entrer les sciences en démocratie*. La Découverte, Paris.

Loncle, P. (2003). *L'action publique malgré les jeunes*. L'Harmattan, Paris.

Loncle-Moriceau, P. (2022). Les jeunes et le politique : au-delà du vote, des formes d'engagement multiples. *The Conversation* [Online]. Available at: https://theconversation.com/les-jeunes-et-le-politique-au-dela-du-vote-des-formes-dengagement-multiples-177648.

ParlonsClimat (2025). Les Français parlent climat en 2025 : quelles perspectives dans un contexte nouveau, étude quantitative. *ParlonsClimat* [Online]. Available at: https://www.parlonsclimat.org/les-francais-parlent-climat-2025.

Petit, O., Buchs, A., Calvo-Mendieta, I., Audouit, C. (2024). L'enseignement supérieur et la recherche française en transition *? Développement durable et territoires*, 15(3). doi: 10.4000/13cq9.

Pope Francis (2015). *Lettre encyclique Loué sois-tu ! (laudato si') du pape François sur la sauvegarde de la maison commune*. Éditions Salvator, Paris.

Pope Francis (2023). *Laudate Deum. Exhortation apostolique à toutes les personnes de bonne volonté sur la crise climatique*. Éditions Première Partie, Paris.

Renouard, C. and Beau, R. (2021). *Responsabilité et environnement. L'enseignement et la formation dans la transition écologique et sociétale*. Institut Mines-Télécom, Palaiseau.

Renouard, C., Beau, R., Goupil, C., Koenig, C. (2020). *Manuel de la grande transition : former pour transformer*. Les Liens qui Libèrent, Paris.

Robyns, B., Lenglet, C., Barry, H., Bozzo-Rey, M. (2023). *Smart Users for Energy and Societal Transition*. ISTE Ltd, London, and John Wiley & Sons, New York.

Thévenot, L. (2001). Constituer l'environnement en chose publique. Une comparaison franco-américaine. In *Cadre de vie, environnement et dynamiques associatives*, Jean-Paul, B., Gillio, C., Jacques, I. (eds). PUC, Paris.

Thévenot, L. (2006). *L'action au pluriel. Sociologie des régimes d'engagement*. La Découverte, Paris.

Walsh, I. and Renaud, A. (2010). La théorie de la traduction revisitée ou la conduite du changement traduit. Application à un cas de fusion-acquisition nécessitant un changement de Système d'Information. *Management & Avenir*, 39(9), 283–302.

22

What User Involvement in Sustainable Campuses? An Analysis of the Decision-making Processes of University Smart Buildings in the Lille Region (France)

22.1. Introduction

As places of work, learning and exchange, universities can play a key role in the environmental, energy and social transition by mobilizing a wide range of stakeholders (administrative and technical staff, teachers, students, local residents). Universities in the Lille region have taken up these challenges and developed eco-transition programs accompanied by concrete achievements such as demonstration buildings, positive energy buildings, smart buildings and HQE buildings (for High Environmental Quality, a French label). More generally, these projects are part of the Rev3 regional initiative, a strategic project for a third industrial revolution in the Hauts-de-France region, launched in 2013 with the aim of inventing a new low-carbon economic model. It is now a collaborative effort, with more than 800 projects being monitored, and public and private investment of around €500 million per year.

This ambitious regional dynamic is reflected in particular in the signing on June 18, 2021, of the Lille Low Carbon Pact, which aims to drastically reduce greenhouse gas emissions linked to construction and urban development. It was signed by 100 construction, development and renovation partners, including five member institutions

---

Chapter written by Laure DOBIGNY, Mathilde SZUBA, Hervé BARRY and Anouk GERME-BARTIER.

For a color version of all figures in this chapter, see www.iste.co.uk/dobigny/rethinking.zip. All figures in this chapter are produced and designed by Margaux d'Hont.

of the Catholic University of Lille. The pact is based on ambitious operational commitments in favor of the climate, the environment and health and applies to all residential and tertiary, public and private, development, construction and renovation projects. However, French universities must also comply with the Elan law of November 23, 2018, which sets the principle of a 40% reduction in the energy consumption of tertiary buildings by 2030. At the same time, the issue of inclusion of vulnerable people in universities is increasingly being addressed, whether through the introduction of a Disability Master Plan at the request of the Ministry of Higher Education, Research and Innovation, or by putting it on the political agenda. On May 12, 2021, the Minister for Higher Education, Research and Innovation and the Secretary of State for People with Disabilities chaired the first National Inclusive University Monitoring Committee. The committee agreed to set up two working groups, one of which is dedicated to global accessibility. This issue is also central to Lille's universities, in terms of institutional policies, research (ethics, medicine), teaching at the Catholic University of Lille and through the construction of a European university alliance for inclusion, InclusU (the European University for Inclusiveness) at the Lille University.

Although most universities and schools deal with the issues of sustainable development and corporate social responsibility within the same department[1], is the ecological transition within universities truly inclusive? What consideration has been given to the involvement of users (staff, students, researchers and teachers)? What vision of ecological transition do technical choices related to building construction or refurbishment stem from? Also, what vision of users and their role in this transition? How is this expressed in the choice of equipment and design of efficient and smart buildings, Positive Energy Buildings (BEPOS) or HQE buildings? If smart buildings require "smart users" (Strengers 2013; Robyns et al. 2023; Dorthe and Dobigny 2025), are users not as a whole becoming vulnerable within a technical and managerial vision of the transition? The originality and the scientific challenge of the INCLUNIV project (2023–2025)[2] was to question the ecological transition of universities from the point of view of the involvement and mobilization of users and their technical vulnerabilities within a smart technical system.

22.2. Methodology

The project compares four higher education buildings in the Lille metropolitan area. Three of these were built or renovated as smart buildings (the IUT Roubaix building, the Rizomm and the Michel Falise buildings at the Catholic University of

1 Sustainable Development and Social Responsibility (SD & RS) or Corporate Social Responsibility (CSR), depending on the institution and its status.

2 Funding by the French Agency for Ecological Transition.

Lille), while the Science Po Lille building was renovated to HQE standards and incorporates a building management system (BMS). Using a socio-anthropological methodology, the aim was to examine the genesis of the university transition programs and the way in which the involvement of users was thought through, particularly in the technical choices (decision-makers, designers, engineers, technicians, etc.).

The results presented in this chapter are based on a monographic study of each institution, using socio-anthropological research methods (ethnographic interviews and observations)[3]. During this first phase of the project, 53 semi-directive interviews were conducted with decision-makers, designers, technicians, staff (administrative, technical, teachers and researchers) and students in the four buildings studied. In addition, observations and participant observations were conducted within the steering, orientation, strategic and technical committees of the universities' ecological transition programs.

22.3. Case studies

22.3.1. *Rizomm (Catholic University of Lille): a desire for co-design surpassed by a desire for innovation*

The Rizomm building of the Catholic University of Lille is a building that has undergone several transformations. In the 1950s, it was used as the chemistry building for the School of Advanced Engineering Studies (École des Hautes Études d'Ingénieur – HEI). In the 1960s, it was extended to house the Lille Higher Institute of Agriculture (Institut Supérieur d'Agriculture de Lille – ISA) and then raised in the 1980s. One consequence of these successive transformations was the difference in levels between the different parts of the building, requiring staircases to compensate.

From the 2000s onward, the need arose to renovate the building to bring it up to standard, particularly in terms of accessibility. The renovation was also a response to the need for visibility, influence and prestige, which the building was not providing due to its dilapidated state. As the former dean pointed out, "it was the building that nobody wanted at the time […] It was a thermal and visual wart".

3 Four students carried out their master's internship as part of the project. Two students took part in gathering ethnographic data (interviews) and two students produced a monograph (Germe-Bartier 2023; Maria 2023). The aim was twofold: on the one hand, to train students through research and, on the other hand, to include students in the research work and integrate their situated analysis and perception of the inclusion of students in the decision-making processes relating to the ecological transition of universities.

As early as 2012, work began on renovating the building, in particular by organizing co-design workshops with users (staff and teachers) to gather their needs and ideas, with a view to planning the future building. The project was strongly influenced by the desire for coherence between the teaching provided (science and engineering) and the building: the building must reflect and embody this teaching at the cutting edge of scientific progress and be both a showcase and a place for experimentation, with both educational and scientific aims. In this way, the renovation project is part of a strong imaginary of innovation and modernity.

Three initiatives, launched in 2013, will bring the environmental issue into the project, and in particular, the search for environmental innovation: the "Third Industrial Revolution" (Rev3) program, launched by the Hauts-de-France region[4]; the Live TREE[5] program for ecological and energy transition at the Catholic University of Lille (see Chapter 20 in this book) and its goal of a zero-carbon campus by 2035.

Finally, the restructuring of the faculties, and the merger of the science and economics/business faculties, which will occupy the building, will once again change the direction of the project.

The project's architect worked with the building's future users (mainly staff and lecturers[6]) on a co-design project. This was part of a strong desire on the part of the rector's team in office at the time to "co-develop" the project. The president-rector at the time, an engineer by training who came from the world of business and innovation, was to bring a new vision and strategy for investment, teaching and work. In other words, a different vision of technical appropriation: "We need technologies in Lego form: everyone can put the bricks together, combine them and disassociate them" (former rector of the Catholic University of Lille).

In this respect, the choice of name for the building is significant: as a source of energy for the plant, the rhizome connects plants that evolve in networks. The rhizome is thus the bearer of another organizational and decision-making vision: that of a horizontal model in constant evolution, as opposed to the subordination, hierarchization and centralization represented in the tree by the trunk and its roots (Deleuze and Guattari 1980).

---

4 Inspired and supported by Jeremy Rifkin (2011), this program aims to reduce energy consumption, increase the production of decentralized renewable energy and develop activities and employment.

5 Lille Vauban-Esquermes in energy, ecological and economic transition.

6 The co-design workshops organized in 2013–2014 brought together a third of the staff (around 30 out of the 100 employees involved), but few students.

Although the building was to become an object of cooperation and joint reflection between the faculties that were to merge, this “co-elaborative” vision of Rizomm did not come to fruition, despite the will of the rector and the choice of an architect open to this process. In 2014, it was decided that the building would become a “demonstrator”: a demonstrator of environmental innovations, Rev3 and Live TREE. One of the specific features of the decision-making process for a university building should be emphasized: teacher–researchers are present on the decision-making bodies and are the prescribers. These researchers are not only familiar with the latest innovations but are also curious and keen to put them to the test as research objects. Added to this is the rectoral team’s desire to develop a “univerCity”, i.e. to create a learning city and place the student experience at the heart of teaching. In this vision, buildings must become educational objects.

These factors led to a change in direction for the project. The co-design project with the original architect was scrapped, and an engineering consultancy office specializing in energy and environmental performance of buildings was appointed. From that point on, the consultancy office would have the upper hand, meaning that the technical dimension would take precedence over the architect (in a negotiating position with the consultancy office) and the users. The desire for innovation and technicality in the building industry is therefore leading to a reversal of the professional logic at work in the building industry, where the architect decides and the consultancy office adapts to find technical solutions in line with the architect’s plan. The choice of innovative techniques will mean difficulties in finding the contractors to implement them and a long building adjustment phase (two years), leading to incomprehension on the part of users when faced with malfunctions in a building that is supposed to be smart:

> It’s not a car you buy in series, which is supposed to work when it leaves the factory. It’s a unique model (former dean).

These new technical systems are not only new for users, but also for the university’s technical services. They will be trained by the consultancy office, but the need will soon arise to create a new job of “smart building manager” dedicated to the (specific) technical management of the building and to mediating between the building and its users (staff and students). Despite the genuine desire to “get people involved” (the former dean) and co-design the building with its users, this objective was going to come up against the parallel objective of technical innovation, as well as existing professional cultures, both internal (technical services, support and general resources) and external (architects and consultancy offices). Additionally, it was difficult to mobilise people in the long term due to student and staff turnover in the context of a merger, as well as participants losing interest in a project involving several twists and turns.

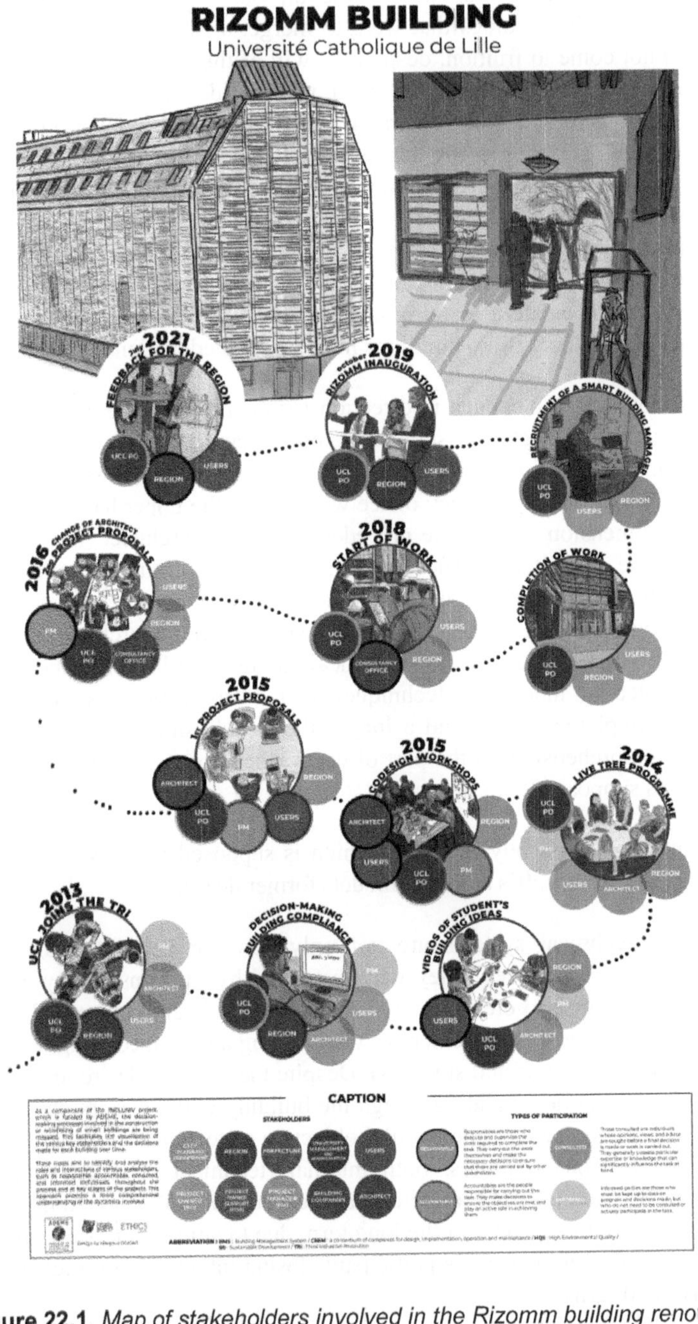

Figure 22.1. *Map of stakeholders involved in the Rizomm building renovation process. Production and illustration by Margaux D'Hont*

22.3.2. *Michel Falise (Catholic University of Lille): the choice of full automation*

Until June 2024, the JUNIA-HEI engineering school (a member of the Catholic University of Lille school federation) was housed in several adjoining buildings, the oldest of which was built in 1885. Since June 2024, this building, now known as Michel Falise, has housed the Faculty of Management, Economics and Science (FGES) at the Catholic University of Lille. After several renovations in the course of the 20th century, the oldest part of the building needed to be brought up to standard again, both technically and in terms of working conditions. This was done in two stages. The first stage of the renovation (2012–2015) was devoted to modernizing the building to meet the requirements of the 2005 Thermal Regulations, with the addition of a Building Management System (BMS) that was never fully operational, and to reorganizing the workspaces, complicated by the introduction of a number of open-plan areas. All this was done without any real consultation with the building's users, apart from a few meetings and occasional consultations with heads of department.

The second renovation (2019–2020) was designed to meet the new priorities of the general management: better control of energy consumption; making the most of in-house expertise in electrical and building engineering; and, finally, being part of the Third Industrial Revolution promoted by the Hauts-de-France Region and inspiring the decarbonization program launched in 2014 by the Catholic University. In the context of this engineering school, the expectations led to the choice of elevating the building to smart building with a socio-technical script based on the exclusion of any human intervention except in cases of excess heat, when the opening of window(s) is tolerated, "because it is less complicated to heat than to cool". A link has been formed between a modernized BMS and the pre-existing network of sensors for monitoring the heating–ventilation–lighting functions. Automatic adjustments are now made to comply with regulatory standards and to take account of foreseeable attendances based on opening times and lesson plans transmitted to the BMS.

This choice for a high degree of automaticity is understandable if we consider the interplay of players influencing the rationality of the technical managers. They have to deal with a general management team that is more demanding when it comes to energy consumption, and occupants whose needs they perceive to be diverse, unstable and therefore elusive, which they summarize in the colorful phrase "otherwise, it's Beirut", when they argue that there is no form of consultation. Of course, they put forward more positive explanations for their choice, in particular by referring to the occupants' freedom to produce comfort in the workplace at their own expense: "We want to be able to respond to needs".

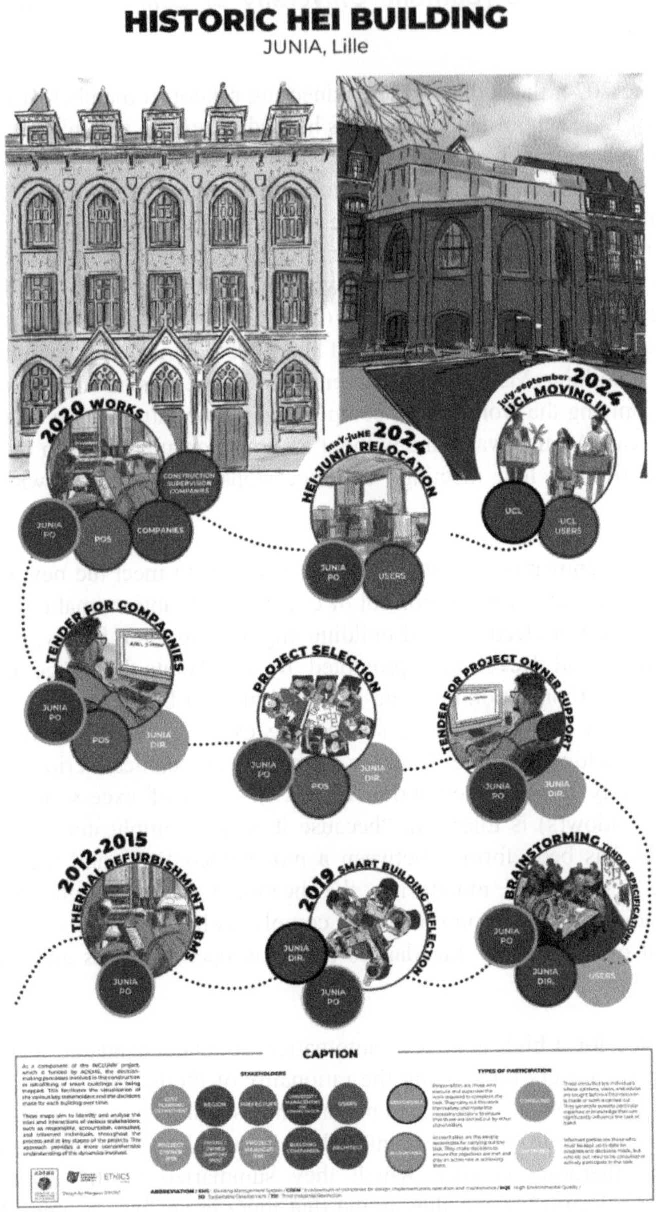

Figure 22.2. *Map of stakeholders involved in the Michel Falise building (ex-JUNIA-HEI building) renovation process. Production and illustration by Margaux D'Hont*

However, make no mistake about it. The real intention of the technical managers is rather to create the conditions for freedom of action in order to test settings that can achieve the best compromise between moderating consumption and satisfying occupants. We can therefore understand that the advent of a context for experimenting with solutions that are acceptable to the main stakeholders leads to a "demonstrator building" in the full sense of the term: a tool where it is possible to "try out lots of things" thanks to the neutralization of the human factor by automatic systems, rather than a building whose overall smooth operation produces globally satisfactory results.

The strategic path chosen by the technical managers also leads to the observation that a degree of involvement (or inclusion) of the occupants in the building's socio-technical script has been restored. This is because experimenting with settings, initially in small areas before being rolled out across the board, is tantamount to giving occupants a warning function in the building's socio-technical script; it is the occurrence of complaints that indicates how far not to go in reducing energy input. In this way, the "fully automatic" scenario shifts the occupants' contribution from action to reaction. This is all the truer in the case of the Michel Falise building, as the technical managers show a clear lack of interest in analyzing the data produced by monitoring the installations (BMS, sensors, etc.).

22.3.3. *IUT Roubaix (University of Lille): users distanced by the technical expertise of designers*

Unlike the other three buildings studied, this building is a new construction. The project to build a new University Institute of Technology (IUT) in Roubaix was launched in 2015. It is part of a practical reflection on the need to accommodate staff and students in a healthy building:

> The very start of the project was the fact that the IUT was located in the old Roubaix post office building, which had a lot of health and safety problems. So it was absolutely essential either to carry out major renovations or to relocate the IUT (university executive).

From the outset of the project, the University of Lille 2's presidential team chose to include the design of the IUT in the Rev3 ("third industrial revolution") public policy of the Hauts-de-France region. This dynamic is part of a technical vision of the energy transition through decentralized production of renewable energy, the deployment of smart buildings as energy production nodes and their networking via smart grids. The construction of the IUT has therefore been designed in line with these criteria: designing a smart, energy-efficient building that (self)produces renewable energy.

The second wish of the presidential team was to stay in Roubaix and build a new building as part of the city's renovation project: the Alma-Gare. The IUT will therefore be built at the same time as another university building to form a campus in Roubaix.

The next step was to publish a call for projects requiring a CREM (a consortium of companies for design, implementation, operation and maintenance), i.e. all of the players involved in the construction of the building set a project up and commit to the performance of this building, defined by the energy performance contract (CPE).

These contracts were democratized in France following the Grenelle Act of 2006, and their necessity was reaffirmed in 2012 by the European Union (Chong et al. 2013). The CREM is one of the versions of the "legal vehicle" for energy performance contracting and commits service providers to meeting targets on pain of financial compensation.

By opting for a CREM, the university's contracting authority was able to anticipate the building's consumption and therefore its operating costs. These aspects were particularly important for the contracting authority, as the public university's finances are severely constrained (Calviac 2019).

The second requirement of the call for projects was to construct a building that complies with the HQE and BEPOS (positive energy building) labels, in order to meet the technical criteria and thus qualify for Rev3 public policy funding.

The Passivhaus (low-energy consumption) label will be proposed by CREM. The CREM is made up of various players in the building industry: an architect, a technical design office, an energy and environmental engineering office, an operator–maintainer, a BIM management company and a builder. Apart from the choice of the CREM, which will be made by the users, they have little involvement in the decision-making process and the technical choices (see Figure 22.3).

For the building's designers, the construction of the IUT enabled them to showcase their expertise, promote their know-how and their organization and give meaning to their work: "It was a dream come true! It's a formula one, it's a Ferrari" (university executive).

The novelty and innovation required to create this building was a recurring theme in interviews with university managers. As the building required technical expertise that appeared to be new, its development was accompanied by a prestigious challenge. Innovation is also linked to another issue: risk (Alter 2013). Designing a building using techniques that are new and little-used means taking a risk for the building's designers. This risk-taking implied by innovation will be perceived as stimulating.

Attached to this representation of innovation is the idea of a certain technical difficulty:

> I really like the building for its ... technical complexity. There are things, the intelligence that manages it all in terms of design and the grey matter that went into it. It's a real gem (university executive).

The prestige associated with the building's design depends on the technical equipment, which requires detailed, cutting-edge expertise. The designers will therefore maintain the image of an innovative building, making their expertise prestigious. This emphasis on expertise will counter the ignorance of users and put them at a distance. Users are said not to have the knowledge they need to understand technology, because "they don't care, it's not their job". This representation of the user will be used to justify the decision to automate technical equipment: "There is an inflationary side to demand that needs to be moderated". As the user would not be able to grasp the technical tools, limits have to be placed on their demands and their requests have to be kept at a distance.

This idea is sometimes so pervasive that it distances not just the user but also the use of the building:

> We could afford to go down to 16°C, so all the calculations were made at 16°C. Except that as the atrium is open to the classrooms, the teachers inevitably leave the classroom open from time to time and so there is a heat exchange between the atrium at 16°C and the classroom [...] So in theory everything worked fine as the doors were closed and in reality they weren't (project management assistance).

The building thus becomes independent of its use: it functions autonomously, without its occupants. While the difficulties encountered by users in using efficient buildings have been studied in the sociological literature (Garabuau-Moussaoui 2015; Zélem 2018), the decision-making and sociological processes leading to these technical choices and the processes by which these techniques distance users are less well known.

The case of the Roubaix IUT is instructive from this point of view, especially as it requires us to consider the user as multiple: students, teachers and staff do not face the same difficulties, to which can be added inequalities and vulnerabilities that complicate the relationship of users to the technologies deployed in the building.

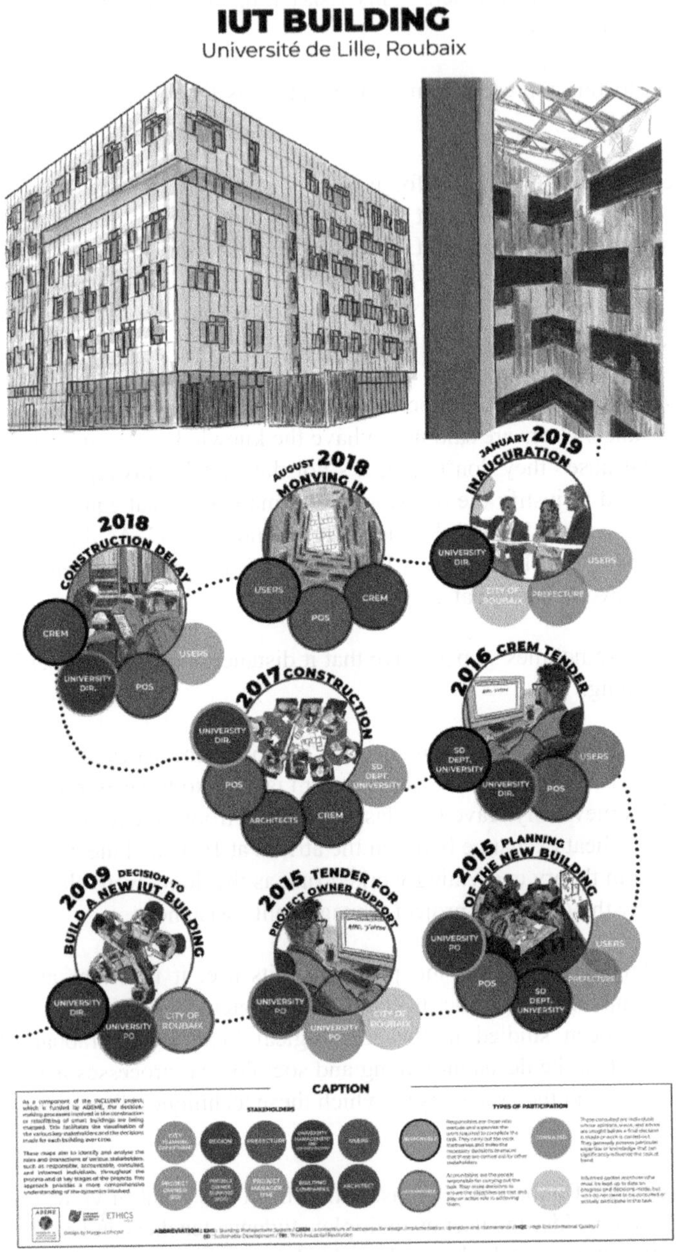

Figure 22.3. *Map of the stakeholders involved in the construction process of the Roubaix IUT. Production and illustration by Margaux D'Hont*

22.3.4. *Sciences Po Lille: involvement users but limited decision-making power*

Sciences Po Lille is a state-run *Grande école*, with competitive entrance exams. The 2,000 students enrolled follow a wide range of multidisciplinary courses in the social sciences. The school was founded in 1991. After occupying several premises over the course of its history, since 2017, it has been based in Lille city center (Angellier street).

The premises occupied by Sciences Po Lille since this latest move are two buildings constructed in 1890 to house the university library and the Faculty of Law and Literature, respectively, emblematic buildings of what was once Lille's "Latin Quarter".

Since their construction, these buildings have undergone several extensions and renovations. The most recent extension took place in 2015–2016, to accommodate Sciences Po Lille, which had previously been housed in a former textile factory (Trévise street) in the Moulins district. With this project and the move, the school will benefit from 30% more floor space in renovated premises of a much higher standard than that offered by the former Moulins factory (a district classified as a disadvantaged neighborhood, drug trafficking, on the outskirts of the city, etc.).

This renovation project (2015–2016) and the rehousing process that accompanied it were the subject of a monograph (Maria 2023) aimed at highlighting the various implications and intentions of the players at work, as well as the reception by users of these newly delivered buildings.

Compared with the other three university buildings studied in the INCLUNIV project, Sciences Po Lille stands out because it is not considered a smart building, and the decision-makers had no intention of moving in that direction. Similarly, the ecological intentions are initially very limited. The site will nevertheless be HQE compliant, although the ecological transition is not a primary objective.

Finally, on the energy front, the diagnosis is less clear-cut: a BMS will be installed to control energy flows, and the quest for energy performance is asserted by some of the players at the helm of the project, but in the end, the ambition is not taken very far, and the result will be several disappointments.

Nevertheless, the case study is instructive when compared with the other three sites, particularly on two points. Firstly, because the ecological and energy ambitions will ultimately come later, after the move, and will take concrete form from 2018 onward in a series of actions and decisions that will lead to a re-examination of the way in which these issues were dealt with, or not dealt with, during the building work. Secondly, because from the outset of the project, the school's management has shown a strong desire to include staff and students. Workshops, visits, working groups and consultations are being organized before and during the works. Interviews conducted in 2023 with people who had taken part in the project at the time clearly underlined this commitment, even as they recounted their disappointment with the outcome.

It should be pointed out that the Hauts-de-France region is the main funder of this, and as such will be the contracting authority. The school management does not have control over the major decisions. This limits the scope for consultation with future users of the building, whether staff or students. Added to this is a difficulty noted in the other case studies: the timetable for a project of this scale is necessarily spread over several years, which in normal circumstances is already an obstacle to the mobilization and involvement of future users, but is particularly acute when it involves students who know from the outset; this is because their course is of limited duration and the future premises will not concern them.

Even though it is not as pronounced and apparent as in the case of smart buildings, the socio-technical imagination of the decision-makers who presided over the construction and relocation of the buildings is evident in several striking features. Firstly, contrary to a choice frequently made in the case of smart buildings where energy consumption is tightly controlled, at Sciences Po Lille, users remain relatively free to open windows and leave doors open, and now also to adjust the radiators in each room. The latter is a recent decision (started in 2021) taken by the head of property and logistics, which was accompanied by a communication campaign aimed at students to encourage them to control their heating consumption. This approach is the opposite of the automatic switch-ons and switch-offs that are generally seen as a sign of mistrust toward users.

The building's BMS presents another interesting and revealing case. As the property and logistics manager recounted in an interview:

> In fact, Automate is Sauter's proprietary software, so it's code developed specifically for the site, by a person who is no longer there, who can no longer be contacted. Just to tell you, it took me a year to get the administrator codes for the controllers, to really be able to enter and modify the IP addresses, etc., things like that. Because it wasn't planned.

This is another category of user whose control over the building has been limited, and who has been prevented from appropriating the very instrument that should enable them to control flow consumption.

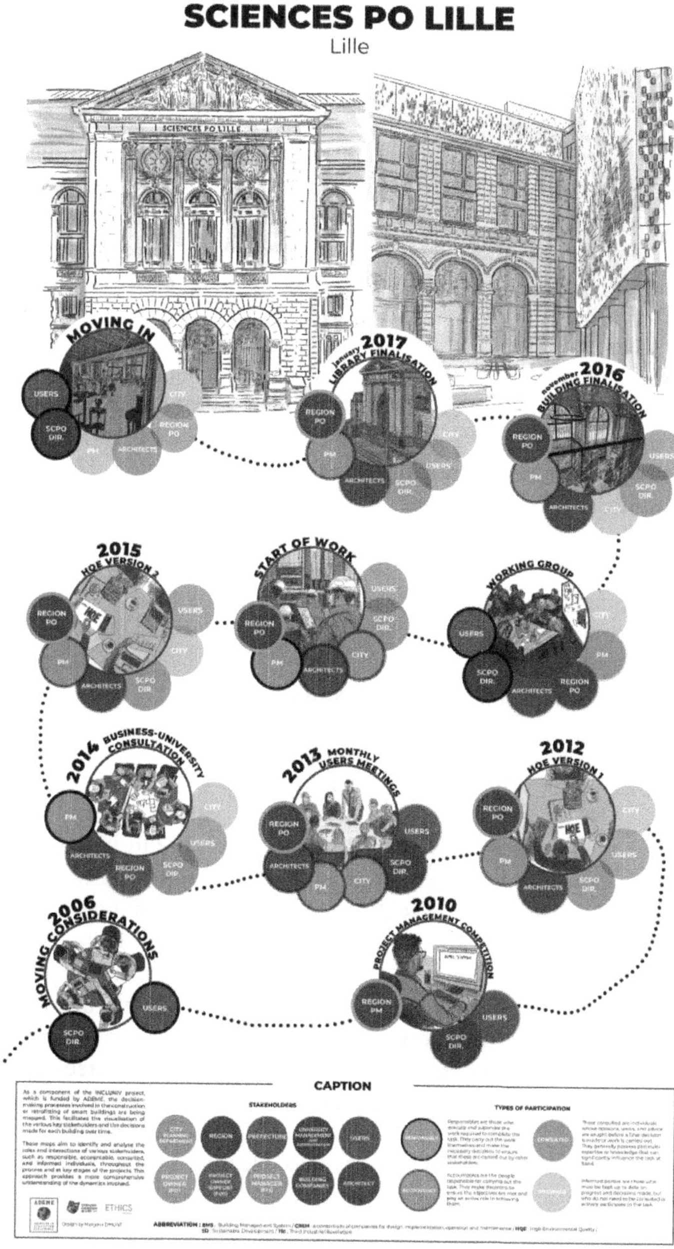

Figure 22.4. *Map of stakeholders in the renovation process at Sciences Po Lille. Production and illustration by Margaux D'Hont*

22.4. Findings and discussion

22.4.1. *University buildings: showcases for their institutions and imaginary of innovation*

The study of these four cases from Lille shows the importance of the building's influence, image and suitability for the educational institution in the renovation or construction decisions. Buildings are seen by decision-makers as showcases for their institutions, which must reflect their prestige and the courses they teach, in the interests of attractiveness and competition between universities.

However, these buildings are also showcases for the region, which is helping to finance them. The launch of the Rev3 initiative in the Hauts-de-France region has played a decisive role in the deployment of demonstration and smart buildings (Junia, Rizomm, IUT). The Rev3 initiative has conveyed a powerful vision of energy and environmental transition through technical innovation. The three "smart" buildings studied (Rizomm, IUT, Junia) were intended to be genuine "demonstrators" of the region's "third industrial revolution" and innovation laboratories. Innovation will be sought for its own sake (symbolic, educational, R&D, etc.) and not necessarily for ecological or economic gain. This regional context of environmental innovation is central to understanding the technical choices made by these different university institutions. Also, the inclusion of users will conflict with this quest for innovation and performance.

22.4.2. *Include: who, when, how and why?*

Although in three of the cases studied, a genuine desire to involve users in the decision-making process was clearly expressed and implemented, it was faced with several difficulties. First, there was the difficulty of "recruiting" participants, of arousing the interest of users in taking part, and then the difficulty of maintaining participation over the long term, as building projects are spread over several years. But above all, it is the difficulty of projecting needs that is proving to be a major obstacle to user inclusion. It is very difficult for users to project themselves into future uses, and into a building that will be occupied, at best, in three- or five-years' time, if they have not changed jobs or services by then.

The socio-anthropology of technology has clearly shown that forgetting the technique is inherent in the technical phenomenon: it is the absence or breakdown that reveals (or reminds) the user of the presence of the technical object. Technique reminds the user of itself through use: it is because we want to plug in our computer or my lamp that we realize there are not enough sockets or that they are badly located. It is therefore very difficult for users to anticipate their future needs in an abstract, theoretical or speculative way, to think about the need *before* the use.

This is compounded in the building sector by a highly standardized phasing (programming, design, construction and operation), where consultation of needs is limited to the programming phase. Data from the field shows that it is not enough to involve users in this phase alone, and that there is a need for co-design and adjustments in the calibration and operation phase. However, it is probably not the same users who need to be involved at these different phases.

Also, as Janda (2011, p. 20) analyzes, architects have a significant role to play in involving users:

> Building users play a critical but poorly understood and often overlooked role in the built environment [...] With climate reduction targets set at 80% of the 1990 levels, designers need to work with users to deliver comprehensive energy reductions. Preparing the public for this interactive role is a job in itself. To fully address the task ahead, the article suggested that either an existing professional group should adapt its jurisdiction to include public education on building literacy, or a new professional group should arise to claim this role. Some architects have the skills and experience to take on this challenge, but the field as a whole would need to develop professional expertise and seek ways of integrating user involvement in building performance to fully succeed.

22.4.3. *A question of temporality*

Including users in the decision-making process (programming and design) requires finding the right timeframe. If the timeframe of the project is too short, it makes it difficult to implement consultation and participation, and the urgency of the decision-making does not allow users – who are not specialists – to understand the issues, study alternatives, etc. (IUT building). However, if the timeframe is too long, users lose interest in participating (Rizomm). All of the cases studied also highlighted a specific feature of university buildings: a large proportion of users are students, and their turnover linked to the length of their course is unsuited to the timeframe of construction and renovation projects. Finally, both construction and renovation projects place the client in a paradoxical timeframe: there is both the perception of a very long process (different phasing, delays in construction and delivery, etc.) and at the same time, an ongoing urgency in decision-making for the project owner. Decisions have to be taken urgently, with decisive consequences for the use of the building, at the various stages of the project (urgency generated by the project management team, which is bound by deadlines for the schedule, execution, delivery, etc.).

22.4.4. *The smart building: an intelligent building instead of users*

The various case studies clearly illustrate the technical paradigm behind the choice of smart building or BMS (Building Management System), i.e. the automation of flows (ventilation, lighting, heating, etc.), which reflects both a mistrust of users and a desire to achieve technical, energy and environmental performance.

It is a line of thinking that is being promoted first and foremost by the building industry:

> When we launched the calls for tender, a lot of companies came to see us. I was a bit involved in the analysis, and they said, not quite like that, but almost, "don't have any confidence in students and young people. In any case, they have very little ecological awareness. So the building has to be intelligent for them". That's pretty much how it was put. In other words, everything has to be closed, there are no skylights, you can't open the windows, the lights have to be switched off automatically. And that's one of the messages. The hyper-technologicalization of systems and automation is a real debate, a real question of accountability (former rector of the Catholic University of Lille).

22.5. Conclusion: inclusion by sensor or the automation in question

What kind of inclusion for the user does this technical paradigm propose? How can users take ownership of the technology that surrounds them on a day-to-day basis, when they have no control over lighting, heating, ventilation, etc?

In this technical approach, sensors are used to take into account the presence of users and their needs (Dorthe 2020). This is the technical artifact that, in a way, includes the user. The user acts without knowing it (their presence triggers the light, heating, etc.), but they perceive the result (heat, light, etc.).

The absence of interaction with the technical system contributes to an even greater invisibility of energy, whereas the visibility of energy is a central issue in achieving low-energy practices (Dobigny 2024). Also, what about environmental citizenship? Education and awareness of environmental issues also involve action and the body.

There is therefore a paradox between the objective of reducing energy consumption through technical mediation (energy efficiency and sufficiency), and the fact that it is impossible to achieve precisely because the user is completely disengaged from the technical system: without any perception and representation of their consumption, how can users adopt practices of sufficiency (Dobigny 2024)?

Through automation, a dual process of derealization and delegation of energy consumption takes place. However, it is possible to design technical systems that allow savings to be made (the system switches off without user intervention) while retaining user control, but without automation (switch, thermostat, etc.). Co-designing technical systems deployed in university buildings with engineers, designers, social scientists and users could make it possible to develop technologies that integrate both user control and savings on usage.

Finally, these concrete issues also question the way in which ecological policies are invested and implemented by universities and their stakeholders, as well as the place they are given. To train "smart users" or ecological citizens, should we raise awareness, integrate and involve them, in what way, how, and over what timeframe? These questions highlight the tension in the imperative of public participation in the ecological transition: universities are predominantly places of education designed from a top-down perspective. As a result, ecological issues bring to the fore the issue of institutional governance in the projects they seek to implement. To address these issues as effectively as possible, it is essential to provide spaces where reflection is shared and encouraged by institutional mechanisms.

22.6. Acknowledgements

The INCLUNIV project (2023–2025) was carried out with financial support from the French Agency for Ecological Transition (ADEME). We would particularly like to thank Pablo Maria for his master's thesis work and Marie Chavoix and Victor Toutain for their work collecting ethnographic data as part of their master's internship with the INCLUNIV project.

22.7. References

Alter, N. (2013). *L'innovation ordinaire*. Presses Universitaires de France, Paris.

Aulanier, A. and Dobigny, L. (2025). Une transition écologique sans expérience. Les émotions des usagers des smart-buildings universitaires, ou l'histoire d'un rendez-vous manqué avec les enjeux écologiques. *Émulations, Revue de sciences sociales*, 50.

Calviac, S. (2019). Le financement des universités : évolutions et enjeux. *Revue française d'administration publique*, 169(1), 51–68.

Chong, E., Le Lannier, A., Staropoli, C. (2013). La rénovation énergétique des bâtiments : les collectivités territoriales au milieu du gué. In *Collectivités territoriales et énergie : ambitions et contradictions*. Le Moniteur, Antony, 155–169.

Deleuze, G. and Guattari, F. (1980). Introduction. In *Mille plateaux*. Les Éditions de Minuit, Paris.

Dobigny, L. (2024). Smart-city, smart-building, smart-user: “Smart” imaginaries and its dead-ends. In *Smart Grids and Buildings for Energy and Societal Transition*, Robyns, B., Dobigny, L., Abbes, D., Durillon, B., Barry, H., Saudemont, C. (eds). ISTE Ltd, London, and Wiley & Sons, New York.

Dorthe, G. (2020). Smart building et pilotage : quelles métaphores pour quels usagers ? Explorateurs de la Transition – Blog [Online]. Available at: https://blog.explorateurs-transition.fr/2020/07/03/smart-building-et-pilotage-quelles-metaphores-pour-quels-usagers/.

Dorthe, G. and Dobigny, L. (2025). From smart buildings to smart users? Energy transition to the test of parasitic humans. In *The Smartification of Everything: Critical Perspectives in Science, Art and Society*, Gugganig, M., Bronson, K., Mirza, V. (eds). University of Toronto Press, Toronto.

Garabuau-Moussaoui, I. (2015). Les occupants de bâtiments performants en énergie en sont-ils les usagers ? In *Énergie et transformations sociales. Enquêtes sur les interfaces énergétiques*, Cihuelo J. (ed.). Lavoisier, Paris, 33–49.

Germe-Bartier, A. (2023). “Ne touchez à rien !” : travailler dans un smart building. Master’s Thesis, Université Lumière Lyon, Lyon.

Janda, K.B. (2011). Buildings don’t use energy: People do. *Architectural Science Review*. 54(1), 15–22.

Maria, P. (2023). Smart building ou smart user ? L’inclusion des publics dans la transition écologique. Étude du projet de relogement de Sciences Po Lille. Research Thesis, Sciences Po Lille, Lille.

Rifkin, J. (2011). *The Third Industrial Revolution: How Lateral Power Is Transforming Energy, the Economy, and the World*. Palgrave Macmillan, London.

Robyns, B., Lenglet, C., Barry, H., Bozzo-Rey, M. (2023). *Smart Users for Energy and Societal Transition*. ISTE Ltd, London and Wiley & Sons, New York.

Strengers, Y. (2013). *Smart Energy Technologies in Everyday Life: Smart Utopia?* Palgrave Macmillan, London.

Zélem, M.-C. (2018). Économies d’énergie : le bâtiment confronté à ses occupants. *Annales des mines – responsabilité et environnement*, 90(2), 26–34.

List of Authors

Dhaker ABBES
L2EP
JUNIA
Lille
France

Hervé BARRY
LITL
Université Catholique de Lille
France

Baptiste BEGUINET
AMPERE
INSA Lyon
France

Sadia BENAMROUZ-VANNESTE
LITL
Université Catholique de Lille
France

Mariana BENITEZ
Laboratorio Nacional de Ciencias de la Sostentabilidad
Instituto de Ecología
Universidad Nacional Autónoma de México
Mexico City
Mexico

Hanene BETTAIEB
L2EP
JUNIA
Lille
France

Rossano BOLPAGNI
Dept. Chemistry, Life, Sciences and Environmental Sustainability
Parma University
Italy

Adrien BOSSU
L2EP
JUNIA
Lille
France

Alessandro BOTTACCI
Società Italiana di Restauro Forestale (SIRF c/o)
Dipartimento di Scienze Agrarie e Forestali (DAFNE)
Università della Tuscia
Viterbo
Italy

Richard BULL
Nottingham Trent University
UK

Sébastien CARCELLE
ART-Dev
CIRAD
Montpellier
France

Kevin CIANFAGLIONE
LITL
Université Catholique de Lille
France

Jonathan COIGNARD
G2Elab
CNRS, Grenoble INP
Université Grenoble Alpes
France

Arnaud DAVIGNY
L2EP
JUNIA
Lille
France

Emmanuel D. DELOCADO
Ateneo Institute of Sustainability
and
Ateneo Biodiversity Research Laboratory
Department of Biology
School of Science and Engineering
Ateneo de Manila University
Quezon City
Philippines

Amira DHORBANI
L2EP
JUNIA
Lille
France

Marcello DI BONITO
School of Animal, Rural and Environmental Sciences
Nottingham Trent University
UK
and
Department of Agricultural and Food Sciences
Alma Mater Studiorum
University of Bologna
Italy

Laure DOBIGNY
ETH+ ETHICS EA 7446
Université Catholique de Lille
France

Hector DOCARRAGAL MONTERO
École Nationale Supérieure d'Architecture de Paris-La Villette (ENSAPLV)
and
École Nationale Supérieure d'Architecture et de Paysage de Bordeaux (ENSAPBx)
France

Ana Rita DOMINGUES
Nottingham Trent University
UK

Agathe DOUCHET
LITL
Université Catholique de Lille
France

Benoit DURILLON
L2EP
JUNIA
Lille
France

Alice FOUILLOUZE
Université Catholique de Lille
France

Anouk GERME-BARTIER
ETH+ ETHICS EA 7446
Université Catholique de Lille
France

Kahina HASSAM
JUNIA
Lille
France

Ishan JALAN
Nottingham Trent University
UK

Florian KLETTY
LITL
Université Catholique de Lille
France

Ma. Margarita Christina A. LACDAO-UMALI
Ateneo Institute of Sustainability
Ateneo de Manila University
Quezon City
Philippines

Silvano LANDI
Società Italiana di Restauro Forestale (SIRF c/o)
Dipartimento di Scienze Agrarie e Forestali (DAFNE)
Università della Tuscia
Viterbo
Italy

Cristina LONGO
LUMEN
Université de Lille
France

Lucas LOPEZ
CEMMC
and
Chaire RESET
Université Bordeaux Montaigne
France

Muhammad MAZHAR
Nottingham Trent University
UK

Franco PEDROTTI
Università di Camerino
Italy

Benoît ROBYNS
L2EP
JUNIA
and
Université Catholique de Lille
France

Christophe SAUDEMONT
L2EP
JUNIA
Lille
France

Bartolomeo SCHIRONE
Società Italiana di Restauro Forestale (SIRF c/o)
Dipartimento di Scienze Agrarie e Forestali (DAFNE)
Università della Tuscia
Viterbo
Italy

Kokou Kouzouahin SOMABE
LITL
Université Catholique de Lille
France

Mathilde SZUBA
CERAPS
Sciences Po Lille
France

Philip Arnold P. TUAÑO
Department of Economics
Rosita G. Leong School of Social Sciences
Ateneo de Manila University
and
Philippine Institute for Development Studies
Quezon City
Philippines

Albin WAGENER
GERTS – ETHICS EA 7446
ESSLIL
Université Catholique de Lille
France

Matthew O'Meara WALLIS
Nottingham Business School
Nottingham Trent University
UK

Gamze YAKAR-PRITCHARD
Nottingham Trent University
UK

Jessica ZAPHIROPOULO
PACTE
Université Grenoble Alpes
France

Index

U, V, W

Other titles from

iSTE

in

Energy

2024

BENALLOU Abdelhanine
Heat Exchangers: Design and Sizing Algorithms
(Energy Engineering Set – Volume 6)

FAVENNEC Jean-Pierre
Geopolitics and Energy Transition 1: The Basics

MORY Mathieu
Mechanical Energy Conversion: Exercises for Scaling Renewable Energy Systems

ROBYNS Benoît, DOBIGNY Laure, ABBES Dhaker, DURILLON Benoit, BARRY Hervé, SAUDEMONT Christophe
Smart Grids and Buildings for Energy and Societal Transition

2023

ALVAREZ-HERAULT Marie-Cécile, GOUIN Victor, CHARDIN-SEGUI Trinidad, MALOT Alain, COIGNARD Jonathan, RAISON Bertrand, COULET Jérôme
Distribution System Planning: Evolution of Methodologies and Digital Tools for Energy Transition

BEN KHADER BOUZID Monia, CHAMPENOIS Gérard
Electrical and Mechanical Fault Diagnosis in Wind Energy Conversion Systems

JANNOT Yves, MOYNE Christian, DEGIOVANNI Alain
Heat Transfer 1: Conduction and Convection
Heat Transfer 2: Radiation and Exchangers

ROBYNS Benoît, LENGLET Claude, BARRY Hervé, BOZZO-REY Malik
Smart Users for Energy and Societal Transition

2022

ABOUELATTA Mohamed, SHAKER Ahmed, GONTRAND Christian
Smart Power Integration

BOUTAUD Benoit
Energy Autonomy: From the Notion to the Concepts

GONTRAND Christian
Electromagnetism: Links to Special Relativity

2021

ROBYNS Benoît, DAVIGNY Arnaud, FRANÇOIS Bruno, HENNETON Antoine, SPROOTEN Jonathan
Electricity Production from Renewable Energies (2nd edition)

2020

BOISGIBAULT Louis, AL KABBANI Fahad
Energy Transition in Metropolises, Rural Areas and Deserts

SOUALHI Abdenour, RAZIK Hubert
Electrical Systems 1: From Diagnosis to Prognosis
Electrical Systems 2: From Diagnosis to Prognosis

2019

BENALLOU Abdelhanine
Energy Transfers by Convection
(Energy Engineering Set – Volume 3)
Energy Transfers by Radiation
(Energy Engineering Set – Volume 4)
Mass Transfers and Physical Data Estimation
(Energy Engineering Set – Volume 5)

LACHAL Bernard
Energy Transition

2018

BENALLOU Abdelhanine
Energy and Mass Transfers: Balance Sheet Approach and Basic Concepts
(Energy Engineering Set – Volume 1)
Energy Transfers by Conduction
(Energy Engineering Set – Volume 2)

JEMEÏ Samir
Hybridization, Diagnostic and Prognostic of Proton Exchange Membrane Fuel Cells: Durability and Reliability

RUFFINE Livio, BROSETA Daniel, DESMEDT Arnaud
Gas Hydrates 2: Geoscience Issues and Potential Industrial Applications

VIALLET Virginie, FLEUTOT Benoit
Inorganic Massive Batteries
(Energy Storage – Batteries and Supercapacitors Set – Volume 4)

2017

BROSETA Daniel, RUFFINE Livio, DESMEDT Arnaud
Gas Hydrates 1: Fundamentals, Characterization and Modeling

LA SCALA Massimo
From Smart Grids to Smart Cities: New Challenges in Optimizing Energy Grids
(Advanced Smart Grids Set – Volume 2)

MOLINA Géraldine, MUSY Marjorie, LEFRANC Margot
Building Professionals Facing the Energy Efficiency Challenge

SIMON Patrice, BROUSSE Thierry, FAVIER Frédéric
Supercapacitors Based on Carbon or Pseudocapacitive Materials
(Energy Storage – Batteries and Supercapacitors Set – Volume 3)

2016

ALLARD Bruno
Power Systems-on-Chip: Practical Aspects of Design

ANDRÉ Michel, SAMARAS Zissis
Energy and Environment
(Research for Innovative Transports Set – Volume 1)

DUFOUR Anthony
Thermochemical Conversion of Biomass for the Production of Energy and Chemicals

SOUSTELLE Michel
Phase Transformations
(Chemical Thermodynamics Set – Volume 5)
Thermodynamics of Surfaces and Capillary Systems
(Chemical Thermodynamics Set – Volume 7)

2015

MONCONDUIT Laure, CROGUENNEC Laurence, DEDRYVÈRE Rémi
Electrodes for Li-ion Batteries: Materials, Mechanisms and Performance
(Energy Storage – Batteries and Supercapacitors Set – Volume 2)

ROBYNS Benoît, FRANÇOIS Bruno, DELILLE Gauthier, SAUDEMONT Christophe
Energy Storage in Electric Power Grids

ROSSI Carole
Al-based Energetic Nanomaterials: Design, Manufacturing, Properties and Applications
(Nanotechnologies for Energy Recovery Set – Volume 2)

SOUSTELLE Michel
Phases Modeling Tools: Application to Gases
(Chemical Thermodynamics Set – Volume 1)
Modeling of Liquid Phases
(Chemical Thermodynamics Set – Volume 2)
Thermodynamic Modeling of Solid Phases
(Chemical Thermodynamics Set – Volume 3)
Chemical Equilibria
(Chemical Thermodynamics Set – Volume 4)
Ionic and Electrochemical Equilibria
(Chemical Thermodynamics Set – Volume 6)

TARASCON Jean-Marie, SIMON Patrice
Electrochemical Energy Storage
(Energy Storage – Batteries and Supercapacitors Set – Volume 1)

2014

DE LARMINAT Philippe
Climate Change: Identification and Projections

2012

BECKERS Benoit
Solar Energy at Urban Scale

FOULADGER Javad
Electrothermics

2011

GAO Fei, BLUNIER Benjamin, MIRAOUI Abdellatif
Proton Exchange Membrane Fuel Cell Modeling

MONMASSON Eric
Power Electronic Converters: PWM Strategies and Current Control Techniques

MULTON Bernard
Marine Renewable Energy Handbook

RAZIK Hubert
Handbook of Asynchronous Machine with Variable Speed

2010

BRUNET Yves
Energy Storage

JUFER Marcel
Electric Drives: Design Methodology

2009

SABONNADIÈRE Jean-Claude
Low Emission Power Generation Technologies and Energy Management
Renewable Energy Technologies

2008

MEUNIER Gérard
The Finite Element Method for Electromagnetic Modeling

Printed and bound by CPI Group (UK) Ltd, Croydon, CR0 4YY

13/09/2026

14961011-0007